POPULATION BIOLOGY

THOMAS C. EMMEL
UNIVERSITY OF FLORIDA

HARPER & ROW, PUBLISHERS
New York Hagerstown San Francisco London

Sponsoring Editor: Joe Ingram
Project Editor: Richard T. Viggiano
Designer: Rita Naughton
Production Supervisor: Will C. Jomarrón
Compositor: Progressive Typographers, Inc.
Printer and Binder: Halliday Lithograph Corporation
Art Studio: Danmark & Michaels Inc.

POPULATION BIOLOGY

Library of Congress Cataloging in Publication Data

Emmel, Thomas C
Population biology.

Bibliography: p.
Includes index.
1. Population biology. I. Title.
QH352.E47 574.5'24 76-13876
ISBN 0-06-041904-0

Contents

Preface

After taking a variety of courses in population biology and diverse fields of ecology, evolution, genetics, statistics, and computer science, all part of the required repertoire of a modern population biologist, one of my graduate students aptly remarked this year that the new breed of population biologists should be called "the neonaturalists." This appellation coined by Steven M. Chambers is not at all off target if one stops to think about the history as well as the current development and vastness of biology.

It was the great biological integrators of the last century—people such as Charles Darwin and Alfred Russell Wallace—who had studied a diversity of topics from geology to mathematics to philosophy to taxonomy to genetics to behavior, and who had looked at the *totality* of the biotic and physical world in their explorations in their home countries as well as around the world, that made the most significant contributions to beginning the now century-old revolution in biological thought that has centered on the concept of evolution. The neo-Darwinists held the center stage in evolutionary biology from the 1920s to the late 1960s with their synthesis of Darwinian natural-selection theory and the theories of the field of population genetics. The efforts of such outstanding men as Sir Ronald Fisher, J. B. S. Haldane, Sewall Wright, Sir Julian Huxley, and Theodosius Dobzhansky began four decades of rapid advancement in evolutionary biology, highlighted by extensive documentation of the mechanisms and actions of selection in increasing organic diversity. Indeed, few young evolutionary biologists during those 40 years dared to transcend the established neo-Darwinian authorities of "the modern synthetic theory of evolution." Finally, a small group of "post-Darwinists," as E. O. Wilson has so appropriately titled them, began to challenge aspects of the central theory by analysis of ecological problems such as competition among species, properties of the niche, and species diversity. The fresh insights and intriguing predictions made by Robert H. MacArthur, Richard Levins, R. C. Lewontin, C. S. Holling, Eric R. Pianka, E. O. Wilson, Paul R. Ehrlich, John R. G. Turner, Gordon H. Orians, Thomas W. Schoener, Peter H. Raven, and others in the past decade have opened up a host of new doors to the expansion of evolutionary biology as an integral part of the new population biology, which aims at nothing less than a set of predictive theories of population processes and phenomena.

To achieve or even attempt to reach that goal, one must be well versed in a diversity of fields in biology and mathematics, and have an intuitive feeling for the complex overall problems of ecology and evolution. And in addition to picking up the Cartesian analytic method of simplification, analysis, and resynthesis that Lewontin rightly stresses, the prospective population biologist must have an ability to perceive and separate, gently but decisively, the significant problems from extraneous background noise in the course of actual field work on natural populations of plants and animals (and laboratory populations where feasible). Few biologists have such a combination of admirable characteristics, and most of us vary away from the median toward one extreme of the spectrum ranging from analytical model builder to purely descriptive biologist. One of the purposes of this book is to attempt to bridge this spectrum for the beginning student in population biology, by presenting a balanced introduction to both the descriptive and analytical facets of modern population biology. Excellent books are already available on broad or specialized portions of this field, such as ecology (Krebs, 1972), ecology and field biology (Smith, 1974), geographical ecology (MacArthur, 1972), evolutionary ecology (Pianka, 1974), and population genetics (Crow and Kimura, 1970).

Part of the problem of writing a book on population biology involves the host of interrelationships among the basic contributing fields of ecology, genetics, evolution, and behavior. Readers are faced by this problem, too, because often they might need to recall part of their background in basic ecology or ideas on natural selection in order to accept a statement dealing with the reasons for exhibition of a particular type of, say, dispersal behavior early in the book. I have attempted to minimize such potential conflicts by organizing the material in what has seemed to be the most satisfactory order of presentation for my students in courses at the University of Florida. Nonetheless, most of the chapters will stand on their own to the well-prepared, advanced undergraduate, and this deliberate arrangement of nearly self-contained units may aid instructors who would prefer to take up certain sections in an alternate order.

I claim no pretense of originality by introducing untested theory or a vast compendium of new and exotic data into the literature with this book. Rather, the consistent aim throughout its writing has been to present a basic overview of the facts and problems known about biology on the population level, an overview that will be interesting to the student, useful in teaching, and perhaps will attract new young biologists of the potential of Charles Darwin or Robert MacArthur to take up a career in this exciting and fascinating field of population biology.

Thomas C. Emmel
University of Florida

Acknowledgments

I am indebted to a great many colleagues, students, and friends who have contributed to the evolution of my interests in population biology and directly or indirectly assisted in the preparation of this book. I particularly wish to acknowledge the important influence of Paul R. Ehrlich, Richard W. Holm, and Peter H. Raven while at Stanford University and, in fact, to the present. Their innovative and often unique conceptual approaches to broad problems in population biology have made signal contributions to the development of this area and have helped to push it to the forefront of modern biology. John F. Emmel, Edward F. Emmel, and Larry E. Gilbert of the University of Texas have also especially helped me to see problems in natural animal populations from new perspectives.

I am very grateful to Boyce A. Drummond III, Steven M. Chambers, Thomas S. Kilduff, Carlos E. Valerio, F. Clifford Johnson, Archie F. Carr, Lewis Berner, James T. Giesel, Jonathan Reiskind, Brian K. McNab, John H. Kaufmann, James E. Dinsmore, and many other colleagues and students at the University of Florida for profitable discussion on various topics in population biology and ecology over the past eight years. I have also greatly appreciated the excellent help of Nancy

P. Drummond in the preparation of the index, and the timely assistance provided by Hal M. Ingman, Jr., during other parts of the project. Ruth F. Smith and Nancy P. Drummond typed several entire drafts of the manuscript with great diligence and accuracy. Errors remaining, of course, are entirely my own. Richard Viggiano of Harper & Row patiently and carefully guided this project through to publication. To all these persons, I express my deepest appreciation.

T.C.E.

1 2 3 4 5 6 7 8 9 10 11

Introduction to Population Biology

Massive migratory herds of wildebeest, zebra, and Thomson's gazelle move across the African savanna each year in a rhythmic and remarkable natural spectacle. These tremendous aggregations of grazing animals appear at first glance to be merely responding, albeit in an unusually dramatic way, to the alternating succession of dry and wet seasons. But studies by population biologists in the Serengeti National Park, an East African area of 5600 square miles of grassland and open woodland (Figure 1–1), have recently shown that these migrations are the manifestation of a rather complex mosaic of interactions between the populations of these organisms, their foodplants, and the physical environment.

The Serengeti is a gently rolling land, and the peaks and depressions affect the distribution of water and grass types. Rain falling on the slopes tends to collect and remain longer in the depressions, and the lower areas also develop a heavier soil with increased clay content. Thus, the grass in the depressions is longer than it is on the slopes and peaks, and this spatial pattern in the quality of the food supply causes major effects on the distribution and movements of the grazing animals. To understand why, we must examine briefly both the nutri-

Figure 1–1 Migrating populations of wildebeest, zebra, and Thomson's gazelle moving across the Serengeti Plains in East Africa. (Photo by Arnold Small.)

tional characteristics of the principal members of the Serengeti grazing herds and the structure of their food supply.

Mammalian herbivores start with the basic difficulty that they cannot digest their own food. Lacking enzymes capable of breaking down the cellulose cell walls of their plant food, they carry bacteria and protozoa that do possess enzymes that dissolve the great quantity of hard cell walls made of this carbohydrate. The gut of the herbivore is then able to utilize the cell-wall products as a source of energy and, of course, can readily absorb the proteins and soluble carbohydrates present in the soft interior cytoplasm. However, the herbivores are faced with the problem of obtaining enough protein from the food supply, which is loaded with huge quantities of refractive carbohydrates. The ruminant mammals, such as wildebeest and gazelle, cope with the problem by being selective in their feeding habits, eating components of the vegetation that have thin cell walls and high concentration of protein, namely the leaves, young shoots, and fruits of grasses. The optimal food supply of the ruminant grazing animals is near the ground, among the leaves of the smaller grasses and their young shoots. These can be processed slowly in the mechanical and chemical cycle of breakdown in the rumen of the gut. The grazing strategy of the ruminant, then, involves selection of high-protein food components and processing of this food at a slow but highly efficient rate. A small ruminant like a Thomson's gazelle has a higher rate of metabolism than does a large ruminant and hence requires relatively more protein and energy per unit of body weight per day. But in terms of absolute food amounts, the smaller gazelle only needs to consume about 20 percent of the amount of food per day that a wildebeest requires. Hence, the smaller species has about five times as much time available to find and eat food as does the wildebeest, and the gazelle can live on a selective diet even under conditions of scarcity that the wildebeest is not equipped to handle.

The nonruminant mammals, such as zebra and elephants, do not have the internal cycle of regurgitation and mastication of their food, as they lack a rumen or similar structure. They ferment their intake of cellulose in the enlarged colon and large intestine, and then extract and assimilate protein in the relatively small and simple stomach. The efficiency of protein utilization is consequently lowered by as much as a third compared to that of the ruminant mammal. But there is no rumen mechanism that limits the speed of passage of material through the gut, and the nonruminant mammal can process perhaps twice as much food in the same period of time. Thus, the zebra can thrive on a diet of grass or herbage that is too low in protein to support a ruminant. The zebra can better tolerate a high level of cell-wall content in its food, but must maintain a high rate of intake to survive on such marginal food, and is therefore less specific in its feeding habits than the ruminant (e.g., wildebeest or gazelle). It can survive on older and taller grasses or shubs which have heavily thickened cell walls in their long stems.

With this background in mind, we can interpret the localized seasonal movements of the Serengeti fauna that graze across the undulating landscape (Bell, 1971). Each depressed area collects runoff water from the slopes, so that the grass in the depressions tends to grow to greater heights than does the grass on the elevated hills (Figure 1–2).

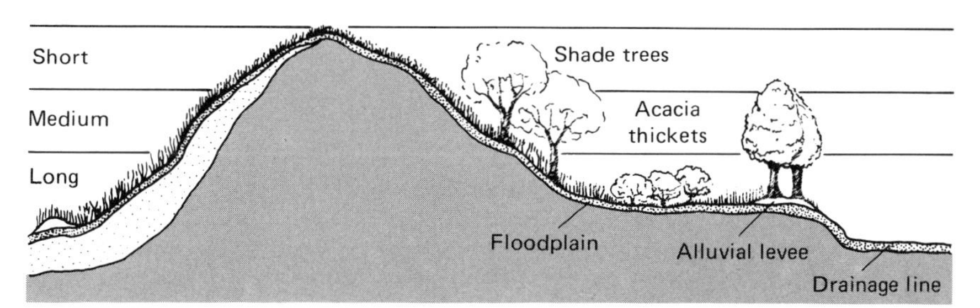

Figure 1–2 The topography and vegetation of a typical undulation of the African savanna on the Serengeti Plains. The grasses grow longer in the depressions where water runoff collects, and some of the grazing species have difficulty reaching the parts of this grass that contain the most food value for them. See text for explanation of resultant effects on ungulate population biology. (Source: R. Bell, 1971, *Scient. Amer., 225:* 86–93.)

There is a topographic succession of grass heights. In the wetter areas especially, the stems above the growing point extend to form the flowering seedheads of the grass, which, like the ordinary stems late in the growing season, contain thick-walled cells and are not as rich in protein as the young leaf growth. Hence, by the end of the rainy season in the longer grass of the depressions several layers of forage are available with varying accessibility to the herbivores. At the lowest level are young shoots and leaves of smaller grasses and herbs, such as clover. In the middle strata are stems and leaves of taller grass. The upper level consists almost entirely of the low-protein flowering culms of the tall grasses.

In the wet season, all grazing mammals concentrate in mixed aggregations on higher slopes where grass is shortest. Trampling and constant cropping by feeding keeps the grass in a short growing stage, which provides maximum protein for animals. This rich food supply also provides a particularly appropriate time for calving and lactation by females.

With the dry season, the grass stops growing and the short grass of the raised areas is increasingly removed. With declining food availability, larger nonruminants like zebras are unable to maintain themselves and they move down into the longer grasses of the depressions. As the grass situation worsens on the elevated areas, even smaller ruminant species like wildebeest and Thomson's gazelle descend. This succession of grazing species corresponds to both body weight and the order of selectivity shown in their dietary preferences. Were it not for the activities of the earlier members of the grazing succession, however, the smaller animals would be unable to reach their preferred lower levels of the savanna herbs. The zebras and buffalo break down the stiff dense stands of grass stems and culms through their grazing and trampling, thereby preparing the structure of the vegetation for the later members. Gazelles and wildebeest must graze on the lowest levels where the highest concentration of protein occurs, and these levels become accessible through their association with the populations of the zebras and buffalo (Gwynne and Bell, 1968). The processes of natural selection have adapted each animal to make maximum use of available habitat and food.

When one looks at the larger geographic pattern of seasonal movements of the migratory grazing populations on the Serengeti Plains, these massive migrations of zebras, wildebeests, and Thomson's gazelle are actually manifestations of the same factors promoting the local movement pattern just examined. There are three zones, arranged in a broad triangle across the Plains (Figure 1–3), which differ in much the same manner as the local topography as far as grass growth is concerned. The Serengeti Plains of the southeastern portion have low rainfall, a dusty volcanic soil, and huge expanses of short

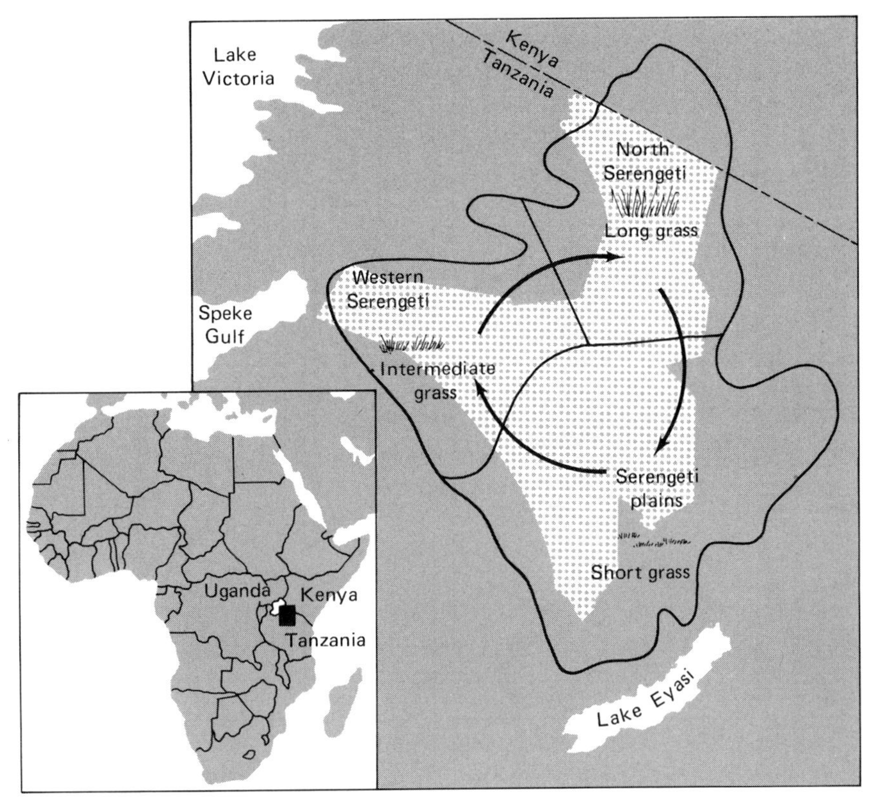

Figure 1–3 The Serengeti National Park (shaded region) contains 5600 square miles of grassland and lies along the northern border of Tanzania within the larger Serengeti Ecological Unit, which covers 9000 square miles. The three ecological divisions and seasonal ranges of migratory populations in this Serengeti region are outlined. The herds spend the wet season (March through May) on the Serengeti Plains, early dry season in the Western Serengeti, and late dry season in the North Serengeti. (Adapted from R. Bell, 1971, *Scient. Amer., 225:* 86–93.)

grass. During the wet season, immense herds of mixed grazing species concentrate here. At the start of the dry season, migrating grazers move into the Western Serengeti, where rainfall is greater and grass grows to an intermediate height. When conditions begin putting stress on the larger nonruminants in the late dry season they move again, this time to the North Serengeti, the location with highest rainfall and longest grasses. Wildebeests follow next, and finally Thomson's gazelles. As in localized seasonal movements, then, these huge migrations result from a grazing succession, ordered ultimately by cellular characteristics of the herbivore's choice of food (to which it adapted through evolutionary change), the pattern of rainfall, and complex interactions among ecologically associated populations (Fig-

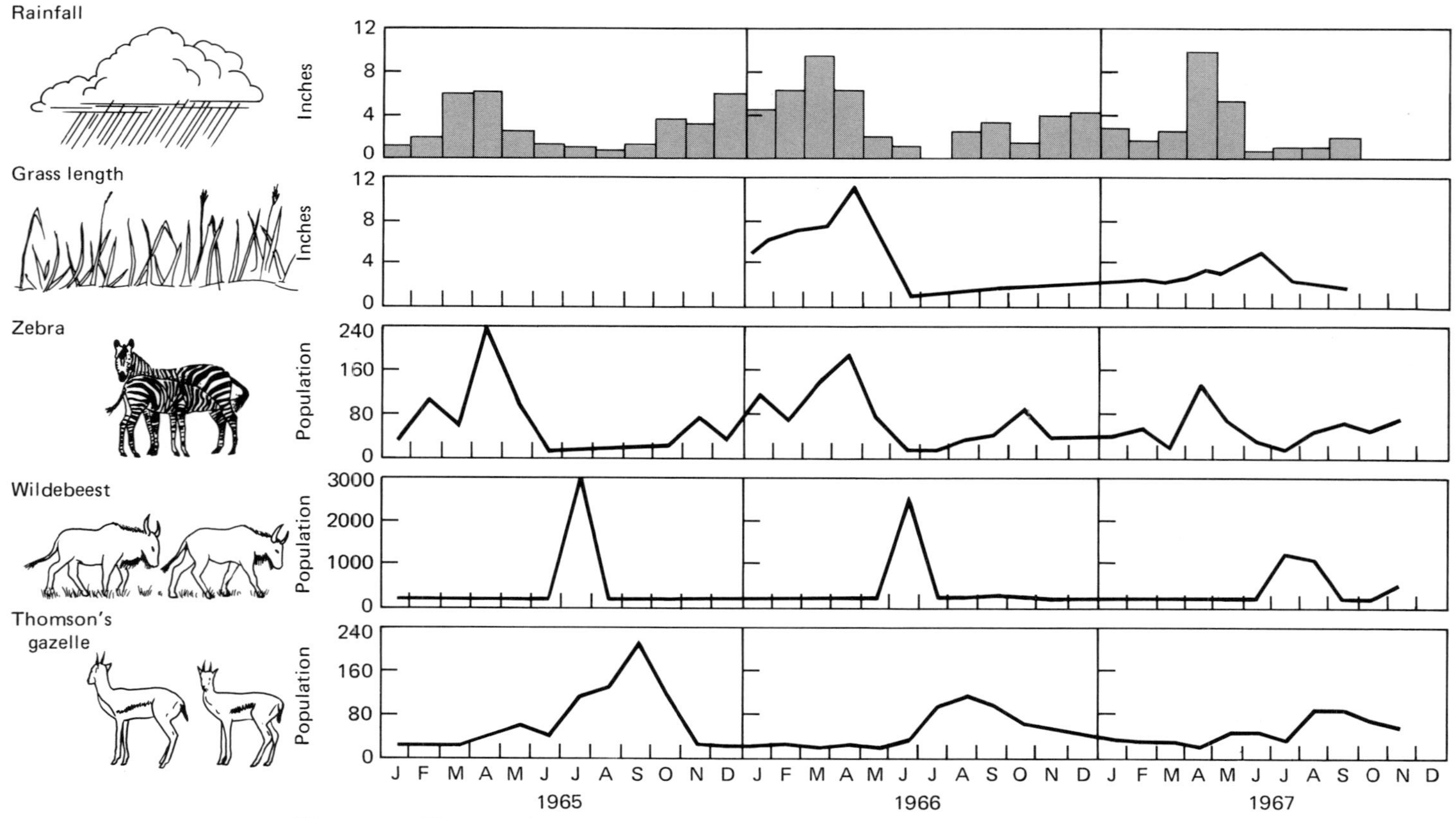

Figure 1–4 The population sizes of migrating species of grazing ungulates in relation to rainfall and height of grass over a three-year period in the Western Serengeti. Daily transect counts were made in a strip 3000 yards wide and half a mile long. The zebra, wildebeest, and Thomson's gazelle visit this western region in a regular grazing succession each year in the early dry season. (Source: R. Bell, 1971, *Scient. Amer., 225:* 86–93.)

ure 1–4), yet reflected in the most readily observable sense in the massive migratory movements of the populations of the Serengeti.

The population biologist is confronted constantly with problems as complex and fascinating as those of wildebeests and the other grazing mammals of the Serengeti. Whether he works with laboratory populations of *Paramecia* and bacteria, counts Antarctic populations of sperm whales, or creates computer models of isolated processes in order to better understand their role in the whole population, his aim is the same: to achieve an interdisciplinary approach to problems of biology at the population level, and from these studies to develop a predictive understanding of the functioning and biological relationships of aggregations of organisms. In the very broadest sense, the field of modern population biology may be defined as including *all aspects of groupings of organisms and organisms in groups* (Ehrlich and Holm, 1962). It is not ecology alone or even population genetics, but the integrated study of the behavior, ecology, genetics, and evolutionary biology of populations and the unique phenomena that they exhibit. The applied aspects of these studies are obvious, ranging from human population problems to pest-control programs. The seemingly inexhaustible richness of the patterns in which organisms are related in time and space will yield an exciting level of growth in basic research on population biology long into the future.

Before we delve into the meat of our subject and begin a discussion of the factors that affect the structure and biological character of populations, let us review some basic principles of ecology and related background factors that we should keep in mind for the study of populations. If you have had a strong basic course in ecology, you may still wish to read the remaining sections of this chapter for an overview and to refresh your memory on certain topics. As we saw in our brief introductory discussion of the population biology of three of the African herbivores on the Serengeti Plains, the population and its environment exist as an inseparable unit. A host of factors need to be considered in analyzing population phenomena, many of which are concerned with the relationships of the organisms to cycles in the physical environment[1] and to the patterns of energy flow in ecosystems. From these considerations, we shall look in later chapters at the importance of the ecological niche concept and the influence of environment and biotic interactions within populations, in controlling patterns of dispersion, dispersal, and general population structure.

THE CIRCULATION OF MATERIALS

The cycling of chemical materials through an *ecosystem*, an ecological unit that includes the physical environment and its component plant

[1] Seasonal cycles of climate and weather phenomena are taken up in Chapter 10.

and animal populations, is part of what Clark et al. (1967) have called the "life system" of a population. A *life system* is composed of a subject population and its *effective* environment, that is, the part of an ecosystem that determines the existence, abundance, and evolution of a particular population. This definition includes biotic as well as abiotic agencies influencing the population. The flow of materials and the flow of energy through the constituent populations (or biological community) of an ecosystem are really inseparable, as the intake of materials coincides with the extraction or use of energy. Nonetheless, there is a fundamental difference in that the flow of materials tends to be *cyclical* with a particular atom being recirculated through the ecosystem many times, whereas the flow of energy is *unidirectional,* and energy is lost to the community once it has passed through its respective parts.

About 40 of the ninety-odd elements that occur naturally on the earth are required by living organisms. In quantity alone, the most important of these elements are carbon, hydrogen, oxygen, nitrogen, and phosphorus, but equally essential are sulfur, calcium, potassium, chlorine, magnesium, iron, iodine, fluorine, boron, zinc, and molybdenum. Some, such as magnesium and iron, are most frequently used as just single ions in the centers of complicated chlorophyll, hemoglobin, or cytochrome molecules. But all are essential for maintaining life. These chemical elements are not destroyed upon the death of an organism, but tend to be used over and over again by flowing from organisms to the environment and back to the organisms again in more or less circular paths. These circulatory pathways are termed *biogeochemical cycles* because they involved the *biosphere* (a term used to refer to all the living organisms on earth), *geological* components such as water, rocks and soil, and simple *chemical* elements moving repeatedly between inorganic geological forms and organic molecules in living creatures. A summary of the overall pathways of biogeochemical cycles on earth, shown in Figure 1–5, points out that they involve atmospheric components as well as soil minerals, and that the cyclical movement of water between the organic and inorganic worlds is more or less tied to the basic cycling of energy and nutrients from the photosynthetic plants to the consumer animals to the decomposer bacteria and other organisms. This representation also reminds us that the entire earth taken as an ecosystem is driven by an outside source of light energy, the sun.

The transfer of elements through a biogeochemical cycle may take place at considerably different rates, depending on the portion of the environment acting as a *reservoir* for the element. The *gaseous cycles* are more nearly "perfect" than other biogeochemical cycles in that the materials in circulation do not become inaccessible to organisms over long periods of time. Gaseous cycles include the *carbon, nitrogen, ox-*

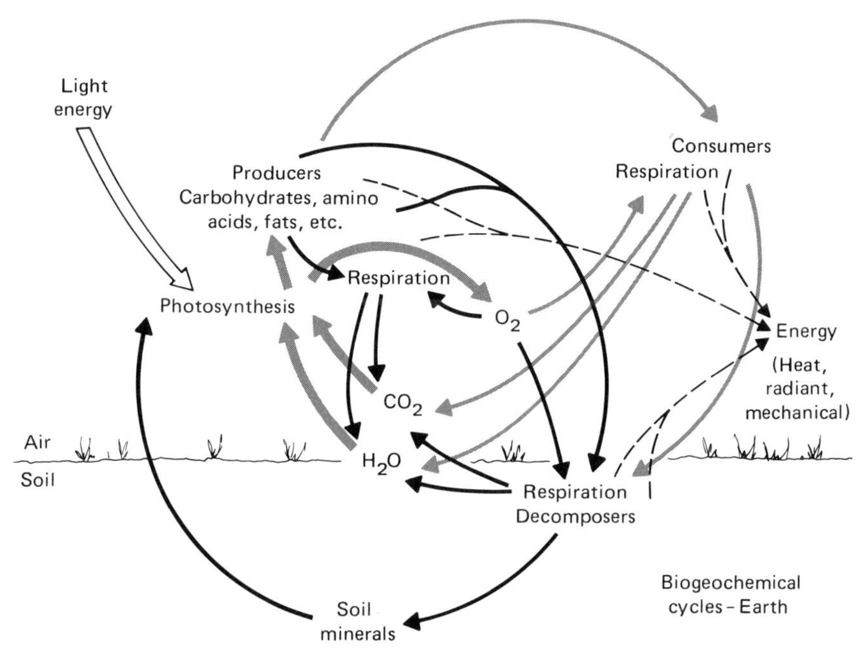

Figure 1–5 Biogeochemical cycles on earth. Cycles of carbon, oxygen, hydrogen, soil minerals, and energy through the living organisms of the earth's biotic communities are indicated (nonbiological transformations are not shown). Solid lines show essential transformations between producers and decomposers. Light lines indicate that consumers are not theoretically essential to a cycling of elements. Dashed lines show energy transfer from input as light energy to expenditure. Forked lines indicate that not all energy is lost through respiratory processes, but that some may enter the external environment via other metabolic and physical functions of chlorophyll-containing plants. Consumers include the animals and many parasitic and saprophytic plants. Decomposers are primarily microorganisms, although it is sometimes technically difficult to draw a line between them and certain consumers. Water is indicated at the soil–air boundary because it usually enters metabolic reactions by first being taken up through plant-root systems. (Source: Frank B. Salisbury, 1962, *Science, 136:* 24. Copyright by the American Association for the Advancement of Science.)

ygen, and *hydrogen cyles.* The earth's atmosphere serves as the principal inorganic storage reservior for these four elements. Because they constitute over 97 percent of the bulk of protoplasm (living matter) and over 99 percent of the atmosphere, these elements must be moved about in tremendous quantities. They cycle quite easily because the reservoir is the gaseous form of the element in the atmosphere.

Sedimentary cycles involve the remainder of the approximately 40 elements vital to the biotic world, and are said to be less perfect cycles because part of the element is retained in inaccessible chemical forms or is buried in geological formations for an extended period of time.

Once these elements have been removed through erosion and leaching from terrestrial ecosystems, they tend to pass literally downhill in a dissolved state, eventually reaching marine sediments where they may be lost to the living world for considerable periods of geologic time.

Let us look at several of these key biogeochemical cycles now, in order to better understand their effects upon plant and animal populations, whether in the relatively simple ecosystems of the savanna or in the extraordinarily complex world of the rain forest.

GASEOUS CYCLES

Carbon cycle

Carbon is found in all organic compounds, and because the carbon-containing carbohydrates and fats are especially important in the transfer of energy from one organism to another in the biotic world, carbon atoms move simultaneously with the flow of energy in ecosystems and populations. The principal inorganic reservoir of carbon is free carbon dioxide in the atmosphere and dissolved carbon dioxide in water (Figure 1–6). Photosynthesis in green plants uses light energy and this carbon dioxide, in combination with oxygen and hydrogen from water, to manufacture carbohydrates, which include carbon atoms as a basic molecular "backbone." The plant can convert these basic sugar compounds into more complicated fats and carbohydrates and eventually into proteins. When the plant respires, some of this stored carbon is released back into the environment as waste carbon dioxide. If the plant is eaten by a herbivore, the stored carbohydrates, fats, and proteins are broken down in the animal's digestive system and body cells, and the carbon atoms are resynthesized into still different carbon compounds. Carnivores feeding on the flesh of herbivores will again break down and assimilate these carbon compounds. These various animals will also release carbon dioxide regularly into the atmosphere as a product of their cellular respiration. Finally, both plants and animals die and decomposer organisms break down the organic molecules of the cells, releasing the protoplasmic carbon into the environment (mostly as CO_2). Under certain conditions, part of the organic carbon is preserved in fossil fuels such as peat, coal, oil, and natural gas, or in limestone and coral reefs. Combustion and weathering return carbon from these deposits to the atmosphere. The key factors throughout the cycle, however, are the actions of living organisms, especially the photosynthetic plants in moving carbon from its atmospheric reservoir to organic compounds in protoplasm, and the actions of decomposers in reversing this process.

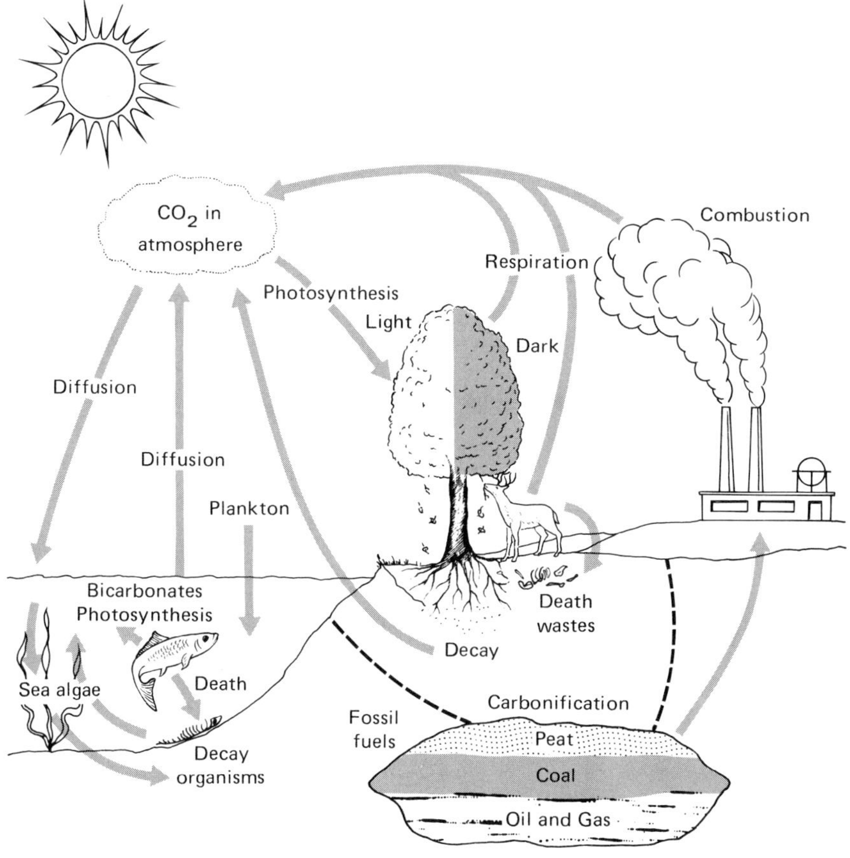

Figure 1–6 The carbon cycle, a gaseous biogeochemical cycle. (Source: Robert M. Chute, 1971, *Environmental Insight*, Harper & Row, New York, p. 45. After R. L. Smith, 1966, *Ecology and Field Biology*, Harper & Row, New York, p. 49.)

Nitrogen cycle

Nitrogen is an essential component of protein and nucleic acid molecules in plants and animals. About 78 percent of the earth's atmosphere is nitrogen gas (N_2), the largest gaseous reservoir of any element. Yet in the gas form nitrogen is useless to most organisms. The transfer of inorganic N_2 to a form usable by green plants and other organisms (Figure 1–7) is accomplished by the nitrogen-fixing bacteria and algae found in soils and aquatic habitats. Nitrogen-fixing bacteria in the genus *Rhizobia* live in the root of nodules of legumes, whereas others such as species of *Clostridium* and *Azotobacter* are free-living in the soil. These microorganisms can convert N_2 into ammonia (NH_3) or nitrates (NO_3), which are taken up by green plants, where the nitrogen is used in amino acid and protein synthesis. Nitrates can

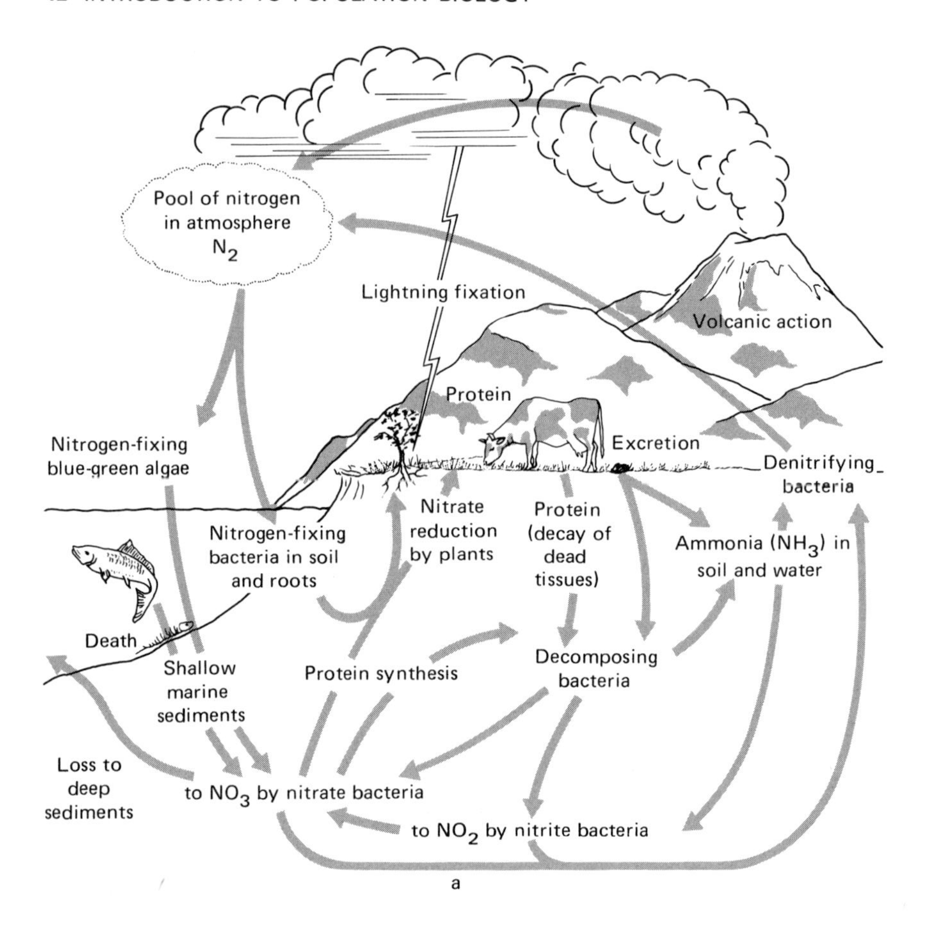

a

also be formed by electrification, where the energy of a lightning bolt passing through the atmosphere binds nitrogen and oxygen together into nitrates, which precipitate onto the soil from the air during thunderstorms.

Once the nitrogen is taken up by a green plant and is utilized in protein synthesis it can be passed along through herbivorous animals to carnivores as in the carbon cycle. Nitrogen is excreted by these animals in the form of ammonia, urea, uric acid, or other nitrogenous compounds, and decomposing bacteria and fungi break down the proteins in the dead bodies of plants and animals to release nitrogen first in the form of amino acids and then ammonia. Nitrite bacteria convert this NH_3 gas into simple nitrite (NO_2) compounds, and nitrate bacteria can add a third oxygen atom to nitrites to produce nitrates (NO_3). Green plants can pull these inorganic nitrogen compounds from the soil again to use in protein synthesis, or denitrifying soil bacteria may return nitrogen directly to the air at this point (Figure 1–7b). Loss of nitrogen from the cycle occurs at the nitrate stage, wherever erosion

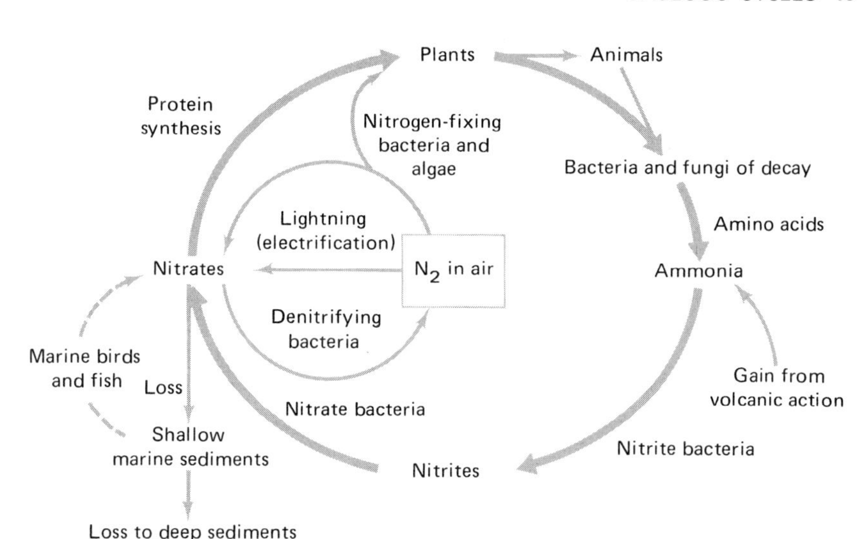

Figure 1–7 The nitrogen biogeochemical cycle. (a) Overview of the biological and environmental components and pathways of this complicated gaseous cycle. (Source: P. R. Ehrlich and A. H. Ehrlich, 1972, *Population, Resources, Environment,* Freeman, San Francisco, p. 200.) (b) The optimal pathways for reciprocal circulation of nitrogen between its reservoir and its various molecular forms, and mode of loss of nitrates to deep marine sediments. (Source: T. C. Emmel, 1973, *An Introduction to Ecology and Population Biology,* Norton, New York, p. 17.)

and leaching of soils by rain carry nitrates into streams and eventually into the oceans. Marine birds and fishes can return these nitrates to land through daily or migratory movements, or commercial harvesting of fish as food or of bird guano deposits as fertilizer.

Phosphorus cycle

Phosphorus is a key element in metabolic energy transfers, the chromosomal hereditary material DNA, and in bones, as well as other tissues and cellular molecules. This element is also a classic example of a sedimentary cycle, and because of its frequent deficiency it may limit biological productivity in some areas. The principal reservoir for the cycle is phosphate rock laid down in sedimentary beds during past geologic ages (Figure 1–8). Water erosion dissolves phosphate out of the sedimentary reservoirs, forming a phosphorus pool in the soil. Here the plants absorb phosphorus through their roots and use it in cellular syntheses. Animals acquire the element from plants, and both groups of organisms return phosphorus to the nutrient pools in the soil via normal excretion of waste products or through decomposition upon death.

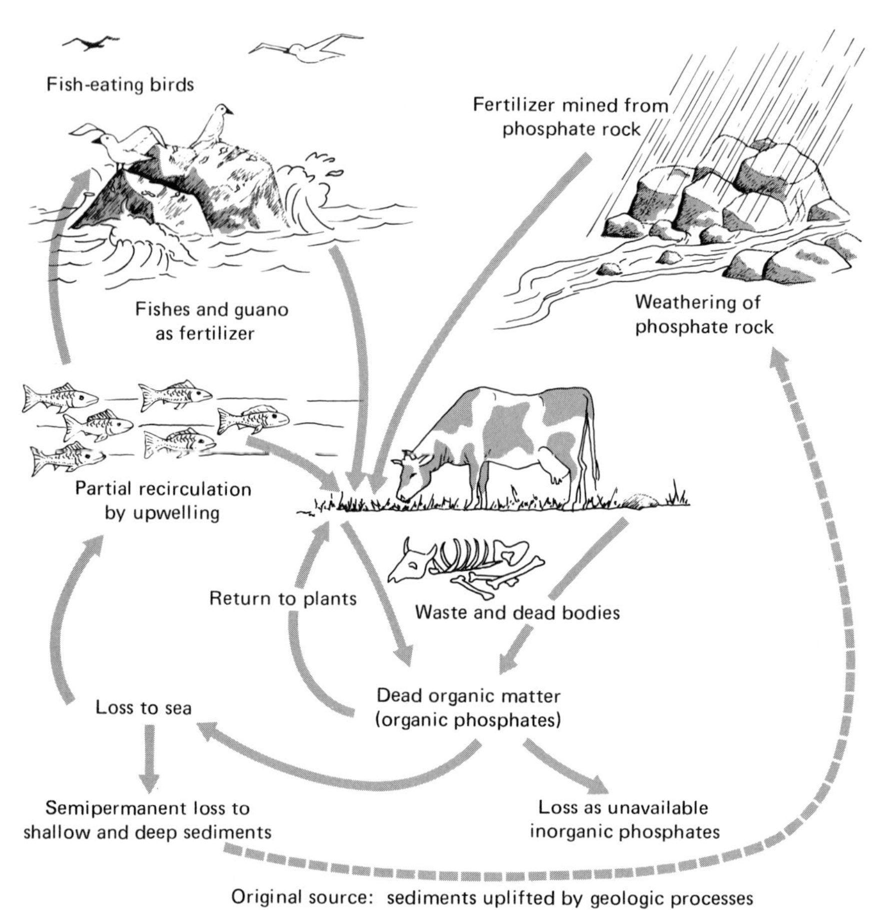

Figure 1–8 The phosphorus biogeochemical cycle, with a sedimentary reservoir. (Source: P. R. Ehrlich and A. H. Ehrlich, 1972. *Population, Resources, Environment,* Freeman, San Francisco, p. 201.)

The principal loss from the cycle comes with downhill transport of dissolved phosphorus into shallow marine sediments. As in the nitrogen cycle, some of this phosphorus is returned to land from the sea by fish-eating birds and migratory fish or by fish brought ashore by man. These marine vertebrates ultimately acquired the phosphorus from the microscopic plankton, which incorporated phosphorus into their bodies from the estuarine waters and shallow coastal sediments where their early development took place. But underwater currents and geological subsidence carry most of the phosphate compounds to deep marine sediments, where the lack of sunlight, low temperatures, and high pressure prevent the extensive growth of plankton. Thus, recycling of phosphorus from the deep ocean floor does not occur until seismic upheaval in a later geologic age.

Calcium cycle

The transfer of calcium through an ecosystem is also an example of a sedimentary cycle. Like phosphorus, calcium is a key element in bones, teeth, and a number of biologically important molecules. It is instrumental in maintaining proper osmotic balance in plant and animal cells, and plays a vital ionic role in neural transmission in animals. The principal reservoir of calcium is in the soil, where the element is released slowly by weathering (Figure 1–9). In small watersheds carefully studied by Likens and Bormann (1972) in Connecticut it was found that about $\frac{1}{14}$ of the available calcium pool is cycled through the vegetation each year. As summarized in Figure 1–9, about 570 kilograms (kg) of calcium per hectare (ha) are held by the vegetation and 1740 kg calcium/ha in organic debris. Some 690 kg/ha are available as nutrients in soil water and on exchange surfaces, and the soil and rock minerals contain about 28,550 kg calcium/ha. The trees and shrubs in the vegetation annually take up

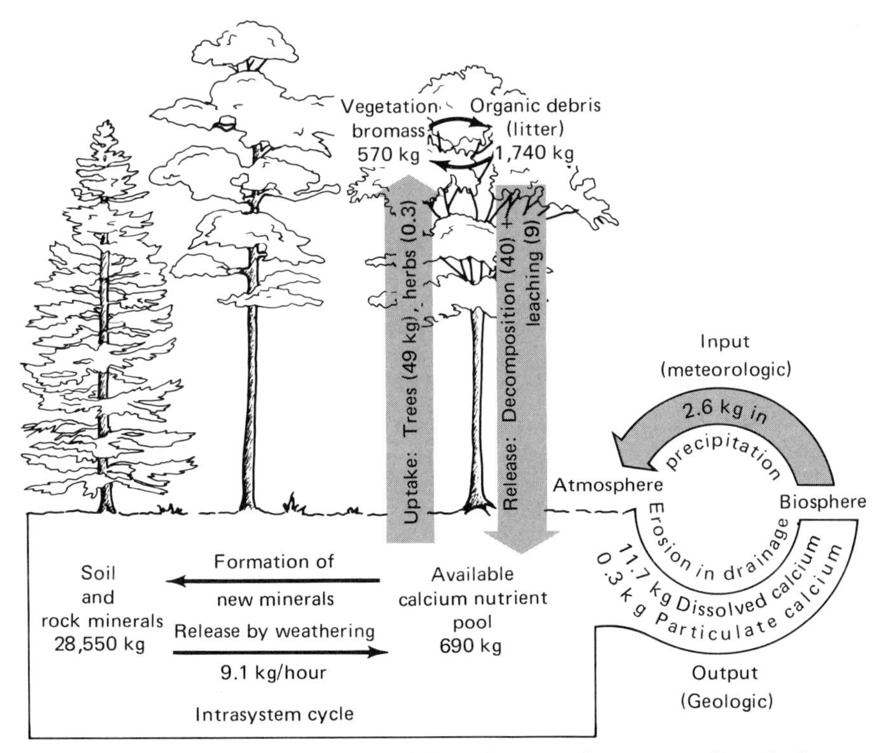

Figure 1–9 The major parameters of the calcium cycle in an undisturbed northern hardwood ecosystem in central New Hampshire. (After G. E. Likens and F. H. Bormann, 1972, in J. Wiens, ed., *Ecosystem Structure and Function,* Oregon State Univ. Press, Corvallis.)

slightly more than 49 kg/ha, whereas about 49 kg/ha are released by decomposition and leaching.

Looking at the relationship of this small watershed ecosystem with the surrounding world, precipitation brings in an average annual input of 2.6 kg/ha; the average annual output of calcium from erosion is 12.0 kg/ha, of which 0.3 kg/ha is lost to the surrounding biosphere as particulate matter and 11.7 kg/ha is lost in dissolved form in the stream water. Thus, this temperate ecosystem is efficient in retaining and circulating calcium, for on the average the net loss caused by weathering is only 9.1 kg calcium/ha per year. This loss represents only about 1.3 percent of the total available calcium (690 kg/ha) in the soil nutrient pool in the ecosystem, and about 18 percent of the amount circulated each year by the vegetation through uptake and release. As in the phosphorus biogeochemical cycle, streams transport

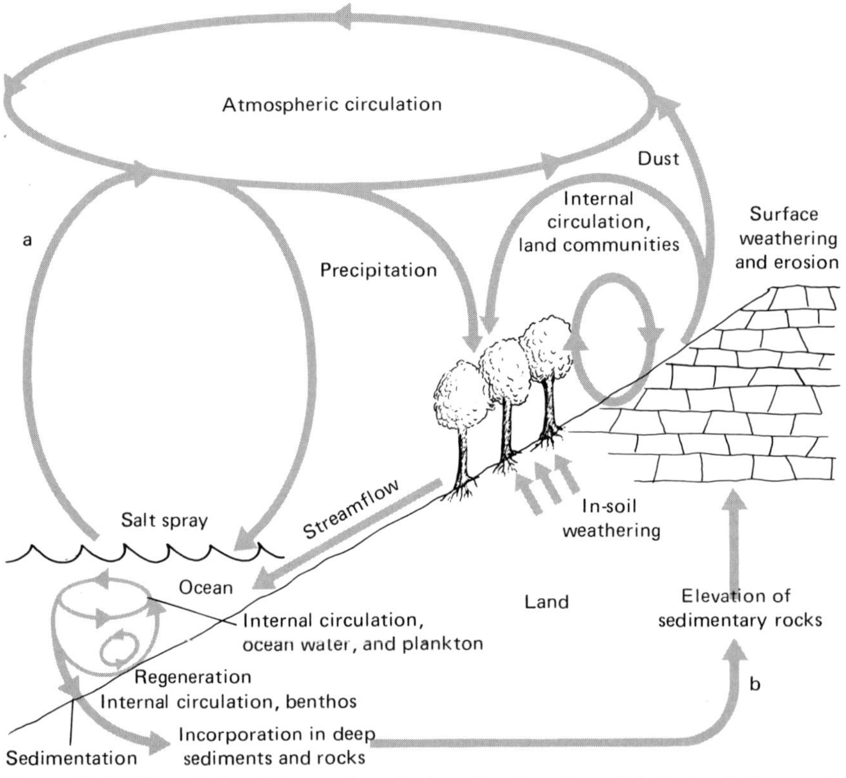

Figure 1–10 The calcium biogeochemical cycle, showing global circulation and circulations within terrestrial and local ecosystems. The shorter term cycles (a) involve salt spray carrying calcium into the atmosphere and back to water or land by preciptation and streamflow. The longer term cycle (b) involves geologic movement of marine sediments and uplift of rocks to exposure by weathering and streamflow again. (Source: Robert H. Whittaker, 1970, *Communities and Ecosystems,* Macmillan, New York, p. 122.)

weathered calcium to the ocean sediments, from which they are very slowly recirculated via future geological uplift (Figure 1–10).

ENERGY FLOW IN ECOSYSTEMS AND POPULATIONS

Just as materials flow in a regular pattern through ecosystems, energy moves in a definite rite of passage. Yet, as previously emphasized, this energy flow is unidirectional and at the end energy is lost from the biological community. In the typical freshwater marsh (Figure 1–11), there is a massive daily input of radiant energy from the sun, varying in quantity somewhat from season to season or with cloudy weather but approximately the same over the years. The green plants capture this energy in chemical bonds through photosynthesis and the potential energy (energy at rest, but capable of performing work) in these bonds is available when the bonds are broken elsewhere in the

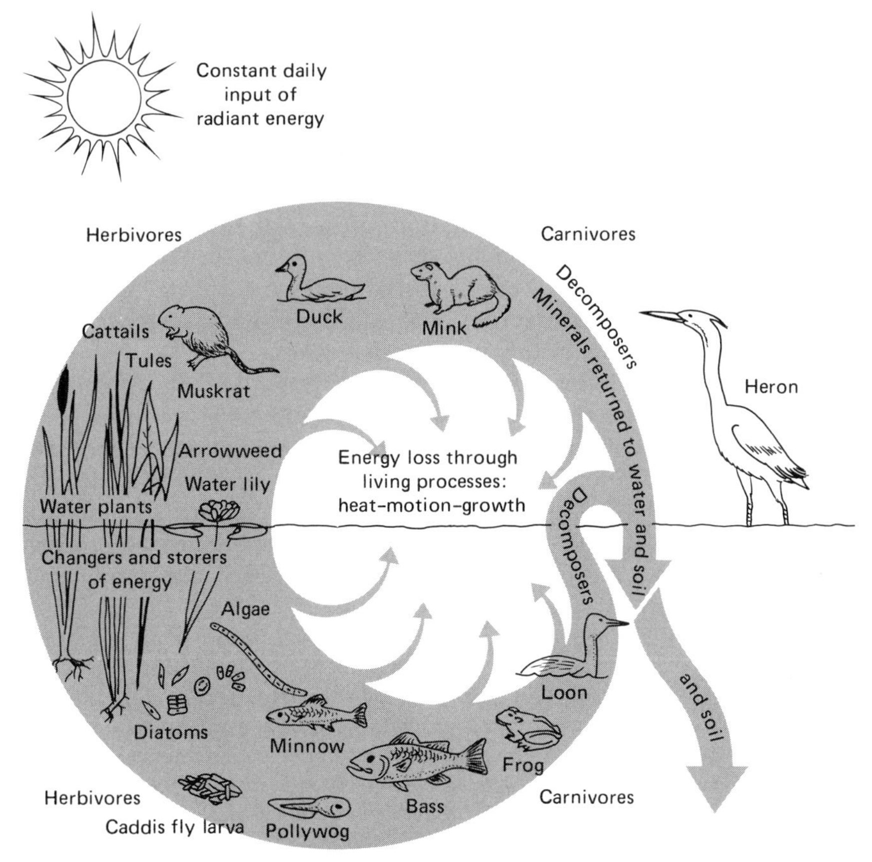

Figure 1–11 Energy flow patterns through a typical freshwater marsh. (Source: Elna Bakker, 1971, *An Island Called California,* Univ. of California Press, Berkeley, p. 143.)

cell or plant body for doing work (then energy is said to exist in its kinetic form, the energy of motion that results in work).

The behavior of energy in the universe is summarized by the two laws of thermodynamics. The first law, which deals with the conservation of matter and energy, states that the amount of energy in the universe is constant and hence that energy cannot be created or destroyed but only changed from one form to another. The amount of kinetic energy released by the breaking of a chemical bond in a glucose sugar molecule equals the potential energy present prior to the attack of an enzyme. When the energy was present in the form of a bond it could be classified as chemical energy; the kinetic energy released by the fracturing of the bond would be heat, another form of energy. The chemical energy was first made by the process of photosynthesis, which involved light, a third form of the same basic energy milieu of the universe.

The second law of thermodynamics emphasizes that this process of energy transfer and transmutation is not perfect. Although the total amount of energy involved in the reaction does not change, some useful energy (energy available for doing work) is converted into heat energy at every transformation of energy. This heat energy escapes to the surrounding environment. Hence, in our illustration (Figure 1–11) of energy flow in a freshwater marsh, a considerable quantity of energy is lost from the plants in the form of heat, the result of chemical reactions in growth and other living processes. At each later stage in the energy flow pattern, heat is lost during the transfer of energy among herbivores, carnivores, and decomposers, whether these feeding relations occur in the air or under water.

Because energy is lose from the biotic system and its constituent populations at every step, the ecosystem could only continue functioning with an outside source of energy. This, of course, is provided in essentially unlimited quantities (for our purposes) by the sun, but terrestrial ecosystems are notoriously inefficient as far as taking full advantage of this "free" solar energy. This occurs because of physical constraints as well as biological factors. Smith (1966) has pointed out that—

> approximately 57 percent of the sun's energy is absorbed in the atmosphere and scattered in space. Some 36 percent is expended to heat water and land areas and to evaporate water. Fortunately, energy released from decay prevents much dissipation of heat into the ground. Approximately 8 percent of light energy strikes plants. Of this 10 to 15 percent is reflected, 5 percent is transmitted, and 80 to 85 percent absorbed. Most of the absorbed energy is lost as heat in the evaporation of water. An average of only 2 percent (0.5 to 3.5 percent) of the total light energy striking a leaf, and this is restricted largely to the red end of the spectrum, is used in photosynthesis.

As we shall see shortly, the efficiency of energy transfer does not improve very much at later stages, either, in the overall pattern of energy flow through the plant and animal populations of an ecosystem.

Given these principles of energy transfer, how can we more precisely describe the pattern of energy flow between the component species populations of an ecosystem? Ecologists have expressed these patterns diagrammatically in the form of *food chains* and *food webs.*

The food web concept and trophic levels

A food web is simply the total set of feeding relationships among and between the species composing a biotic community. That these relationships may achieve considerable complexity is indicated in Figure 1–12, which shows part of an estuarine food web along the coast of Long Island. The bases of this food web are found (to the left in the figure) in organic debris, water plants, marsh plants, and microscopic plankton. A number of herbivorous insects, crustaceans, mollusks, and fish feed on these various sources of energy, and they, in turn, are fed on by organisms higher in the food web (shown on the right side). If one food source for the osprey, such as the blowfish, becomes rare or disappears for awhile, the osprey can feed more on the other prey species (eels, billfish) that are in the same nutritional relationship as the blowfish. Thus, the many interlocking *food chains* in the web, such as

water plant → mud snail → billfish → osprey

or

water plant → eel → osprey

contribute a measure of stability to the ecosystem. With many food chains and cross-connecting links, there is greater opportunity for the prey and predator populations in an ecosystem to adjust to changes. This diagram (Figure 1–12) of the Long Island estuary food web also depicts how the concentration of a persistent insecticide like DDT tends to be increased in food chains. Each species feeds on numerous smaller organisms that have picked up the chlorinated hydrocarbons in their environment, particularly if they are filter-feeders on the bottom and are constantly filtering water around them (contrast the concentration of DDT in the filter-feeding clam and the free-grazing mud snail). The chlorinated hydrocarbons are passed along readily in such food chains because of high solubility in body fat and their low solubility in water, and over an extended period, organisms at the end of a food chain build up huge concentrations relative to those in the surrounding environment.

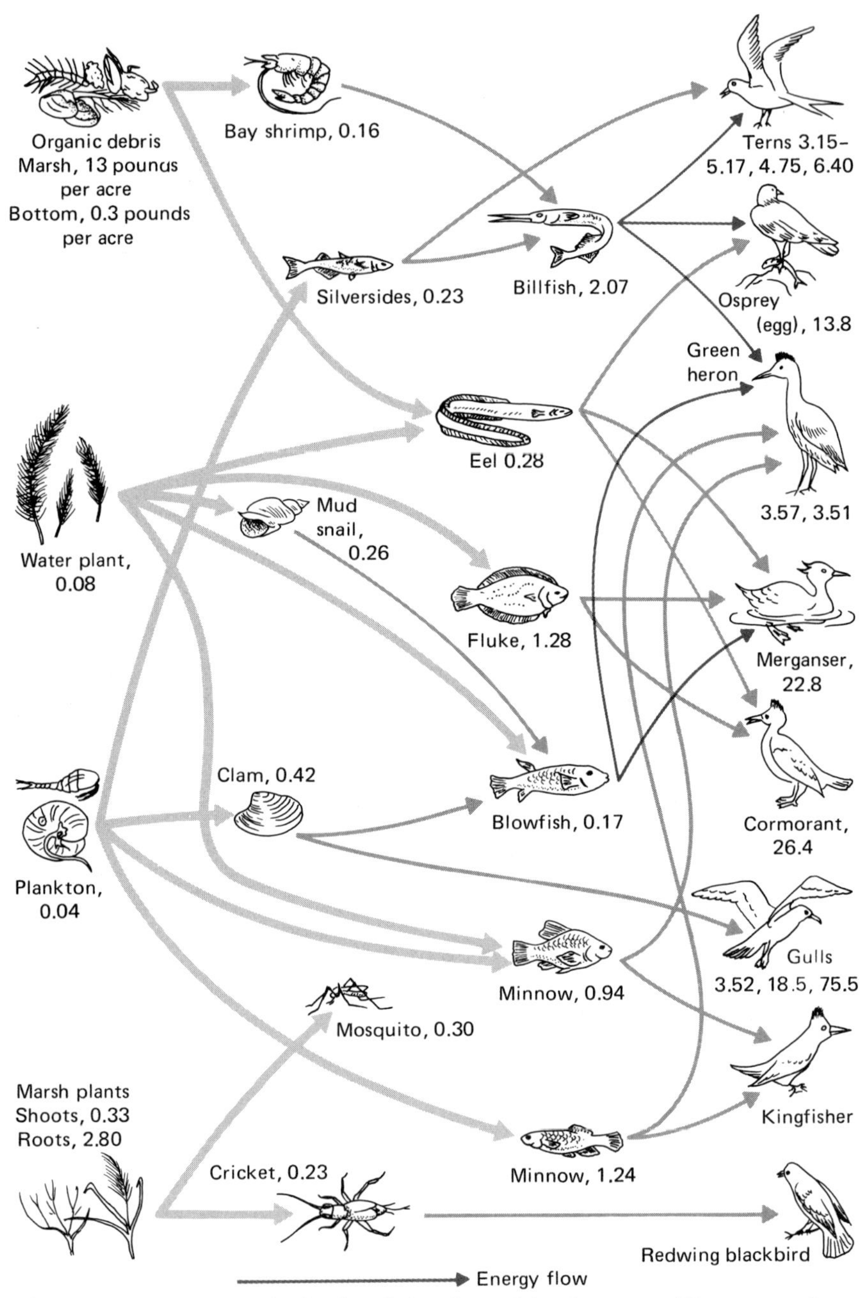

Figure 1–12 A portion of a food web in a Long Island estuary. The arrows depict patterns of energy flow, and the numbers indicate the parts per million of DDT found in each kind of organism in the 1960s by George M. Woodwell and his coworkers. (Source: P. R. Ehrlich and A. H. Ehrlich, 1972, *Population, Resources, Environment,* Freeman, San Francisco. From G. M. Woodwell, 1967, *Scient. Amer.*)

At each step in the food chain, a considerable portion of the potential energy being transferred in the food is lost as heat. Thus, the longer the food chain, the more restricted the amount of energy that will reach the terminal members. As a result, we rarely find food chains of more than four or five steps in natural situations.

All organisms that share the same general source of nutrition are said to be at the same *trophic level.* The feeding level concept implies that these organisms obtain their energy through the same number of steps from the ultimate source—the sun—in a food chain. The first trophic level in ecosystems is represented by green plants and comprises the *producers* (or autotrophs), organisms that convert light energy from the sun into chemical-bond energy via photosynthesis. The remaining heterotrophic components of the biotic community depend on the autotrophic producers to fix light energy and manufacture carbohydrate compounds from simple inorganic substances. The second trophic level is that of the primary consumers, or *herbivores*—organisms that eat plants and convert potential energy in plant tissues into animal tissue. As we have seen in the Serengeti grazing fauna, the herbivores are adapted to handling considerable cellulose in their diet. The *carnivores,* or flesh eaters, make up the third trophic level. These secondary consumers or first-level carnivores feed directly upon the herbivores. If other carnivores feed on these first-level carnivores, like a hawk feeding on a weasel, which in turn has fed upon herbivorous mice, then those carnivores are called second-level carnivores. The terminal carnivore in a food chain represents the *top carnivore,* usually being in the fourth or fifth trophic level (tertiary or quaternary consumers) of a community. Naturally, some animals feed on several different trophic levels. A black bear (*Ursus*) may feed on plant berries and then catch and eat a ground squirrel. These consumers belonging to more than one trophic level are called *omnivores.* The *decomposers,* organisms such as bacteria, yeasts, and fungi that feed on dead organic matter, consume plants and animals from all trophic levels. They are instrumental in breaking down dead plant and animal bodies into soluble organic and inorganic molecules which can be recycled.

The trophic structure of a food web in coral reefs of the Marshall Islands in the Pacific is shown diagrammatically in Figure 1–13. The feeding habits of over 200 species of reef fishes were classified into five trophic groups (Hiatt and Strasburg, 1960), which were rather complexly related to each other and to the plankton and algae producers.

A different emphasis by many ecologists recently has been to quantify energy flow in ecosystems and the ecological efficiencies of these various trophic levels and their constituent populations. Thus, in a pioneering study on an aquatic Florida ecosystem, H. T. Odum

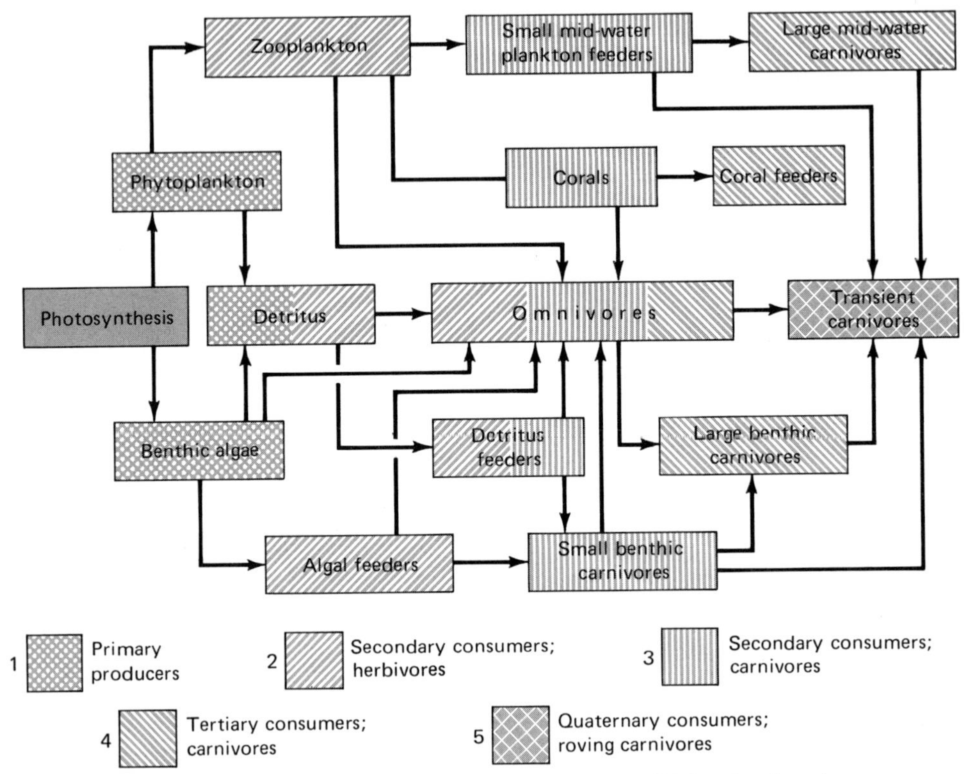

Figure 1–13 The food web of coral reefs in the Marshall Islands, showing the trophic structure in a qualitative manner. (After R. W. Hiatt and D. W. Strasburg, 1960, *Ecol. Monogr., 30:* 65–127. Source: D. R. Stoddart, 1965, *Geography, 50* (228): part 3.)

(1957) measured the quantity of energy captured by each trophic level over a year's time. He then determined the ecological efficiencies of energy transfer by looking at the ratios between energy flow at different points along the food chain of trophic levels. His study is summarized in Figure 1–14. At this particular location in Florida, Odum found that 1,700,000 kilocalories (kcal) (the most common unit of thermal energy used by ecologists[2]) fell on each square meter of surface at Silver Springs, Florida over a 365-day year. Of this potentially available quantity, only 20,810 kcal of energy per year were captured by photosynthesis in the green-plant producers present in each square meter. By definition of ecological efficiency of the producer level, only

[2] The small calorie (cal) is used when small quantities of energy are involved and represents the amount of heat needed to raise 1 g (or 1 ml) of water 1°C. One kilocalorie (kcal) represents the quantity of heat required at a pressure of 1 atmosphere (sea level) to raise the temperature of 1000 g of water 1°C.

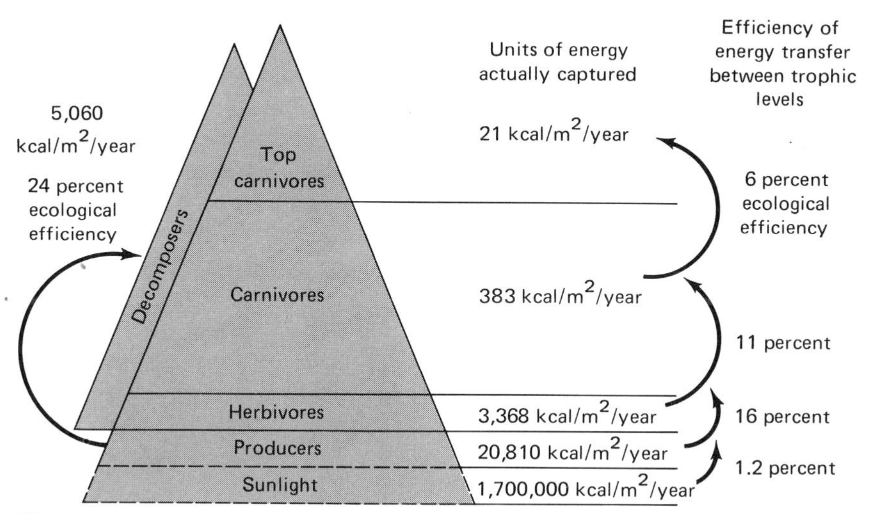

Figure 1–14 The ecological efficiencies of the trophic levels at Silver Springs, Florida. (After T. C. Emmel, 1973, *An Introduction to Ecology and Population Biology*, Norton, New York, p. 34.)

1.2 percent of the solar energy present (20,810/1,700,000 kcal) was actually trapped by chlorophyll molecules and converted into organic compounds. The low efficiency is perhaps not unexpected, as only about one-half of the sun's energy is represented by the usable spectrum, 4000 to 7000 angstroms (Å), and the plants themselves are not organized for total efficiency (leaves or algal bodies not covering all available surface area, light lost between chlorophyll molecules or in cell-wall material, and so forth).

Continuing up through the trophic-level diagram, herbivores consumed and utilized some 3368 kcal of the 20,810 kcal available over the course of a year in the plant material found in an average square meter. This represents a 16 percent efficiency (84 percent was lost to the environment), a partial improvement over producers because herbivores can move around to feed. In the two carnivore levels at Silver Springs, Florida, efficiency of energy utilization remains at a relatively low level, although the searching behavior of predators greatly increases their chances of finding suitable energy (in prey) compared to sessile plants. By comparison, the decomposers, feeding on all trophic levels, are quite efficient and they manage to utilize about 24 percent of the potential energy originally available in the chemical bonds of the producers.

With relatively thorough coverage of most portions of producers by grazing species in the African savanna, the total ungulate population in the Tarangire Game Reserve in Tanzania is suggested by Wiegert

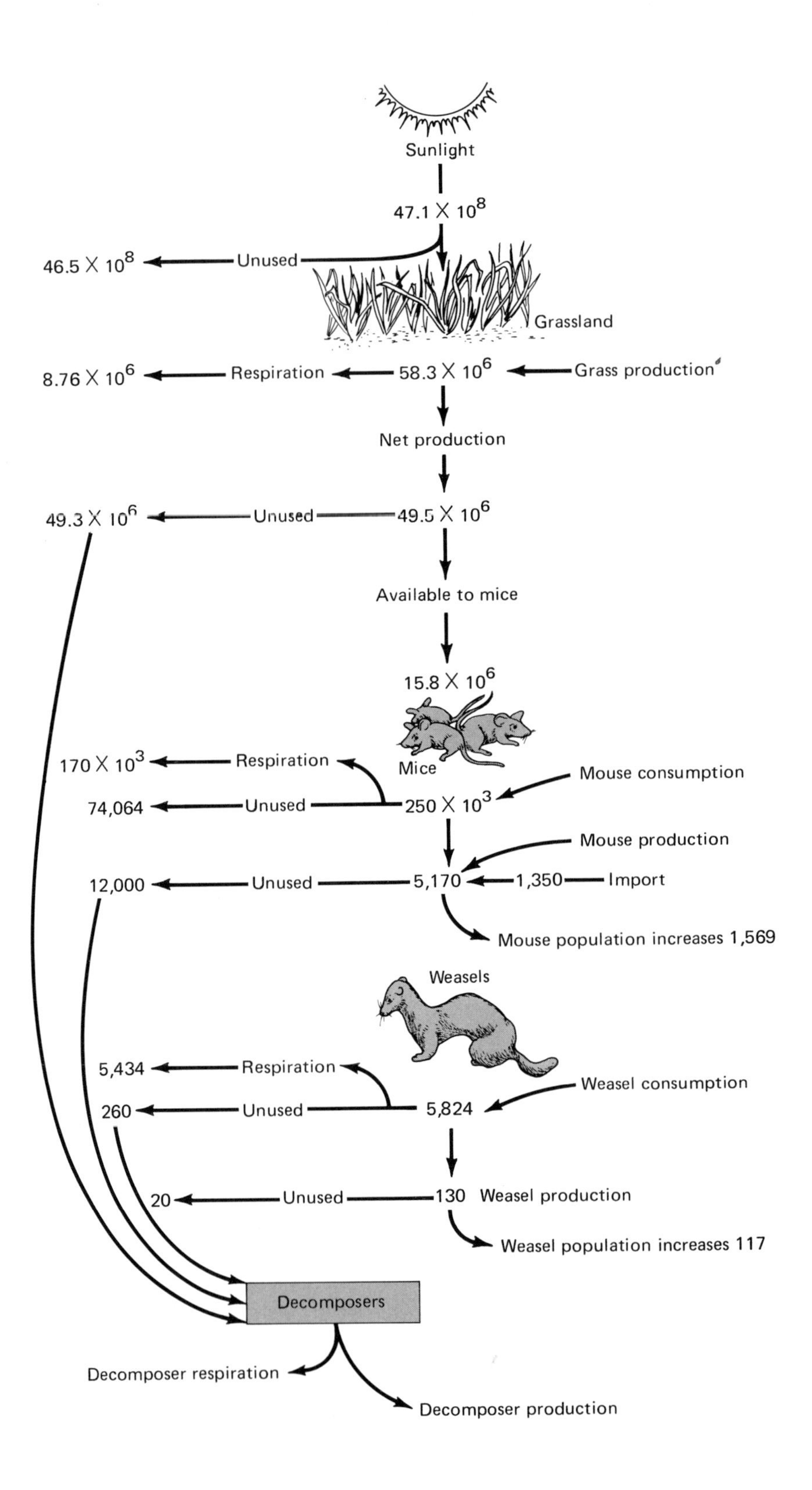
Sunlight
47.1 X 10^8
46.5 X 10^8
Unused
Grassland
8.76 X 10^6
Respiration
58.3 X 10^6
Grass production
Net production
49.3 X 10^6
Unused
49.5 X 10^6
Available to mice
15.8 X 10^6
Mice
170 X 10^3
Respiration
74,064
Unused
250 X 10^3
Mouse consumption
Mouse production
12,000
Unused
5,170
1,350
Import
Mouse population increases 1,569
Weasels
5,434
Respiration
260
Unused
5,824
Weasel consumption
20
Unused
130
Weasel production
Weasel population increases 117
Decomposers
Decomposer respiration
Decomposer production

and Evans (1967) to reach 28 percent efficiency as regards exploitation of primary production. In the Queen Elizabeth Park, Uganda, exploitation efficiency is said to be as high as 60 percent. There is considerable question as to the accuracy of these limits (Bourlière and Hadley, 1970), but they do suggest the possible extent to which stable natural grasslands can be utilized by mixed populations of grazing mammals.

Note again how energy transfers in a single food chain reflect the same types of efficiencies as are found in overall trophic-level exchange, and that this limits the number of steps in a food chain as well as population size in the terminal species. Figure 1–15 shows the energy-flow budget for a simple grass-meadow mouse-weasel food chain in an old-field community in southern Michigan (Golley, 1960). About 1 percent of the total solar energy received (47.1×10^8 kcal/ha) was converted into plant material. Of this 58.3×10^6 kcal of energy that went into production of plant tissue, some (8.76×10^6 kcal) was lost by respiration (15 percent) and much (49.3×10^6 kcal) was unused by the principal vertebrate herbivores in the habitat to be decomposed or eaten by plant-eating insects. About 2 percent of the actual gross grass amount (49.5×10^6 kcal) available to the meadow mice was actually consumed by them (250×10^3 kcal). The bulk of this energy (68 percent) was then lost through respiration or passed to decomposers upon death. Some went into production of new mouse tissue and even new mice. At the third step in the food chain, the weasels, which fed almost exclusively on meadow mice here, utilized 30 percent of the mouse biomass actually available (5824 kcal). Of this amount, fully 93 percent of the intake of energy was used in respiration (5434 kcal), and with unused energy lost to decomposers only 130 kcal (per hectare of weasel population area) went into production of new weasel tissue and new weasels. This would be clearly insufficient to feed and maintain a fourth trophic level (top carnivore) in this habitat, unless such a secondary carnivore could range over a tremendous number of hectares each year.

From the energy standpoint, it is obvious why the number of steps in a food chain is limited, and why the number of organisms involved in the populations through which this energy passes becomes smaller with each new link. The relationship between successive trophic steps can be shown graphically in the form of an ecological pyramid (Figure 1–16). C. E. Elton, a pioneering ecologist, emphasized (Elton, 1927) that the lower end of a food chain usually involved the most abundant species and that successive steps of carnivores contain fewer and fewer individuals, with finally one individual or a pair of top carni-

Figure 1–15 Energy flow through a plant–meadow mouse–weasel food chain in an old field community in southern Michigan. Figures are in kilocalories hectare. (After F. B. Golley, 1960, *Ecol. Monogr., 30:* 187–206.)

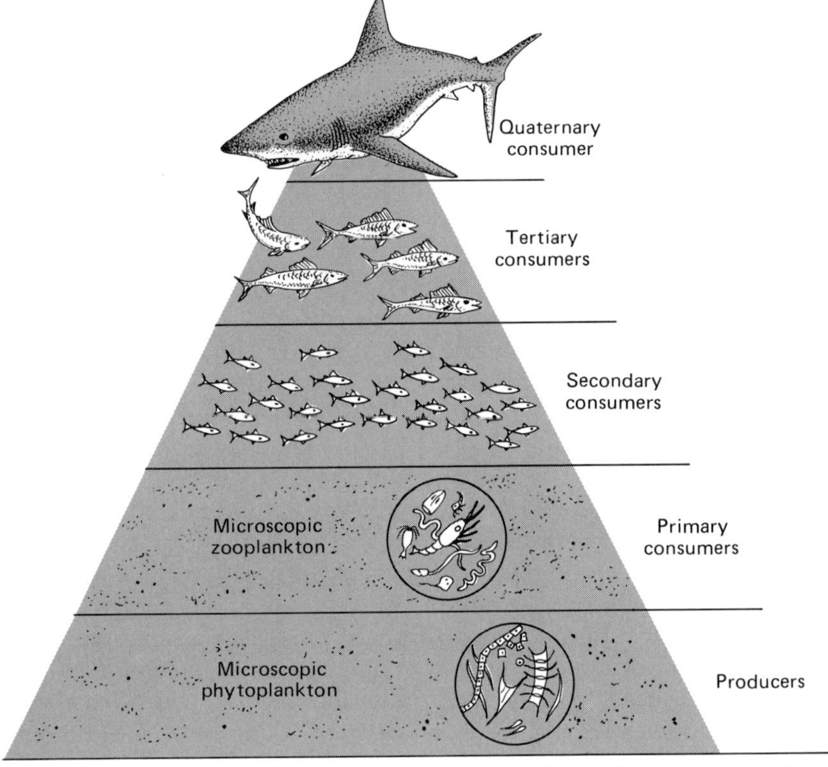

Figure 1–16 An ecological pyramid in the ocean. The higher the step in the food chain composing the pyramid, the fewer the number of individuals and the larger the size. (Source: A. S. Boughey, 1968, *Ecology of Populations*, Macmillan, New York, p. 99.)

vores covering a great area of hunting territory. This pyramid of numbers corresponds in shape with the pyramid of energy flow already examined (Figure 1–15). Pyramids of biomass can also be used to show the total bulk of organisms (by weight or other measure) present at any one time in each link. Thus, on the tropical African savanna, Bourlière and Hadley (1970) cite the biomass of ungulates in some national parks and game reserves as varying from 1.1 to 31.5×10^3 kg/km^2. The unusually high biomasses in this range are not abnormal densities that may occur around water holes during the dry season, but represent yearly averages in areas where about three-quarters of the standing crop biomass is represented by the elephant and the hippopotamus populations—two large, slow-growing and late-maturing species with slow turnover rates. Interestingly, in the rain forest at Barro Colorado Island in Panama, some 70 percent of the total mammalian biomass is maintained by the populations of *arboreal* species, especially those of foliage-eating sloths and howler monkeys (Eisenberg and Thorington, 1973). Determinations of biomass in populations of species that inhabit different areas and exploit different resources at various levels of specificity will undoubtedly contribute much in the future to our understanding of how food consumption and nutrient cycling affect the population structure and ecological success of particular species.

1 2 3 4 5 6 7 8 9 10 11

The Genetic Structure of Populations

The genetic side of population biology is concerned with the behavior of genes among the individuals that compose a population, and how the genetic constitution of the population influences the numbers and other properties of the group. Populations consist of an array of genotypes, and we know that in many species this array changes in both space and time. This characteristic of genetic plasticity makes evolution possible and permits adjustment by the species to cope with ecological changes in its environment. In a genetic sense, the *population* is commonly defined as a group of sexually interbreeding or potentially interbreeding individuals, because the fate of individuals and their genes determines the overall genetic constitution of the population. We must examine all factors contributing to transmission of genes within the population, including population size, the available array of genotypes among individuals in the population, and the abilities to survive and reproduce of individuals carrying specific genes. In this chapter, we shall examine the fundamental principles of population genetics and how gene frequencies, mutation rates, differential migration, chance, and selection are related in determining the genetic structure of populations. For a discussion of the many advanced ramifications of population genetics, the reader is referred to the texts on population genetics by Mettler and Gregg (1969), Li (1955), Wallace (1968), and Crow and Kimura (1970).

GENES IN POPULATIONS

To begin, let us consider a population as an assemblage of genes existing through time, rather than as a collection of individuals reproducing themselves generation after generation. An assembly of genes can be described in terms of the frequencies of their various alleles (gene frequencies). On the other hand, collections of diploid individuals must be described in terms of the frequencies of the various combinations of alleles present in each of them (zygotic frequencies), that is, the combinations of genes or alleles present in fertilized eggs and the resulting adults. It is easier to describe gene frequencies than zygotic frequencies for a population. As Bruce Wallace (1968) has pointed out, it is easy to enumerate the 52 cards of an ordinary playing-card deck. It is far more difficult to write down the 1326 possible pairs of these cards.

Also, of course, the gene frequencies among the haploid gametes will be related to the resulting diploid zygotic frequencies if the combination of gametes is random and left to chance. Genetically speaking, then, populations have two important attributes: *gene frequencies* and a total *gene pool* for the population.

Gene frequencies

Gene frequencies may be defined simply as the proportions of the different alleles of a gene present in a population. To obtain these proportions, we count the total number of organisms with various genotypes in the population and estimate the relative frequencies of the alleles involved. Recall, of course, that we are usually dealing with diploid organisms, and therefore there can be two forms of alleles or genes at any one locus. When both genes at this chromosome locus are the same, the organism is said to be homozygous. If they differ, the organism is said to be heterozygous. In a haploid organism such as a bacterium or alga, we would find only one gene at any particular locus. Were the plant or animal one of the rarer triploid organisms, there would be three genes at any one locus, that is, three copies of each chromosome type would be present, and therefore the genes located on them would be duplicated three times. Higher multiples of chromosomes (polyploidy) are also known in nature.

As an example of how to determine gene frequencies, let us consider a population of snapdragons in a garden. In snapdragons, red flower color R is incompletely dominant over white r with the heterozygous condition being pink, a distinct phenotype. Our population of 100 individuals contains 70 *RR*, 20 *Rr*, and 10 *rr* individuals. So in this population there are a total of 200 genes with respect to the gene locus that controls flower color. Of these, 140 + 20, or 0.8 of the total (160/200), are *R* genes, and 20 + 20, or 0.2 of the total (40/200) are *r* genes.

The same gene frequencies can also be calculated from the frequencies of the three *genotypes*, 0.70 *RR*, 0.20 *Rr*, 0.10 *rr*, according to the following formulas.

Frequency of R = the frequency $RR + \frac{1}{2}$ frequency Rr (i.e., $0.7 + 0.1 = 0.8$)

and

Frequency of r = the frequency $rr + \frac{1}{2}$ frequency Rr (i.e., $0.1 + 0.1 = 0.2$)

Let us now make this the general case for the calculations of gene frequencies based on the samples of individuals themselves. Assume again that both homozygous and heterozygous individuals can be recognized. Let the zygotic frequencies of *AA*, *Aa*, and *aa* individuals be *D*, *H*, and *R* for males and females combined.

The frequency of *A* (designated gene frequency p) is defined as

$$D + \tfrac{1}{2} H$$

and

The frequency of *a* (designated gene frequency q) is defined as

$$\tfrac{1}{2} H + R$$

We assume that *A* and *a* are the only alleles at this locus so that the sum of *D*, *H*, and *R* is 1.00; consequently $(D + \frac{1}{2} H) + (\frac{1}{2} H + R)$ is also equal to 1.00, and we can say $p + q = 1.00$; that is, the frequency of allele *A* plus the frequency of allele *a* equals 100 percent of the alleles for that gene in the population, as they are the only two alleles at the locus.

Gene pool

The gene pool is defined as the sum total of genes in the reproductive gametes of a population. It can be considered a gametic pool, that is, a pool of sperm and eggs, from which samples are drawn at random to form the zygotes of the next generation. Thus, the genetic relationship between an entire parental generation and the subsequent generation is very similar to the genetic relationship between a specific parent and its offspring. So we can now ask the specific question: How will the frequencies of genes in the new generation depend on their frequencies in the old generation? We could almost ask: How are these gene frequencies "inherited"?

Some early geneticists thought, for instance, that with the presence of *dominant* alleles, no matter what frequency you started out with, you would achieve a stable equilibrium frequency of three dominant individuals to one recessive, for this was the Mendelian segregation pattern for these genes. In 1908, Hardy in England and Weinberg in Germany disproved this concept, showing that gene frequencies do not depend on the presence or lack of dominance, but may remain essentially unchanged or be *conserved* from one generation to the next under certain conditions. We shall now examine conservation of gene frequencies in detail.

CONSERVATION OF GENE FREQUENCIES: HARDY-WEINBERG EQUILIBRIUM

The principle disclosed by Hardy and Weinberg may be simply illustrated by an example from human genetics. In man, the difference between those who *can* and *cannot* taste a certain chemical, phenylthiocarbamide (PTC), resides in a single gene difference with two alleles, T and t.

The allele for tasting T is dominant over t, so that heterozygotes Tt are tasters, and only nontasters are tt.

Were we to choose an initial population composed of an arbitrary number of each genotype, we might ask what the frequency of those genes would be after many generations. For example, let us place upon an isolated island a group of teenage children fleeing from a potential atomic world war. They have the genotypic ratio among them of 0.40 TT : 0.40 Tt : 0.20 tt. The gene frequencies in this initial population, are, therefore,

$$0.40 + 0.20 = 0.60\ T$$

and

$$0.20 + 0.20 = 0.40\ t$$

Let us assume, further, that the number of individuals in the population is large, and that tasting or nontasting ability for this particular chemical has *no* effect upon survival (viability), fertility, or attraction between the sexes. Let us also assume that there is an equal number of males and females among these teenagers.

As these children mature in isolation in this tropical paradise, they will undoubtedly choose their mates randomly with respect to their tasting abilities. That is, although some may prefer blondes and others brunettes, they probably will not have any PTC on the island or even have the slightest interest in testing whether the prospective mate can

Table 2–1 Types of random-mating combinations and their relative frequencies in a population of 0.40 *TT*, 0.40 *Tt*, and 0.20 *tt* genotypes.

		MALES		
		TT 0.40	*Tt* 0.40	*tt* 0.20
FEMALES	*TT* 0.40	0.16 ①	0.16 ②	0.08 ③
	Tt 0.40	0.16 ④	0.16 ⑤	0.08 ⑥
	tt 0.20	0.08 ⑦	0.08 ⑧	0.04 ⑨

TT × *TT* (1) = 0.16
TT × *Tt* (2 + 4) = 0.32
TT × *tt* (3 + 7) = 0.16
Tt × *Tt* (5) = 0.16
Tt × *tt* (6 + 8) = 0.16
tt × *tt* (9) = 0.04
1.00

or cannot taste this chemical. Matings between any two genotypes, then, can be predicted solely on the basis of the relative frequencies of these genotypes in the population.

The types of random mating combinations and their relative frequencies in a population such as we have described are predicted in Table 2–1. As shown in this table, nine different types of matings can occur, of which three matings are reciprocals of each other (for example, *TT* × *Tt* = *Tt* × *TT*). Therefore, there are six different mating combinations in all, between these genotypes, which will produce offspring in the next generation in the following ratios: 0.36 *TT*, 0.48 *Tt*, and 0.16 *tt*. (Table 2–2).

Note that although the frequencies of genotypes, that is, zygotic frequencies, have been altered by random mating, the *gene* frequencies have *not* changed. For *T* the gene frequency is equal to

Table 2–2 Relative frequencies of the different kinds of offspring produced by the matings shown in Table 2–1.

PARENTS		OFFSPRING		
TYPE OF MATING	FREQUENCY	*TT*	*Tt*	*tt*
TT × *TT*	0.16	0.16	—	—
TT × *Tt*	0.32	0.16	0.16	—
TT × *tt*	0.16	—	0.16	—
Tt × *Tt*	0.16	0.04	0.08	0.04
Tt × *tt*	0.16	—	0.08	0.08
tt × *tt*	0.04	—	—	0.04
—	—	0.36	0.48	0.16

$0.36 + \frac{1}{2}(0.48) = 0.60$, and the frequency of t is $0.16 + \frac{1}{2}(0.48) = 0.40$, exactly the same as before. Under these conditions, no matter what the initial frequencies of the three genotypes, the gene frequencies of the next generation are the same as those of the parental generation. For example, if the founding population of this island had contained 0.25 *TT*, 0.70 *Tt*, and 0.05 *tt*, the gene frequency for *T* would be $0.25 + \frac{1}{2}(0.70) = 0.60$, and for t it would be $0.05 + \frac{1}{2}(0.70) = 0.40$. However, despite the new frequencies of genotypes, random mating among the offspring again results in the ratio of 0.36 *TT*:0.48 *Tt*:0.16 *tt*, or a gene frequency of 0.60 *T*:0.40 *t*.

In algebraic terms, p = the frequency of *T*, and q = the frequency of its allele *t*. If no other alleles exist, $p + q = 1$. The equilibrium frequencies of the genotypes are given by the binomial expansion of this expression squared. That is,

$$(p + q)^2 = p^2 + 2pq + q^2$$

In this algebraic expression, p^2 is the equilibrium frequency of the *TT* genotype, $2pq$ is the frequency of the *Tt* genotype, and q^2 is the frequency of the *tt* genotype.

Two important conclusions follow from this discussion. First, under conditions of random mating in a large population where all genotypes have an equal chance of survival to the age of reproduction, gene frequencies of a particular generation depend on gene frequencies of the previous generation and not on genotype frequencies. Second, frequencies of different genotypes produced through random mating depend only on gene frequencies of potential mates. This allows us to express the conservation of gene frequencies in the form of a principle or law.

The Hardy–Weinberg equilibrium principle describes the behavior of genes in populations in the following manner. In the absence of outside forces, namely selection, differential migration, and differential mutation of genes, and with random mating between all genotypes, the initial gene frequencies in a population will be maintained from generation to generation. Furthermore, after the first generation, genotype frequencies remain stable, that is, at an equilibrium point. Genetic events in a randomly mating population of this sort can be represented as follows:

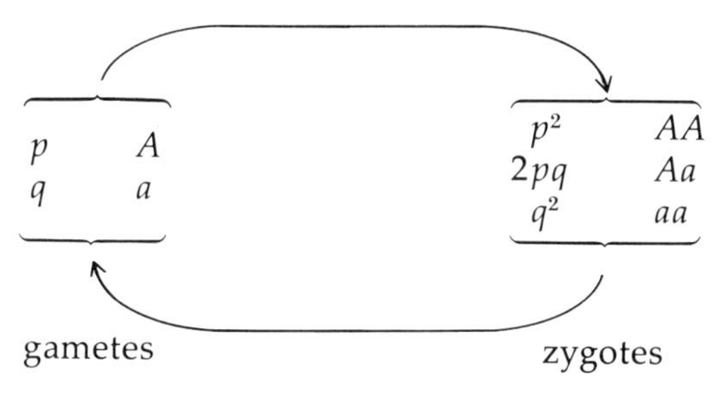

Undisturbed, the cycle shown above will exist indefinitely. Each generation produces gametes that carry the alleles A and a in proportions p and q. These, in turn, unite at random to give rise to zygotes AA, Aa, and aa in the proportions $p^2 : 2pq : q^2$.

ESTIMATION OF EQUILIBRIUM FREQUENCIES IN NATURAL POPULATIONS

Population biologists often want to look at a natural population to determine gene frequencies from observation of phenotypic frequencies in the group. They may also wish to see whether the population is in Hardy–Weinberg equilibrium or whether the gene frequencies differ enough from that idealized standard to cause one to suspect an outside factor such as selection which may be distorting the expected gene frequencies. However, there are several ways to handle phenotypic data depending on the genetic control of traits involved. Let us look at several of these situations.

Codominance in natural populations

When every genotype expresses itself as a different phenotype, that is, a heterozygote differs from either homozygote, gene frequencies can be estimated reliably, and observed genotype frequencies can then be easily compared to their expected equilibrium values. For instance, there exists codominance at the M–N blood-group locus in man. Boyd (1964) classified 104 American Ute Indians into 61 MM, 36 MN, and 7 NN individuals, or genotype frequencies of 0.59 MM, 0.34 MN, and 0.07 NN in the total sample. The gene frequency of M is therefore $0.59 + 0.17 = \mathit{0.76}$, and for N it is $0.07 + 0.17 = \mathit{0.24}$. At equilibrium, therefore, expected genotype frequencies should be

$$(0.76)^2 = MM = 0.58$$
$$2(0.76)\,(0.24) = MN = 0.36$$
$$(0.24)^2 = NN = 0.06$$

The agreement between the observed and expected genotype frequencies is usually tested by the chi-square method. Here the agreement is very good, and it suggests that indeed the individuals in this population of Ute Indians are mating at random (that is, they are panmictic) with respect to M–N blood groups. This seems reasonable because few human beings know their M–N blood group, and fewer still would ever consider this trait in a potential marriage partner.

Dominance in natural populations

When the effect of one allele at a locus is completely dominant over that of another, the heterozygous genotype (e.g., Aa) cannot be phenotypically distinguished from the homozygous dominant (e.g., *AA*). Under these circumstances, gene frequencies cannot be obtained directly as in codominance, as two of the genotype frequencies are unknown and submerged within one phenotypic class.

One method used to estimate gene frequencies in these cases relies upon the only genotype whose frequency is definitely known, the recessive homozygote. Under random mating, we know that the genotype frequencies are

AA	Aa	aa
p^2	$2pq$	q^2

Therefore, the recessive homozygotes are present in a frequency q^2, equal to the square of the recessive gene frequency q. If, say $q^2 = 0.49$, q = the square root of 0.49, which equals *0.70*, and therefore the dominant allele frequency $p = 1 - q$, or *0.30*. The genotypic class, which is homozygous dominant, therefore has the frequency $p^2 = (0.30)^2 = 0.09$, and the heterozygotes have the frequency $2pq = 2(0.30)(0.70) = 0.42$.

By calculating q as the square root of the frequency of the homozygous recessive genotype class, rather than from all observed genotypic classes, we are assuming that Hardy–Weinberg equilibrium has been reached. Thus, we cannot turn around and calculate expected gene frequencies from these data to prove that Hardy–Weinberg equilibrium exists because we will get exactly the same results as with the observed data we used to calculate the gene frequencies in the first place.

We ought to note some general observations on the Hardy–Weinberg equilibrium principle at this point:

We have a yardstick against which to measure *changes* from expected gene frequencies in the population. Thus, we have a sensitive index for assessing evolutionary and ecological phenomena that may be affecting individuals of different genotypes in different ways.

We should note that with knowledge of just the frequency of one allele in the population, such as, for example, the frequency of $A = 40$ percent, we learn that (1) the frequency of the alternative allele (a) is 60 percent, (2) the expected frequency of *AA* individuals is 16 percent, (3) the expected frequency of *Aa* individuals is 48 percent, and (4) the expected frequency of *aa* individuals is 36 percent. So knowing one gene frequency, we know the relative frequencies of the two alleles as well as the relative frequencies of three types of zygotes or genotypes in the population that would be expected if the population is in Hardy–Weinberg equilibrium.

We should also note that the frequency of the heterozygote class cannot exceed 50 percent in a population in which Hardy–Weinberg equilibrium is in operation; that is, this is the value obtained when both p and q equal 0.50. This is the same value as expected when

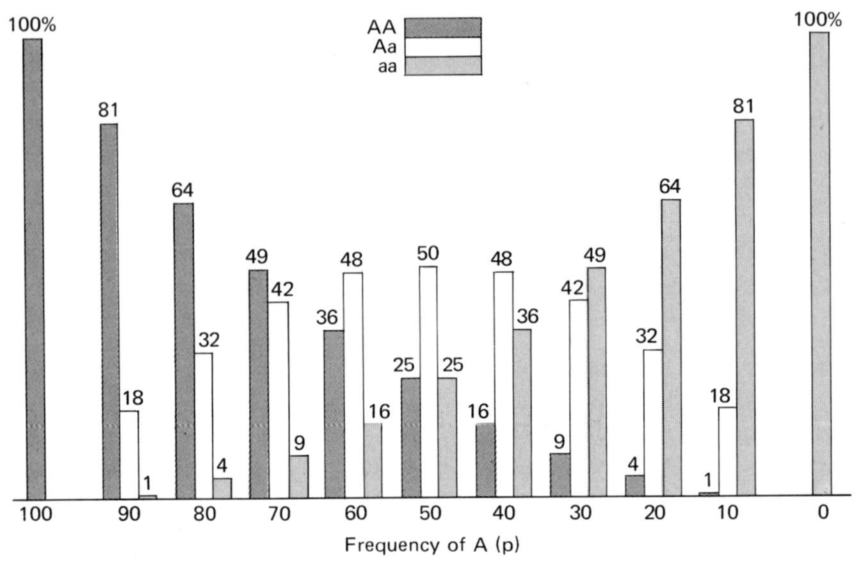

Figure 2–1 The Hardy–Weinberg proportions of homozygotes and heterozygotes that are possible for various gene frequencies. (Source: B. Wallace, 1968, *Topics in Population Genetics,* Norton, New York, p. 61.)

mating F_1 hybrids in order to produce an F_2 generation, of course. This fact is shown graphically in Figure 2–1. In all possible equilibrium populations, the maximum percentage of heterozygotes that may be possessed is 50 percent. It is quite possible, of course, for a population to be in Hardy–Weinberg equilibrium in a trivial sense if the total population consists of only homozygous *AA* or *aa* individuals (that is, $p = 1.00$ or $q = 1.00$) and no heterozygotes.

As an example of a population that is not in Hardy–Weinberg equilibrium with a larger than 50 percent frequency of heterozygotes, let us look at this example. We collect in a population of Florida damselflies; some have a red spot on the wing caused by the *A* allele, whereas damselflies possessing an *a* allele in the recessive homozygous genotype lack the red spot on the males and females. Our sampling on this particular date shows us that we have the following array of genotype frequencies.

AA	*Aa*	*aa*
0.09	0.66	0.25

In this example, $p = 0.09 + 0.33$, or $p = 0.42$. The frequency of the recessive allele is $q = 0.33 + 0.25$, or $q = 0.58$. This is the gene frequency for the present generation we are sampling. In the next generation, however, the array of genotype frequencies that would be produced in the absence of seletion or other nonrandom factors would be:

p^2	$2pq$	q^2
0.18	0.48	0.34

This is quite a change in genotype frequencies, and now the population could go into Hardy–Weinberg equilibrium, with these frequencies remaining stable at these levels if there are no outside forces of nonrandom mating occurring in the population.

Finally, populations with identical gene frequencies may have different genotype frequencies but will reach the same equilibrium state after only one generation of random mating (assuming no selection or outside forces).

Let us sum up what we have learned about Hardy–Weinberg equilibrium in its applicability for studying the genetic structure of populations.

USES OF THE HARDY–WEINBERG EQUILIBRIUM

The Hardy–Weinberg Equilibrium provides a base line for analyzing data obtained by sampling field populations. In other words, using the Hardy–Weinberg equilibrium principle, one can ask, do observed zygotic frequencies agree with observed gametic frequencies? That is, how do the observed and expected proportions of genotypes compare? If they do not agree, the investigator can then look to see which assumption(s) of Hardy–Weinberg equilibrium is (are) not being satisfied in that population. We look at these assumptions in more detail below.

In calculating gene and zygotic frequencies, information is obtained that permits us to compare samples (1) from the same population at different times, or (2) from populations inhabiting different geographic regions. Significant differences between such samples demonstrate microevolutionary changes that occur in the framework of either time or space or both.

LIMITATIONS ON USE OF THE HARDY–WEINBERG EQUILIBRIUM

One pitfall in the use of this simple mathematical formulation is that many people get excited solely over the demonstration that alternative alleles are distributed among members of a population in the manner that the Hardy–Weinberg equilibrium predicts. This equilibrium relationship should be tested in each instance, not because it is the end or the goal of the study but because it is by such test that discrepancies are found (Wallace, 1968). These offer opportunities to examine a population for migration, nonrandom mating, mutation, selection, and all other factors that tend to upset the theoretical expectation. What limitations are there on the use of this equilibrium principle?

Attempting to extract more information than contained in the sample

As an example of this common problem, it is perfectly legitimate to take a sample of individuals from a population and after determining the genotypes to compute the gene frequencies. Then with these frequencies, you can compare the observed frequency of genotypes to the expected distribution of genotypes.

However, if these observed and expected frequencies do not match up, for example more than 50 percent heterozygotes appear in the sample, we cannot immediately assume, for instance, that there is strong selection (say nearly complete lethality) against the two homozygotes. We have to take more than one sample in time and see if the frequency distribution is constant or if it is changing toward an equilibrium point.

The insensitivity of the Hardy–Weinberg Equilibrium

As an example of this problem, let us say that the investigator has trouble identifying the heterozygote in a sample of animals, and he misclassifies 50 percent of *Aa* individuals as *AA*. Wallace (1968) offers this apt analysis of the problem.

Genotype	*AA*	*Aa*	*aa*
True zygotic frequency *A* equals 0.900, *a* equals 0.100	0.81	0.18	0.01
Apparent zygotic frequency Apparent gene frequencies are thus: *A* equals 0.945; *a* equals 0.055	0.90	0.09 50% misclassified	0.01
"Expected" zygotic frequencies	0.893	0.104	0.003

You will note that in the apparent zygotic frequency line, 50 percent of the real heterozygotes have been misclassified as dominant homozygotes, halving the genotype frequency. In comparing the apparent zygotic frequency with the expected zygotic frequencies, there is remarkably good agreement. It would take a huge sample to show that the 90:9:1 apparent frequency (the one erroneously tabulated) does in fact differ from the Hardy–Weinberg distribution expected from the apparent gene frequencies. So you cannot expect the Hardy–Weinberg equilibrium calculations to show up an error of this sort. It is only as good as the data you put into it.

The pooling of data

In too many papers, a number of collections are frequently lumped into a single large test of the Hardy–Weinberg distribution. This is an error because data pooled in this way are systematically biased. If the gene frequencies in the various samples differ, the pooled data possess fewer than the "expected" number of heterozygous individuals (Wallace, 1968).

An extreme example can be given in support of this claim. Suppose that a collector of butterflies unknowingly samples an area that really includes two populations in neighboring localities which are, in fact, effectively isolated from one another by unsuitable habitat or some other factors. Let us say, too, that the frequency of A in one population is 100 percent, whereas its frequency in the other is zero percent (100 percent a):

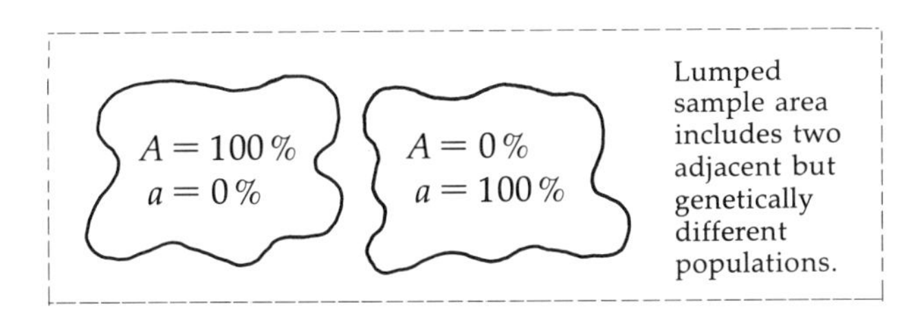

If the two groups composing his lumped sample were of equal size, the collector will calculate gene frequencies as $p = q = 0.50$, from which he will conclude that the frequency of heterozygotes Aa should be 50 percent.

Actually there will be *no* heterozygotes in the sample because there were none in either of the two component populations.

Combining data from partially isolated populations, or from several generations for a population in which gene frequencies are changing, always leads to "expected" frequencies of heterozygotes that are greater than those which actually exist. This kind of error may help to explain unexpected deficiencies in field data, as Wallace (1968) emphasized. Cunha (1949) studied wild populations of a polymorphic species of fruitfly, *Drosophila polymorpha*. In this fly, a single pair of alleles controls light, dark, or intermediate phases of color of the abdomen. The two alleles, E and e, do not show dominance. Cunha collected 8070 individuals in the state of Paraná, Brazil, which had the following genotypes: 3969 EE, 3174 Ee, and 927 ee flies. Hence, the frequencies of the alleles E and e were 0.69 and 0.31, respectively. The expected numbers of the three genotypic classes are 3841, 3454, and 775. Thus, there is a clear excess of homozygotes of both sorts and a deficiency of heterozygotes according to these data. This is very unexpected, for there seems to be no heterosis effect enhancing the viability of the het-

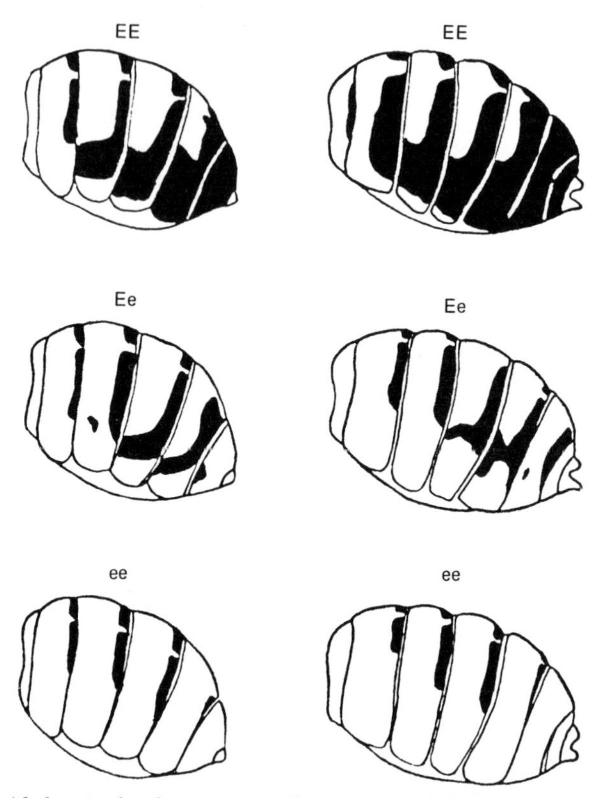

Figure 2–2 Abdominal color-pattern phenotypes of dark (EE), intermediate (Ee), and light (ee) color types in *Drosophila polymorpha* in Brazil. Male left, female right. (Source: A. Brito da Cunha, 1949, *Evolution, 3:* 239–251.)

erozygote in the wild, yet heterozygotes were superior in laboratory population cage experiments. Dobzhansky (1951, p. 131) and Ford (1964, p. 131), were also surprised at the lack of wild heterozygotes to match expected numbers. In contrast, as Wallace (1968) points out, the original data show that the frequencies of *E* and *e* vary from season to season and from place to place. Furthermore, *Drosophila* populations are often extremely localized in distribution, so at least *part* of the apparent deficiency might have come from pooling samples from many small populations in this Brazilian area. Also, Heed has described a third allele at the *e* locus of this species, which if present in Cunha's populations (and unrecognized) would have led to a false deficiency of heterozygotes (Figure 2–2).

MULTIPLE ALLELES AND THE HARDY–WEINBERG EQUILIBRIUM

If the Hardy–Weinberg calculations were restricted to gene loci occupied by no more than two alleles, we would have a relatively limited application to natural situations. However, this equilibrium principle can be extended to cases in which three or more alleles exist. The basic principle for calculating the gene frequency of a particular allele is the same as in the two-allele locus situation. In the case of multiple alleles, one calculates the frequency of a particular allele as the sum of the frequency of the homozygous class plus one-half the frequency of every heterozygous class carrying that particular allele. As an example, if we were dealing with the ABO blood groups, we could use the following expressions to calculate the frequency of the *A*, *B*, and *O* alleles of this locus:

$$\text{Frequency of } A = p = AA + \tfrac{1}{2}AO + \tfrac{1}{2}AB$$
$$\text{Frequency of } B = q = BB + \tfrac{1}{2}AB = \tfrac{1}{2}BO$$
$$\text{Frequency of } O = r = \mathrm{OO} + 1/2\,AO + 1/2\,BO.$$

The sum of the gene frequencies for each of the multiple alleles are expressed by the standard symbols $p, q, r, s, \ldots$ equals 1.00. That is, the three alleles with frequencies of p, q, and r, respectively, will have $p + q + r$ equal to one. If we were to express the gene frequencies of multiple alleles from the genotype proportions, where each genotype produces a phenotypically distinguishable characteristic, then we would have, in the general case of three multiple alleles:

$$p = p^2 + pq + pr$$
$$q = q^2 + pq + qr$$
$$r = r^2 + pr + qr$$

Some genes recently discovered in natural populations have as many as 10 or 12 alleles, as in the case of electrophoretically visible esterase enzyme variants in *Colias* butterfly populations (Burns and Johnson, 1967). Sometimes we wish to look at only one of a series of alleles and study its variation between populations, or with time in a single population. A group of multiple alleles can be treated as if they consisted of only a pair of alleles by considering the frequency of one of them as being represented by q and letting the total frequency of all the others be $1 - q$. The Hardy–Weinberg binomial for the group of genotypes then becomes

$$[q + (1 - q)]^2 = q^2 + 2q\,(1 - q) + (1 - q)^2$$

In the general case, to determine the frequency of the genotypes that will exist in a population in equilibrium, one squares the array of the a_1, a_2, $a_3 \ldots$ gene frequencies in the gametes $(p + q + r + \ldots)^2$. For a three-allele system, the six possible genotypes will exist in a population in ratios of p^2 (genotype $a_1 a_1$) $+ 2pq$ $(a_1 a_2)$ $+ 2pr$ $(a_1 a_3)$ $+ q^2$ $(a_2 a_2)$ $+ 2qr$ $(a_2 a_3)$ $+ r^2$ $(a_3 a_3)$. An example involving three alleles with $p = 0.3$, $q = 0.5$, and $r = 0.2$ is presented in Figure 2–3. As in the simpler case of two alleles examined earlier, equilibrium is

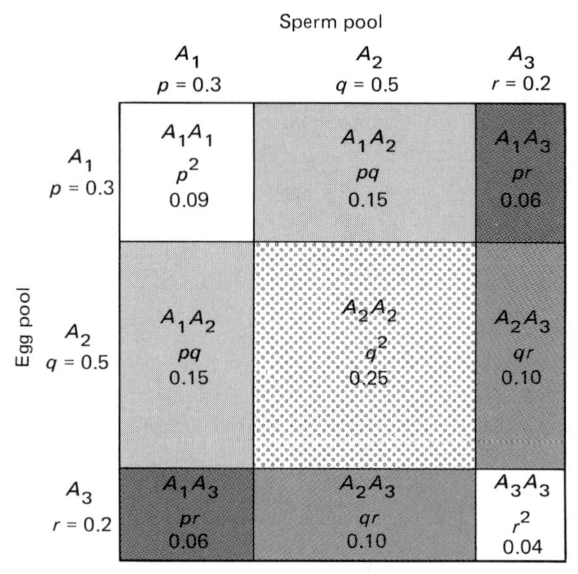

Figure 2–3 Multiple alleles in an equilibrium population. The three alleles, A_1, A_2, and A_3, occur with frequencies $p = 0.3$, $q = 0.5$, and $r = 0.2$ in each gamete pool. The genotype frequencies are given as the areas within the square, which are formed by geometrically squaring the gametic array. (Source: L. E. Mettler and T. G. Gregg, 1969, *Population Genetics and Evolution*, Prentice-Hall, Englewood Cliffs, N.J., p. 44.)

established in one generation through random mating and random union of gametes.

In a four-allele system where p, q, r, and s are the frequencies of the alleles a_1, a_2, a_3, and a_4, the binomial expansion provides the following expected zygotic frequencies:

$$\begin{array}{cccccccc} a_1\,a_1 & a_1\,a_2 & a_1\,a_3 & a_1\,a_4 & \ldots & a_2\,a_2 & a_2\,a_3 & a_2\,a_4 \quad \ldots \\ p^2 & 2pq & 2pr & 2ps & \ldots & q^2 & 2qr & 2qs \end{array}$$

$$\begin{array}{cccc} a_3\,a_3 & a_3\,a_4 & \ldots & a_4\,a_4 \\ r^2 & 2rs & \ldots & s^2 \end{array}$$

These calculations can also be extended to cases involving sex-linked genes, autopolyploids, multiple loci, and other situations. Such cases are considered in courses in population genetics, but are beyond our scope here.

FACTORS UPSETTING THE HARDY–WEINBERG EQUILIBRIUM

Random mating

Using the Hardy–Weinberg calculations, we assume that the frequencies of different sorts of mating depend solely on relative frequencies of different types of individuals in that population. With respect to hidden genetic traits such as blood-type systems, matings are usually random, but unfortunately for the general applicability of the Hardy–Weinberg principle, visible traits usually influence mating patterns. We should emphasize the important point that nonrandom mating patterns do *not* normally alter *gene frequencies* except through some associated process of differential reproductive success (Wallace, 1968). Selective mating patterns will affect the relative frequencies of different *genotypes,* however. For instance, self-fertilization and matings between close relatives will increase the frequency of homozygous individuals at the expense of heterozygotes. Matings between dissimilar individuals, on the contrary, tend to exaggerate heterozygote frequencies.

Migration

The Hardy–Weinberg equilibrium principle is applicable to *closed* populations in which all members of one generation were derived from parents who were members of the same population. If immigrants coming into a population have different gene and zygotic frequencies from that population, the gene frequency and zygotic frequencies in the recipient population will not remain constant. Likewise, if immigrants leaving a population are not a random (unbiased) sample of various genotypic classes, the gene frequency among remaining individuals will change.

Selection

The Hardy–Weinberg equilibrium assumes that individuals of different genotypes contribute equally to the next generation. However, if some are preferentially surviving or are being eliminated, then zygotic frequencies will change and gene frequencies need not remain constant from generation to generation. We explore the importance of selective factors in Chapter 3.

Mutation

The equilibrium in gene frequencies may be upset by differential mutation rates. One allele may change into another allelic form owing to spontaneous or induced mutation. Mutational changes of this sort, unless they occur equally frequently in opposite directions, lead to changes in gene frequencies in the population. That is, unless the rate of U going to u matches the natural rate of u going back to U, the relative proportion of each allele type in the population will change.

Sampling error

The Hardy–Weinberg calculations assume a population of infinite size so that gene frequencies will be constants with no sampling error. But all populations are finite in size, and consequently zygotic frequencies as well as the gene frequencies to which they give rise are not constant but rather fluctuate from generation to generation. We should note that if the Hardy–Weinberg equilibrium is displaced, it tends not to return to its original value by inertia or conservation of momentum like a physical object—it is "neutrally" stable. In fact, in a small population, the sampling fluctuations for adults that actually reproduce each generation can continue until, by accident, one allele or another has been lost from a population. This process is known as *random genetic drift*. Because of its importance, we explore drift in the following section.

GENETIC DRIFT

The previous forces we considered such as mutation, selection, and migration act in a *directional* way to change gene frequencies progressively from one value to another. Various *nondirectional* forces are changes that are *not* tied to the gene frequencies involved, and hence change unpredictably from one generation to the next. One of the most important—random genetic drift—is caused by the "sampling error" discussed above. Random genetic drift is defined as a variable sampling of the gene pool at each generation. Genetic drift is *caused* by a reduction in population size such that gene frequency changes occur because of sampling errors. This result is easy to see if we consider that when the number of parents of a population is consistently large each generation, there is always a strong likelihood of obtaining a good sample of the genes of the previous generation as long as directional forces are not acting to change them. In contrast, if only a few parents are chosen to begin a new generation, such a small sample of genes may *deviate widely* from gene frequency of the previous generation.

We can measure mathematically the extent of the deviation from the last generation in both cases by the standard deviation of a proportion:

$$\sigma = \sqrt{pq/N}$$

where p is the frequency of one allele, q is the frequency of the other allele, and N is the number of genes sampled. For diploid parents, each carrying two genes,

$$\sigma = \sqrt{pq/2N}$$

where N is the number of actual parents. As an example, let us look at the following two situations.

Consider a large diploid population, where $p = q = 0.5$. We will continue this population using 5000 parents as active breeding stock each generation. Then

$$\sigma = \sqrt{(0.5)(0.5)/10{,}000} = \sqrt{0.000025} = 0.005$$

The values of the gene frequencies in such populations will therefore fluctuate mostly around 0.5 ± 0.005, or between 0.495 and 0.505.

In contrast, consider a small diploid population, where we have only two parents as "founders." Here the gene frequency will have a standard deviation of

$$\sigma = \sqrt{(0.5)(0.5)/4} = \sqrt{0.0625} = 0.25$$

or values of 0.50 ± 0.25 (range of 0.25 to 0.75 in one generation). The next generation of two adult parents could start with a gene frequency of as low as 0.25 to as high as 0.75, and their offspring could have a gene frequency as low as 0.03 or as high as 0.97.

In other words, sampling accidents because of smaller population size will easily yield, in a short term, gene frequencies that are either zero or one. As gene frequency limits are also zero or one, sampling accidents because of small population size may easily fix one or the other of the alleles in all of the members of the population (Figure 2–4). If such small sizes are continued through subsequent generations, the likelihood increases that such a population will eventually reach fixation for an allele. This *change* in gene frequency can thus arise in the absence of *any* of the directional forces previously considered. Note that each small population in the range of a species is independent as regards this drift process. Hence, on the average, about 50 percent of any group of small populations will go to fixation for one allele and about 50 percent will go the fixation for the other allele. One would predict, then, that in assessing allele frequencies in a series of small populations, loci that are monomorphic in any one population might be monomorphic for another allele of the same locus in the neighboring population. In fact, for multiple allele systems, a series of monomorphic populations might each be fixed at a different allele in proportion to the total number of alleles available for that locus.

THE FOUNDER PRINCIPLE

An extreme case of genetic drift in a natural population may occur when a population sends forth only a few "founder" individuals to begin a new population. They carry with them a biased sample of alleles, for whatever genes they take in their gametes, whether they are detrimental or beneficial alleles, stand a good chance of becoming es-

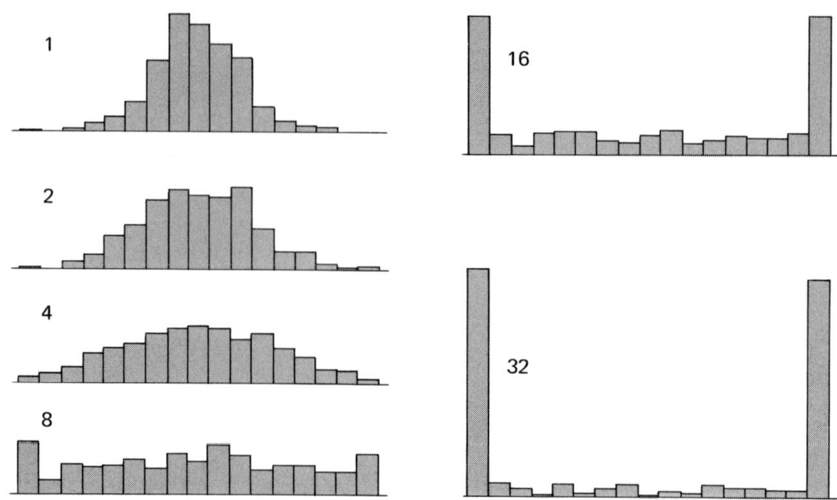

Figure 2–4 Dispersal of gene frequencies among 400 hypothetical populations ("Monte Carlo" simulation on a high-speed digital computer). The following conditions were made for the genetic model: (1) each population consisted of only eight diploid individuals per generation—four randomly formed pairs; (2) each individual mated but once, and the number of offspring produced by each mating varied—a Poisson distribution with a mean of two; (3) selection and mutation were absent; and (4) each population was started with two AA, four Aa, and two aa individuals, so that there was an initial gene frequency of 0.5 for each allele at a single autosomal locus. Four hundred such populations were simulated over 32 generations. The figure depicts the populations classed according to gene frequencies in generations 1, 2, 4, 8, 16, and 32. Because of chance variation in such small populations, there is an increasing spread in gene frequencies toward fixation of one allele or the other in most of the populations. Data courtesy of Henry Schaffer. (Source: L. E. Mettler and T. G. Gregg, 1969, *Population Genetics and Evolution,* Prentice-Hall, Englewood Cliffs, N.J., p. 53.)

tablished in the new population because of the sampling accident. Thus, species that are good travelers or migrants may "found" island populations that will eventually have quite different gene frequencies from their source populations.

GENETIC VARIATION AND CHANGE IN NATURAL POPULATIONS

The fundamental principles of population genetics that we have reviewed here have their greatest importance in population biology in their role of providing the hereditary mechanism for adaptive change and dynamic evolution of the subject population. The mutation process is constantly generating a supply of mutant alleles in populations. Most mutants that arise in this manner are decidedly harmful to the organism, as the animal or plant is already a well-balanced functioning unit and any small change will usually be for the worse. With the mutation process introducing these new mutants into the gene pools of population, there must be some opposition to the accumulation of these harmful mutants, or populations would soon be full of genetically deficient parents, and reproduction would severely diminish or cease. Darwin's theory of natural selection, of course, is based on the obvious fact that only a portion (and usually a small portion) of the progeny of any species survive and become parents of the next generation. Genetically different parents will produce different numbers of surviving offspring. Hence, the better adapted variants will tend to constitute a greater proportion of the survivals than will the less well-adapted ones, and the incidence of the genes carried by those better-adapted variants will increase in succeeding generations. The evolutionary aspects of this process in population biology will be taken up in Chapter 3.

We should note here, however, that genetic elasticity and phenotypic plasticity are two ways in which organisms are adapted to survive and multiply in a wide range of environments or are adapted to cope with a change in environment (Birch, 1960). The ability for a population of organisms to fluctuate between certain values of gene frequencies or even drift in time to different gene frequencies en-

Figure 2–5 The Stephen Island wren, *Xenicus lyalli*, once inhabited a small island near New Zealand. A very small population size, which limited genetic elasticity and phenotypic (especially behavioral) plasticity, did not allow this flightless wren to cope with a change in environment—the introduction of one domestic cat. The history of this species has been told by the ornithologist W. R. B. Oliver: "In 1894 the lighthouse keeper's cat brought in eleven specimens, which came into the hands of H. H. Travers. . . . A few more captures made and duly reported by the cat and then no more birds were brought in. It is evident, therefore, that the cat which discovered the species also immediately exterminated it." (Source: S. Carlquist, 1965, *Island Life*, Natural History Press, New York, p. 349.)

hances the ability of the individual genotypes to survive and multiply in a wide range of environments (Figure 2–5). Thus, the presence of mutations in creating new alleles on which selection can operate is not necessarily a disadvantage. Adaptation through genotypic plasticity is only possible when considerable genetic variability is available at all times:

> The source of such variability is twofold: mutation and, in sexual organisms, recombination of genes. Recurrent mutation is the main source of genetic variability in bacteria, algae, and protozoa, but mutation rates are, for the most part, so low that the effectiveness of mutation on its own has been questioned for all but such organisms as these which multiply at a prodigious rate. A high rate of increase combined with natural selection can lead to rapid change despite relatively low mutation rates. With sexual organisms mutation rate is reinforced with the element of stored variability which means that there is a reserve of variability over and above that which mutation alone can provide (Birch, 1960).

The three main genetic mechanisms involved in stored or concealed variability are described as follows:

Genes that have no phenotypic expression such as recessive alleles concealed by dominant alleles will be stored in the gene pools, as it is the phenotype that is exposed to selection.

Polygenic inheritance, that is, the control of a character by a number of genes, favors stored variability (Mather, 1941, 1943). Crossing over releases this variability from the linked polygene complexes, and such released variability can be of adaptive value when the environment changes, in addition to the higher immediate fitness of most heterozygotes.

Heterosis or superiority of the heterozygote over the homozygotes creates a mechanism for storing variability in the population. In fact, as we shall see, this is the essential condition for establishing balanced polymorphism in Mendelian populations. As Birch (1960) points out, if a mutant produces a heterotic heterozygote, natural selection will retain this mutant in the population even if the homozygote is lethal. Dobzhansky and his co-workers have found among *Drosophila* populations, that heterosis is the mechanism that maintains a diversity of chromosomal inversions in populations. This diversity adapts a species to a wide variety of environments (Dobzhansky et al., 1950). We shall be studying this aspect of the effects of stored variability in Chapter 3.

When a population inhabits an environment that is constant in its features, genetic plasticity and diversity of genotype will be of no advantage. Hence, a balance between flexibility and stability of the genetic composition of a population is often maintained through a variety of genetic mechanisms. Chromosomal inversions, for instance, tie up blocks of genes which have been proved to be of adaptive value in selection.

> Inversions suppress crossing over with consequent reduction in variability. This would be advantageous in some environments, but where the environment is constantly providing new challenges, such as on the periphery of the distribution of a species, greater genetic flexibility may be necessary for survival. This is born out by the finding of a decreased number of inversions at the periphery of the distribution as compared with the center in certain populations of *Drosophila* (Birch, 1960).

The alternation of sexual and asexual generations is another mechanism for maintenance of a balance between stability and flexibility of the genotypic composition of a population. In the asexual generation, genotypes are kept stable. The change from asexual to sexual phase is related to change in environment. The sexual generation occurs when the environment becomes unfavorable and so provides the individuals in a population at this stage with an increase in genetic variability (Birch, 1960; Lewontin, 1957).

1 2 3 4 5 6 7 8 9 10 11

Evolution at the Population Level

The evolutionary aspects of population biology pervade the ecological, genetic, and behavioral interactions within and between animal populations. Evolution at the population level necessitates consideration of the nature of selection and its various manifestations, including group selection and kin selection. We must look at factors that reduce genetic variation in populations as well as those that store and protect genetic variation and even increase it. The unity of the genotype must be considered along with the role of isolating mechanisms and hybridization in maintaining or breaking up a useful genotype. The evolution of dominance in populations, together with geographic variation, speciation, and the relative significance of gene flow between populations versus selection in holding species together as a series of phenotypically similar populations or in promoting differentiation between the populations, must be carefully considered. In studying evolution at the population level, then, we must concentrate on th- adaptive features of organisms in populations. Because this becomes our overall concern when studying population biology, in this chapter we are really synthesizing an overview of the evolutionary aspects of population biology. Topics treated in subsequent chapters, of course, ultimately involve selection and evolution as well.

THE NATURE OF SELECTION

The term *natural selection* is a metaphorical expression for the net effect of relative reproductive success among organisms (the total expectation of offspring or "fitness") living together in the same environment. Selection is often considered in terms of the result of differential reproductive success of genetically different individuals (Wallace, 1968). Selection may alter the gene frequencies in a population just as do the processes of migration and mutation, but it does not necessarily do so. Fitnesses of different genotypes may differ but in a way that opposing tendencies balance and the gene frequency remains unchanged (Crow and Kimura, 1970).

What does the term *fitness* really mean? In genetics, fitness refers to relative reproductive success or adaptive value. As long as one genotype can produce more offspring than another in the environment, its fitness is said to be superior. One would like to be able to measure the magnitude or intensity of a differential reproductive capacity. This may be done in several ways.

In its simplest form, this measurement commonly consists of counting the number of offspring produced by one genotype as compared to those produced by another. For example, if individuals of genotype *AA* produce an average of 100 offspring that reach maturity, whereas genotype *aa* individuals produce only 90 offspring in the same environment, the reproductive success of *aa* is reduced by 10 offspring or by the fraction $10/100 = 0.1$. Looked at in terms of fitness or adaptive value, *AA* may therefore be considered to possess a superior adaptive value relative to *aa*. If we designate the adaptive value (fitness) of a genotype as W, which will equal the proportion of offspring that survive to maturity relative to those of other genotypes, W can be chosen so as to fall in a range between 1.00 for the most productive genotypes and 0 for lethals. In the case just given, therefore, the adaptive value or fitness of $A = W_A = 1.00$, and that for $a = W_a = 0.90$. The natural selection acting on each genotype to reduce its adaptive value is defined by the *selection coefficient s*. For the above example, $s = 0$ for A and $s = 0.1$ for a. Therefore, the relationship between W and s is simply $W = 1 - s$, or $s = 1 - W$.

In many instances, one genotype may be arbitrarily designated at the beginning of an evolutionary study as having a fitness of 1.00, and the fitness of competing genotypes evaluated relatively. In such cases, adaptive values may exceed the 1.00 value. For instance, in the study of sickle cell anemia, heterozygotes in malarial regions of East Africa turned out to have a relative fitness value of about 1.26 compared to homozygous dominant normal individuals with normal hemoglobins (1.00 for survival in areas where they are not being compared with the survival of heterozygotes under the selection of malaria).

The term "selection intensity" has been used by Simpson (1953) in the same sense as for selection coefficients, that is, in terms of the disadvantage of one genotype relative to another, which he limits to ± 1. If a lethal has a -1 "selective intensity" value, the normal allele may have an upper limit of fitness of $+1$, that is, completely favorable selection, whereas at 0, there would be no fitness differential and all genotypes would have equal reproductive success.

Animal breeders define selection intensity as the amount of culling necessary to make progress toward a goal. More commonly, it is defined in terms of a practiced differential between the average of those selected to be parents and the average of the population in which they were born. But this definition involving changes in mean values toward predetermined goals is not always relevant to natural selection, although it is found in stabilizing as opposed to directional selection. Haldane (1954) states that natural selection, obviously, is often directed toward stabilization: maintenance of the mean value and removal of extreme variants. However, if this result occurs in an animal breeder's project (i.e., if it turns out that the mean value of a quantitative character is the same in the survivors as in the original population), then the breeder puts the intensity of selection at zero! The pertinent example of extremely heavy artificial or natural selection on a pure line may be cited. If we only breed from the heaviest 1 percent of members of a pure line of pigs, this artificial "selection" has no effect on the distribution of weights in the next generation. All have the same genotype. It is not *selection* unless there is heritability of selected traits. So Haldane claims that we cannot judge intensity of natural selection by its effect on the next generation. Instead, we have to compare the actual parents of the next generation with the population of which they are a sample (and a sample biased by the result of selection). Therefore, in the measurement of natural selection, Haldane considers selection by differential mortality within a population and at a certain part of the life cycle, and he is principally concerned with continuous-variable metrical characters (such as weight), considered to be the commonest (and multifactorial) phenotypic traits of the genotype.

Haldane measures the intensity of selection as the natural logarithm of the ratio between the survival frequency of the optimal phenotype and the survival frequency of the total population. For example, take a population of butterflies with several phenotypes present, each controlled by a different genotype. Over the period considered, should 10 percent of the population die, the intensity of natural selection reasonably cannot exceed 10 percent, or if the logarithmic scale is used,

$-\log_e (0.9)$ or 0.1054. Three cases may result: (1) If *all* phenotypes and genotypes have the same mortality (i.e., 10 out of every 100 of each phenotype die), the deaths are entirely accidental and there is *no* natural selection. (2) If one phenotype has no mortality at all (i.e., 0 out of 100 die), but all deaths occur in other phenotypes, we say that the intensity of selection is 10 percent, or 0.1054 on the natural log scale. (3) If we can distinguish several phenotypes, there is an optimal phenotype, or a set of phenotypes, of which a fraction s_o survives, whereas of the total population, a smaller fraction S survives. That is, s_o is the *optimal* phenotype's survival frequency (the phenotype that is surviving at a higher rate than the whole population), and s is the total population's survival frequency. If *all* the population had belonged to the optimal phenotype, then a fraction s_o of it would have survived. The extra mortality resulting from natural selection is $s_o - S$; that is, the difference between the survival frequence of the optimal phenotype and the total population. Thus, Haldane (1954) defines the *intensity* of natural selection as

$$I = \log_e s_o - \log_e S$$

or

$$I = \log_e s_o \, S$$

For example, if $S = 0.90$ and $s_o = 0.95$, that is to say 10 percent of the whole population dies, but only 5 percent of the optimal phenotype dies, then $s_o - S = 0.05$ (the extra mortality attributed to natural selection) and $I = 0.0541$ (the intensity of natural selection). When death rates in a population are low, these two measures are nearly the same. In practice, one generally considers only one measurable character x at a time, say weight at birth. One can then roughly determine the values of x for which mortality is minimal, that is, the optimal phenotype. This is often close to the mean value, and not usually, if ever, an extreme value.

Haldane's method of measuring the intensity of natural selection for a metrical character is quite different from that of animal breeders. If the mean value of x is the same in the survivors as in the original population, animal breeders put the intensity of selection at zero. This may be justifiable when artificial selection is being measured. It is not so in the case of natural selection, which may have had no effect on the mean, while considerably reducing the variance. Haldane offers an actual example of natural selection which approaches this theoretical case. Karn and Pinrose (1951) recorded the birth weights of 13,730 babies and the fractions in various weight groups that survived the hazards of birth and of the first 28 days of life. Haldane considers only their data on the 6693 females (the mortality rates for males were slightly higher).

The overall survival S	$= 0.959 \pm 0.002$	or	$95.9 \pm 0.2\%$
The survival for			
babies of 7.5–8.5 lb	$= 0.985 \pm 0.003$	or	$98.5 \pm 0.3\%$
babies under 4.5 lb	$= 0.414 \pm 0.037$	or	$41.4 \pm 3.7\%$
babies over 10 lb	$= 0.905 \pm 0.064$	or	$90.5 \pm 6.4\%$

Although the latter figure does not differ significantly from the optimal survival of 98.5 percent, there is no doubt as to the lower viability of heavier babies. When the various rates are plotted against weights at birth, the optimal survival s_0 can be more accurately calculated. Thus, Haldane found that $s_0 = 0.9828$, or 98.3 percent, or very close to the survival frequency of the optimal group found by more casual observation of the data. That is to say, only 1.7 percent of the babies would have died if all had been of the optimal weight, whereas in fact 4.1 percent did die (overall survival S was 95.9 percent). Thus, $s_0 - S = 4.1 - 1.7 = 2.4$, and 2.4 divided by $4.1 = 58$ percent. On the sole criterion of weight, then, 58 percent of all deaths were selective, and taking the natural logs of $s_0 - S$, the intensity of natural selection for weight was $I = 0.0240 \pm 0.004$.

The effect of this natural selection on the population was to increase the mean birth weight from 7.06 to 7.13 lb, that is to say, by 1 percent, but to decrease the standard deviation of birth weights from 1.22 to 1.10, that is to say, by 10 percent. The effect of natural selection in reducing the variance was therefore far greater than its effect in increasing the mean.

This selection against extremes appears to be true in every case in which natural selection for a metrical character has been observed; that is, selection reduces the variance and could be called stabilizing or normalizing selection. Haldane's paper gives a rather easily applied method of estimating the *intensity* or *magnitude* of this and other types of natural selection. We should reemphasize here that in artificial selection, which is usually *directional,* at least as practiced by plant and animal breeders, overall fitness may not be as important as meat or milk or grain production, etc., whereas in natural selection, the greater general fitness of some types means their increased survival. Human breeding programs make fitness literally proportional to meat or milk yield; the individual's genes are not preserved unless it "makes the grade" in that particular characteristic.

When selection occurs because one genotype leaves a different number of progeny than another, this may be the result of differences in *survival, mating,* or *fertility*. We should note that *pure assortative mating* does not change the gene frequencies in a population, so it is not selection (Crow and Kimura, 1970). It could change gene frequencies if some did not mate or if there were variation in mating success. In pure assortative mating, all genotypes make the same average contribution as a product of their assortative mating. Where there are selective differences in mating or fertility, fewer offspring are left as products of genotypes in particular matings, but not in others. Ignoring mating, then, as a cause of differential reproductive success, we see further that selection involving mortality and fertility is almost always complicated. One consequence of differential *mortality* is that a population counted at any stage except the zygotic stage will usually depart from Hardy–Weinberg ratios even when mating is random. This would suggest that the proper time to census a population would be as soon as possible after fertilization. In contrast, from the standpoint of assessing the effects of random gene frequency drift, it is more meaningful to count adults at the beginning of the reproductive period (Wright, 1931; Fisher, 1939). When probability of mating or fertility is being considered, it may be more meaningful to measure fertility of mating pairs than that of individuals (Bodmer, 1965).

TYPES OF SELECTION

The three basic modes of selection in a population can be described as (1) stabilizing selection, (2) directional selection, and (3) disruptive selection. The expected effects of each of these types of selection on genetic variation in a population are shown diagrammatically in Figure 3–1.

Stabilizing selection has been defined as discrimination against

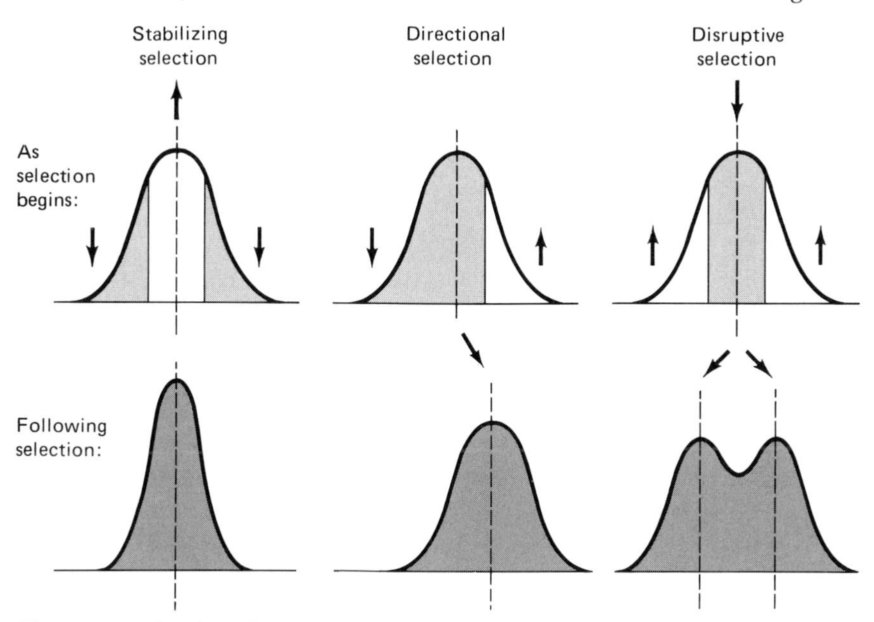

Figure 3–1 The three basic modes of selection and the change in genetic variance expected from each. Adverse (↓) and favorable (↑) selection pressures are exerted on various parts of the population frequency distribution of a variable phenotypic character. The *ordinates* (vertical axes) represent the frequencies of individuals in each population, and the *abscissas* (horizontal axes) show the phenotypic variation.

phenotypically peripheral individuals of a population by natural selection (Mayr, 1963). Selection of this type has also been called centripetal selection by Simpson (1953) or normalizing selection (Waddington, 1957). In this situation, selection operates to maintain an optimal phenotype and discriminates against any deviants that may segregate in the normal course of recombination of genes and alleles. This is a commonly observed type of selection in nature for which many examples could be cited. The classic paper is by A. C. Bumpus (1896) who looked at differential survival in samples of English sparrows rescued from a severe snowstorm in New England. The sparrows that survived were closest to the mean values for this species in many characters, but the same characteristics varied more widely in the dead sparrow samples. Mertens (1947) compared immature samples and adult samples of *Natrix natrix* snakes: 26$\frac{2}{3}$ percent of the snakes in the immature group had unusual variance of the postocular scales while only 13 percent of adult samples had variance at these scales, indicating higher mortality of extreme types during maturation. During winter, extreme variants of the wasp *Vespa vulgaris* are differentially eliminated from hibernating populations in North America, so that population variability in spring is much reduced from that present before hibernation (Thompson et al., 1911).

In directional selection, a shifting environmental factor causes selection to shift gene and genotypic frequencies until the new balance is achieved. As in stabilizing selection, a single optimum is reached, but the favored group during selection is made up of phenotypes other than the modal class. Directional selection is the commonest type of natural selection, and a great many examples have been described. One of the most dramatic of these cases is industrial melanism, which has affected the evolution of at least 80 species of cryptically colored moths in Britain during the past 100 years and at least 40 species of moths in temperate North America. Until heavy industrialization deposited soot on the barks of forest trees, light-colored individuals of these moth species were relatively abundant and had cryptic coloration that made them inconspicuous on light-barked or lichen-covered tree trunks. With increased smoke pollution, soot was deposited on tree trunks downwind from industrial areas in England and eastern North America, which made light-colored individuals highly conspicuous on darkened tree trunks. The very rare melanic forms of these moths, which appeared from time to time by spontaneous mutation prior to the industrial revolution, now possessed a selective advantage, for they are relatively inconspicuous on soot-covered tree trunks and consequently are overlooked by birds (Figure 3–2). With the light-colored moths suffering a high mortality and the dark-colored forms surviving at a high rate, the populations of these moths in industrial regions have become virtually completely melanic in phenotype (Kettlewell, 1956, 1973).

Figure 3–2 Industrial melanism in the English moth *Biston betularia*. (a) The whitish form *typica* and its melanic form *carbonaria* on lichened oak trunk in Dorset; (b) form *typica* and its melanic form *carbonaria* on polluted oak trunk in the Birmingham area. (Photos by H. B. D. Kettlewell. Source: B. Kettlewell, 1973, *The Evolution of Melanism*, Oxford Univ. Press [Clarendon], New York, p. 113.)

One of the earliest experiments showing directional selection was with the praying mantis, *Mantis religiosa*. Cesnola (1904) tethered green and brown individuals of this mantid on patches of green and brown grass. After 19 days of leaving brown and green mantids exposed to natural predation, Cesnola found that 35 out of 45 brown insects had been eaten by birds, whereas none out of 20 green insects had been eaten on green grass. In the brown grass plots, none out of 20 brown mantids was eaten in 19 days, whereas 25 out of 25 green insects had been eaten within 11 days.

In disruptive selection, multiple forces of selection favor two or more diverse phenotypes and discriminate against intermediates. This type of selection leads to discontinuity of variation and ends up in either (1) polymorphism or (2) divergence through isolation or sympatric speciation. In polymorphism, each selective optimum is represented by a distinct morph. The population remains structurally integrated and intact. Batesian mimetic polymorphisms are good examples of the result of disruptive selection where the population remains an effective unit. These must be maintained by strong selection to have them work efficiently with predators. In Batesian mimicry, an edible species mimics a distasteful or poisonous species and predators, confusing the mimic with the model, will avoid both. However, the mimic must not become too common in relation to the model or predators will eat a number of mimics before they ever taste a noxious model. This fact would tend to restrict the population size of the mimic species to conform to the abundance of the model species. However, in tropical areas, some mimic species circumvent this problem by being polymorphic, with a *series* of mimic color patterns, each phenotype resembling a different model species in the general area. For instance, a number of female morphs in the African swallowtail butterfly *Papilio dardanus*, each mimicking a different model, have developed in many regions of Africa. This results in the achievement of maximum efficiency of mimetic polymorphism and high population levels. Otherwise, the population will become fixed for one morph. If that happens, a lower number of adults would be required because of the selective necessity of maintaining a low mimic-to-model ratio.

In divergence with subsequent isolation or sympatric speciation, the population splits up, and the segregates often develop habitat preferences or seasonal and temporal differences in activity. Reproductive isolation results. Let us examine how divergence may occur as a result of disruptive selection, with divergence occurring either in isolation or with sympatric populations.

Isolation by disruptive selection was demonstrated in a classic series of experiments by Thoday and Gibson (1962) with fruit flies, *Drosophila melanogaster.* Although the flies were allowed to choose their own mates in this experiment, reproductive isolation resulted because of disruptive selection. A typical experiment was started by counting the number of sternopleural bristles on the thoraxes of 80 males and 80 females, and selecting from them the eight males and eight virgin females with the highest number of bristles, together with the eight males and eight virgin females with the lowest number of bristles. These 32 flies are placed together in a small vial for 24 hr. After this period, the males are discarded and the females are separated into "highs" and "lows" once more. The females are then placed in culture bottles.

When the progeny flies hatch, 40 males and 40 virgin females are collected from the "high" cultures whereas similar numbers are collected from the "low" ones. The eight highest bristle-number males and females of the "high" line and the eight lowest males and females of the "low" line are again placed together in the small mating vial for 24 hr. Males are again discarded; females are separated into "highs" and "lows" and are placed in culture bottles. The cycle is repeated each generation (Wallace, 1968).

Results of the typical experiment are shown in Figure 3–3. Within 12 generations of selection, there has been a noticeable divergence between the "high" and "low" lines of *Drosophila,* despite the opportunity for random mating between flies of both types when they were grouped as virgins in a single mating chamber. Apparently this result comes from our selection of the highest and lowest offspring, when we are undoubtedly selecting offspring from high × high and low × low matings. We are exercising disruptive selection here on the offspring of particular matings. So selection for flies of different bristle numbers will also represent selection for sexual isolation, assuming there is a genetic basis for sexual preference between flies with similar numbers

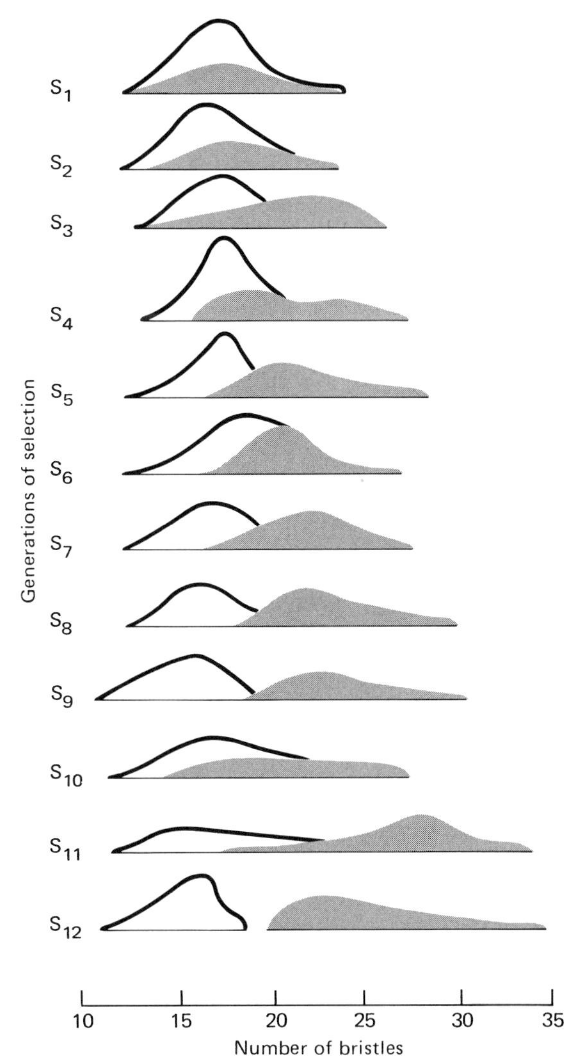

Figure 3–3 Divergence of "high" and "low" lines of *Drosophila melanogaster* obtained by Thoday and Gibson, despite the opportunity for the occurrence of random mating provided by placing males and females with high and low bristle numbers together as virgins in a single mating chamber. (Source: B. Wallace, 1968, *Topics in Population Genetics,* Norton, New York, p. 394. After J. M. Thoday and J. B. Gibson, 1962, *Nature, 193:* 1164–1166.)

of sternopleural bristles. In our experiment, females with high bristle numbers that accept males with low bristle numbers will not contribute to the selected progeny. When Thoday and Gibson did individual mating tests, it was indeed shown that disruptive selection operated through the development of sexual isolation. We should also note that the accumulation of different sets of modifier genes is involved by our selection of high and low bristle-number offspring.

A situation in nature that has apparently resulted in sympatric speciation through disruptive selection has been recorded for the "frit" flies in the genus *Rhagoletis,* where foodplant races develop sympatrically and disruptive selection results in speciation (Bush, 1969). Larvae of these flies feed within developing fruits of many plant species such as cherries, blueberries, apples, currants, walnuts, and tomatoes. Courtship and mating in *Rhagoletis* occur on the larval host plant; thus, there is a direct correlation between mate and host selection. Bush (1969) presents compelling evidence that this characteristic coupled with other biological attributes of these flies has likely led to some members of certain groups of sibling species (that is, closely related but distinct species) having evolved in sympatry as a result of minor alternations in genes associated with host plant selection. Of particular interest in the *Rhagoletis pomonella* species group is a new host race that was established on introduced apples a little over a hundred years ago from the original hawthorn-infesting form in the Hudson River Valley. Today, sympatric populations of the apple and hawthorn races have slight but significant differences in morphological characters sufficient to establish them as at least distinct races. The initial infestation of apples apparently occurred around 1865, and this racial attribute spread rapidly through the rest of the Northeast by 1916. A striking difference is apparent in the emergence times of these two races. Both races emerge from the pupal stage at a time when their

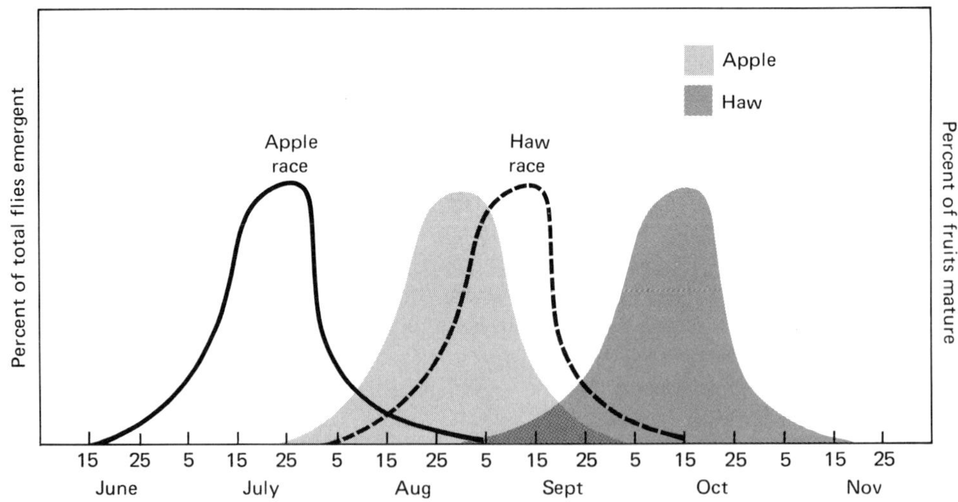

Figure 3–4 Emergence period of the apple and hawthorn races of *Rhagoletis pomonella* in the northeastern United States. The cross-hatched curve represents the approximate period of fruit maturation when larvae are leaving the fruit and pupating in the soil. (Source: G. L. Bush, 1969, *Evolution, 23:* 237–251.)

respective host fruits are in the appropriate condition for oviposition. Thus, the apple race emerges and begins oviposition approximately four to five weeks before the hawthorn race (Figure 3–4). The two populations are isolated seasonally from one another. In addition to the separation caused by using different fruits for mating and oviposition substrates, this seasonal difference effectively minimizes gene flow between them. Reproductive isolation between host races and their parental populations are documented for a number of other cases. Bush (1969) concludes that such factors as isolation on related hosts with different fruiting times, disruptive selection, and conditioning as to choice of food, may considerably reduce gene flow between host races and lead to the rapid sympatric evolution of host races and sibling species.

Storage and protection of genetic variation in balanced polymorphisms is of sufficient importance to our discussion of evolution at the population level that we next consider these subjects at length.

BALANCED POLYMORPHISM

Ford (1940) defined genetic polymorphism as "the occurrence together in one habitat of two or more forms of a species in such proportions that the rarest cannot be accounted for by recurrent mutation." Hence, in many populations, we find naturally occurring distinct variations which have relatively simple genetic inheritance. If these genetic polymorphisms persist in a population, one generally assumes that they are maintained by selection in favor of the heterozygote or by opposing selective forces which are in balance. Thus, in human populations, we have various blood group systems such as the *ABO*, *MN*, and *Rh* allelic systems, or polymorphisms in body biochemistry such as the ability and inability to taste PTC. At least three alternatives come to mind in attempting to explain the presence of these alleles.

Chance

Explaining the existence of a polymorphism by chance would suggest that the several common alleles at a given locus that exist within that population are there because these alleles arose by mutation and persisted because they are selectively neutral. In case after case, this explanation has been advanced and then later shown to be wrong. A selective difference has always been found. Even if there were selective neutrality of alleles at a particular gene locus, the most likely result would be loss from the population of all but one allele for each locus. There would be no selection at all favoring a new allele at that locus, and because perhaps one in 10,000 individuals would possess that mutant initially, it is extremely unlikely that chance would establish a new allele in that population.

Selection in favor of heterozygotes

This is the most common type of selection that results in retention of two or more alleles in a population. Here the heterozygote is selectively or adaptively superior to both homozygotes. Thus, in the malarial areas of East Africa, a heterozygote with the sickle cell anemia allele has adaptive superiority because of his resistance to malaria and his lack of acute anemia. A person with normal hemoglobin is subject to malaria, whereas the person who is a recessive homozygote for the sickle cell anemia allele will normally die before reproductive age of sickle cell anemia. The selection in favor of the heterozygote preserves the deleterious sickle cell allele in these populations. In North American black populations where malarial protection is not currently of selective importance, the sickle cell allele is gradually being selected out and is decreasing in frequency. The incidence of the trait in the slave population about 300 to 350 years ago was likely at least 22 percent (Allison, 1954) and in about 12 generations it has fallen to around 9 percent (Neel, 1951).

Selection in favor of a rare allele

In certain types of selective situations, rare alleles are favored, and this will perpetuate genetic variation within populations. Two types of selection for rare alleles are found in polymorphisms for self-sterility alleles and in apostatic selection.

In many plant species, among them various species of the genus *Oenothera,* new alleles that confer self-sterility on a plant are favored, and thus an extensive polymorphism for these alleles may be established. Pollen carrying a given allele at a particular locus (the self-incompatibility locus) cannot germinate on the female stigma or grow down the style of a plant whose genotype includes the same allele. Thus, no a_1 pollen can fertilize eggs of $a_1\ a_x$ plants, where a_x stands for any other allele at the *a* locus. The advantage of self-incompatibility systems of this sort seems to lie in the protection they afford against inbreeding in populations. A plant species must have at least three alleles in the population to utilize a self-sterility system, for it is only in such cases that pollen of a given sort (a_1, for example) encounters a plant ($a_2\ a_3$) upon which it can germinate and send a sperm cell down the style. Hence, every new self-incompatibility allele will have an advantage over the previously well-established ones in a population, because it can function on every common genotype among that group of plants. However, the more alleles that enter the population, the less the advantage of the new allele as compared to the old ones. Wright (1964) found 45 self-incompatibility alleles in a population of *Oenothera organensis* numbering no more than 1000 and probably fewer then 500 individuals. This system, then, is capable of maintaining a considerable number of alleles in a population in a balanced polymorphism.

With reference to predator selection on a prey population, a rather similar situation develops with rare alleles affecting color or other visible traits in a population. This has been termed *apostatic selection* by Clarke (1962). In this type of selective situation, predators exert frequency—dependent selection on their prey species. That is, the commonest genotypes are most often selected by predators; hence, rare genotypes are favored by selection. Predators apparently develop a searching image on one or a few common varieties of prey that they frequently encounter, and then tend to overlook rarer forms even if they may be obviously colored. Rare varieties, then, largely escape predation; thus, Clarke describes this type of selection as "apostatic" because it involves the advantage of rare phenotypes that stand out from the norm. It may be an important factor in promoting the evolution of diversity within species (in the form of polymorphisms) and between species also, according to R. A. Fisher and Clark. Considerable evidence has accumulated that selection of this sort may be quite unrelated to the cryptic or visual properties of the morphs involved. Hence, in the land snail, *Cepaea nemoralis,* Clark reports a study by Arnold who took two morphs and offered different frequencies of these morphs to wild thrushes on a uniform background of leaf litter in beech woods. Arnold found that the likelihood of predation for a particular morph was negatively related to its frequency; that is, the more abundant prey formed a search image right away for the thrushes, and they tended to overlook the rarer form. Apparently no protective coloration was involved, for these snails looked equally conspicuous to human eyes, and of course the frequency experiments described above showed no effects of crypsis—the predator almost always took the more abundant form.

Alternative selective explanations for balanced polymorphism in these *Cepaea* snails include natural selection against noncryptically colored snails, and physiological selection. In many areas, snails with particular color patterns matching their environment seem to escape predation more than their fellows, and in other areas physiological differences in morphs with different color patterns seem to result in differential survival. How can we reconcile these various alternative explanations? First, there is no particular reason why one factor cannot be operating at one locality and not at another. That is, in areas where bird predation may not be extensive or where soil type frequencies or habitat environmental differences are great, physiological selection may be most important. Also, in areas of uniform habitat where no morph has a particular cryptic advantage, apostatic selection may be more important than either of the other two types. A further important clue is provided by a very interesting study done in Africa by D. F. Owen (1963) on polymorphism and population density in the African land snail, *Limicolaria martensiana*. Owen's data suggest that the fitness of several color forms may vary with the density of the population. *Limicolaria* is very abundant in Uganda and occurs in well-defined isolated populations. Also, it is highly sedentary. It lives on or near the ground, feeding on both living and rotting vegetation. It is polymorphic in shell pattern, and Owen found that he could easily distinguish four forms (streaked and three different pallid forms). He then determined the relative frequency of these color forms in four populations, all within a 1-mile radius in the Kampala area.

Comparing the relative frequency of color forms with the population density in these four populations (Table 3–1) we see that in the areas of highest density, the greatest relative frequency of pallid morphs occurs. In the area of lowest density, there are no pallid morphs, yet all these populations occurred within a mile of each other. We could explain this distribution in several ways. The area in which the most snails exist might represent the area most favorable to the species, and perhaps many different genotypes would thus be able to exist there as compared to a less favorable area. Alternatively, we could suggest that apostatic selection might be affecting the polymorphism and depend in part upon population density. Owen's data support this second explanation more than the first.

The streaked form of the snail is cryptically colored and is difficult to see (at least to the human eye) when in natural situations, but the pallid forms do not seem at all cryptic; in fact, they are even conspicuous. A great many predators eat these snails so one would expect crypsis to be of value. What are the possible effects of apostatic selection in this situation? Predators form a search image for the common color form, and they probably form this image much faster when the population density of the prey is high than when it is low. Owen suggests, then, that the streaked form may be at a selective advantage

Table 3–1 Relative frequency of color forms and population density in four populations of *Limicolaria martensiana*.

POPULATION	DENSITY (PER m^2)	PERCENTAGE				SAMPLE SIZE (*N*)
		STREAKED	PALLID 1	PALLID 2	PALLID 3	
1	>100	61.4	16.2	19.2	2.7	1594
2	15.3	68.4	15.2	12.9	3.5	3455
3	8.0	80.0	10.0	10.0	—	428
4	<1	100.0	—	—	—	382

SOURCE: Owen (1963).

because of its cryptic coloration as long as the density of the population does not exceed a certain critical level. At that point, contrastingly noncryptic forms may receive an advantage because the predator's search image does not include them. Thus, the larger the population, the more morphs because each new allele producing a new morph will be at an advantage in that population until it reaches an equilibrium level. This study by Owen offers evidence that the controlling selective explanation in at least some polymorphisms in nature should depend on population density; that is, at low population densities, selection would occur primarily on a cryptic basis, whereas at a high population density a greater priority would be placed on apostatic selection.

Polymorphic systems are remarkably stable in natural populations. Their advantage seems to lie in providing a versatile biochemical or developmental system involving, for example, two or more gene products, two or more times of action for a given locus to act, or two or more sites of action on membranes within the cells. Polymorphism seems to be the most likely solution to arise in a population within a short time period, at least in cases in which the heterozygote is selectively superior. An inefficient and long-term process that could create the same effect would be by gene duplication; that is, with duplication of a segment in the chromosomes we could add several different alleles to the same member (homologue) of a pair of chromosomes. But this is a long-term process occurring by chance and one which is unable to account for rapidly rising polymorphisms.

Many examples of the long-term persistence of balanced polymorphism exist in nature. Geographic distributional patterns of gene arrangements (inversions) in the chromosomes of *Drosophila pseudoobscura,* studied by Dobzhansky and his colleagues during the past 40 years in the southwestern United States, have remained essentially constant. No formerly polymorphic population has become monomorphic. In contrast, local changes such as in areas downwind from the smog-producing areas of Los Angeles show adjustments in gene-arrangement frequencies. Populations of Mexican platyfish (*Xiphophorus maculatus*) in various streams across northern Mexico vary in frequency of spotting morphs. Gordon (1947) showed that the alleles controlling spotting patterns in these fish have remained at approximately the same frequency in various localities in Mexico across a period of some 70 years, based on preserved samples in museums and modern samples. The polymorphism in shell color pattern in the African land snail, *Limicolaria martensiana,* shows the same stability over time. Owen (1966) found a fossil site in Western Uganda where shells of this species have been preserved in volcanic ash strata. The volcanic eruptions that covered these snails occurred about 8000 to 10,000 years ago, so the fossil snail shells are apparently that old. Owen found that colors and patterns of these fossil shells were quite adequately preserved and hence he could collect a sample of 1277 shells and classify them, and compare the frequencies

Table 3–2 Comparison of frequencies of shell patterns in fossil and living population samples of *Limicolaria martensiana* in Western Uganda.

	FOSSILS	LIVING		
COLOR FORM	KABAZIMA ISLAND ($N = 1277$)	KAYANJA ($N = 2840$)	ISHASHNA RD. ($N = 882$)	RWENSHAMA ($N = 841$)
Streaked	61.0	54.6	40.6	33.7
Broken-streaked	5.2	8.9	4.2	9.8
Pallid 1	3.9	24.0	7.0	1.9
Pallid 2	28.3	11.9	43.2	37.3
Pallid 3	1.6	0.6	5.0	17.3

SOURCE: Owen (1966).

to those of living populations. Several points may be noted from Owen's comparisons in Table 3–2. These fossils show that the five listed forms of *L. martensiana* are at least 8000 to 10,000 years old. Also, the fossil forms do not include any that are not found today. Finally, the overall proportions of morphs in the fossil population are much like those found in living populations of the same species in adjacent areas. This suggests that after the basic polymorphism arose, conditions have remained relatively static in a balanced polymorphism and that, in Owen's words, "Evolutionary trends in the past 8000 to 10,000 years have undoubtedly been in repeated adjustments of the frequencies of the forms to local conditions with occasional extermination or spread of a form into a new population and with periods of relative stability."

Genetic variation may be extraordinarily reduced in animal species in populations that are approaching extinction. Blood samples from northern elephant seals (*Mirounga angustirostris*) representing five breeding colonies in California and Mexico were recently surveyed electrophoretically for protein variation reflecting underlying genetic differences (Bonnell and Selander, 1973). No polymorphisms were found among 21 proteins encoded by 24 loci. This uniform homozygosity may be a consequence of fixation of alleles brought about by the decimation of this species by sealers in the last century. Because the northern elephant seal now lacks a pool of variability with which to adapt to changing conditions, it may be especially vulnerable to environmental modification. The absence of protein polymorphisms at these selected loci in this animal, however, indicates that genic variability is not necessarily essential for the continuing existence of animal species. Of course, these seals may well be polymorphic at other gene loci that were not tested.

COADAPTATION AND DOMINANCE: GENETIC CHANGES ASSOCIATED WITH POLYMORPHISMS

The storage and protection of genetic variation in polymorphisms involves more than mere retention of two or more alternative forms of a given gene in a population. Interactions among the genes on a chromosome evoke selective changes at other loci, and the selection may affect an entire network of gene-controlled reactions. Two outcomes of selective changes that take place within the network are *coadaptation* and *dominance*.

Genes are said to be coadapted if high fitness for the individual depends on specific interactions between these genes (Wallace, 1968). Individuals in the Mexican populations of the platyfish *Xiphophorus maculatus* may possess a variety of black markings including various spots and stripes (Figure 3–5). The purpose of these markings in nature is unknown; they may serve as camouflage, as warning flashes, or as recognition signals during courtship. Whatever the reason, the entire genetic system of morphs seems to be one stabilized by selection because there are a variety of gene-controlled patterns in all the populations studied (therefore one assumes the polymorphism must be important), and second, there has been long-term stability of the relative proportions of these patterns in the various river basin populations.

Crossing fish from the same locality, Gordon (1957) found that different patterns were caused by different alleles at a main locus—the standard situation for a simple genetic polymorphism. However, when hybridization crosses of fishes from different localities were carried out (using fish with the same morph patterns), results were quite unexpected (Figure 3–6). For instance, crosses involving "dorsal-spotted" individuals from two different river basins would produce F_1

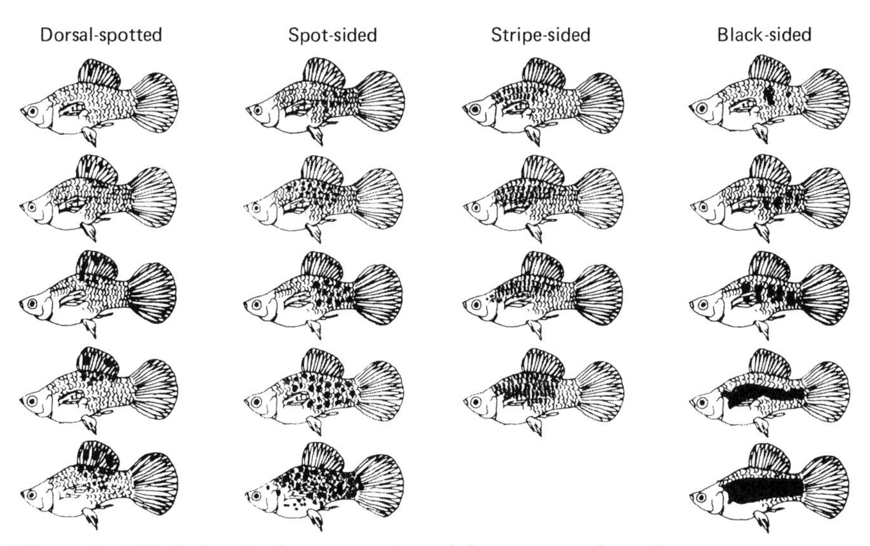

Figure 3–5 Variation in the expression of the macromelanophore pattern genes in the Rio Jamapa and Rio Coatzacoalcos populations of the platyfish, *Platypoecilus maculatus.* (Source: B. Wallace, 1968, *Topics in Population Genetics,* Norton, New York, p. 302. After H. Gordon and M. Gordon, 1957, *J. Genet.,* *53:* 1–44.)

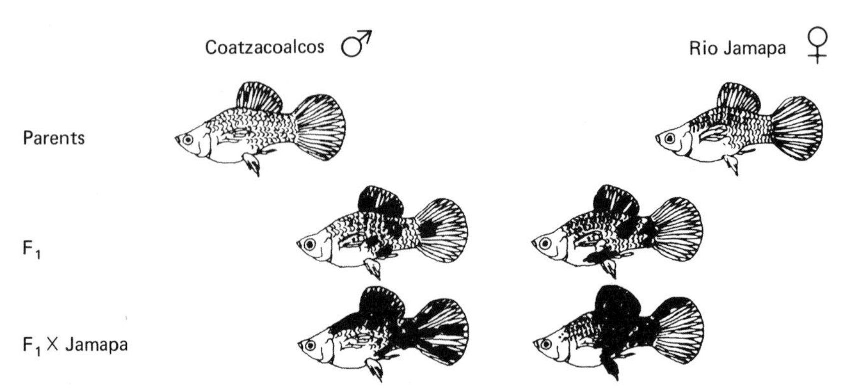

Figure 3–6 Parents, F_1 hybrids, and backcross hybrids of platyfish from two isolated river basins, Rio Jamapa and Rio Coatzacoalcos. (Source: B. Wallace, 1968, *Topics in Population Genetics,* Norton, New York, p. 305.)

hybrids having heavy melanic spots appearing at various unexpected places on the fish's body. If a backcross was done to individuals of the second river basin, the "dorsal-spotted" allele from the first river basin caused enormous numbers of pigmented cells to blacken nearly 50 percent of the fish's body. Clearly, then, within any one population there exists a need for the black patterns formed by macromelanophores and for a precise control over the deposition of these melanic cells. The reasons are unknown. The solution to these two needs lies not entirely with the alleles at the polymorphic locus, but with the interplay of these alleles and the background genotype. If one of these alleles is removed from its proper background and is inserted in another background, potentially dangerous melanotic growths occur. When we have a situation like this and high fitness depends on the proper interactions between genes, we say that the genes in question are coadapted. Coadaptation arises as the result of selection for a well-functioning network of genes whose genetic basis can be transmitted as an interdependent system from one generation to the next.

Similarly, wide geographic crosses in species of the satyrid butterfly genus *Cercyonis,* especially those involving California and Colorado population of *C. pegala boopis,* result in chaotic spotting patterns on the hindwings of F_1 offspring. Crosses done between spotting morphs within either California or Colorado populations result in a relatively simple and regular genetic inheritance of spotting morphs. In contrast, crosses done between Point Reyes, California and Glenwood Springs, Colorado butterflies show no simple spotting outcome but rather a full range of variation, with every butterfly different and some spots misplaced from their normal locations to new areas on the wings (Emmel, unpublished).

Clarke (1969) postulates that the so-called "area effects" found in *Cepaea* and *Partula* land snails may represent areas of coadapted gene complexes in these land snails. Area effects were described over a decade ago among English populations of *Cepaea nemoralis* and *C. hortensis*, in which a few morphs would predominate over wide areas, regardless of habitat or background. Between such areas, the morph frequencies might change violently within distances of 200 meters or less, often in apparently uniform environments. The same sorts of dramatic changes have been found in *Partula* land snail populations over even shorter distances in uniform habitat on Moorea, an island in the South Pacific (e.g., Clarke and Murray, 1969).

Various authors originally suggested that these area effects might be caused by cryptic environmental differences that produce sharp changes of selective forces over short distances. Alternative explanations of genetic drift or the founder principle do not seem tenable. Instead, as Clarke suggests, it may be more logical to assume that an originally uniform series of populations might become differentiated into a series of separate coadapted gene complexes. That is, large regions could be inhabited by a series of identically polymorphic populations, but because of local environmental differences, some alleles would be favored in one part of the region but not in another. Selection would favor other genes that are compatible with an allele that is of primary importance in a particular local environment. These modifier genes would become integrated into a coadapted gene complex with the main allele controlling color pattern and other variations. This process would give rise to regions of comparatively uniform morph frequency but separated from other such regions by narrow zones of transition. In other words, it could give rise to what we call "area effects." The zones of transition would likely involve changes in a large number of genes, and the genetic disturbance caused by hybridization between the coadapted gene complexes might be sufficient to cause genetic or developmental instabilities in these snails (or whatever other organism might be involved).

The evolution of dominance rises from the same general genetic basis as coadaptation of genes. Dominance and recessiveness refer primarily to the presence or absence of a primary gene product, such as a functional enzyme. In other words, dominance is a term that refers to the phenotype, not to the gene itself. To summarize the principal present view of the evolution of dominance, which was originally put forth by R. A. Fisher, dominance is an *acquired* characteristic. New genes acquire dominance, and the old wild-type allele loses it through the modification of gene action by the background genotype. Fisher especially emphasized the effect of selection in heterozygous individuals for modifier genes to enhance the dominance of a particular allele. There are two important series of experimental observations on evolution of dominance: industrial melanism and dominance in mimicry complexes.

During the evolution of industrial melanism in the moth *Biston betularia* since the early nineteenth century, the dominance of the melanic form has been increasingly perfected from the visual standpoint. Early heterozygotes of this moth have been preserved in museum collections and show various white lines on the wings as well as pale patches. In modern forms, these markings have disappeared so that present-day heterozygotes are essentially indistinguishable from homozygous melanics. A considerable portion of this effect has been brought about by modifier genes. Kettlewell (1973) demonstrated this in crossing a dominant melanic allele from a population near industrial Birmingham into a hybrid background of a nonmelanic population of the western, nonindustrialized section of Britain. The result was that the artifically bred heterozygotes had greater numbers of white scales than the normal Birmingham heterozygotes. Crosses of melanic versus normal genotypes in other well-isolated populations of various moth species showed the same results. Crosses within a particular population show the melanic form as a good dominant giving a 3:1 ratio in the F_2 with only a couple of intermediate moths being hard to classify. However, matings involving interpopulation heterozygotes produce many intermediate forms, with typical individuals accounting for one-fourth of the total progeny in the F_2 as expected. The best explanation of these effects is that modifier genes have become established in each population to perfect the dominance of the melanic alleles. When crossed into a different genetic background, the dominance of that allele is no longer expressed as perfectly because of the absence of its coadapted modifier genes.

Dominance in mimicry presents additional evidence for the evolution of the genetic control of dominance through modifier genes in the total gene complex. In Batesian mimicry complexes, a model species that is poisonous or otherwise harmful normally occurs in fair abundance in an area with a mimic that resembles the model and thereby gains protection from its deception of predators. The mimic may not become too common in relation to the model or predators will associate its appearance with a good edible butterfly rather than with the bad taste of the model. Thus, many butterflies that are mimetic have a polymorphism of mimic patterns. In the African swallowtail *Papilio dardanus,* female morphs mimic a number of distasteful danaid butterfly models in a particular area. In areas in which models are quite common relative to mimics, imperfect mimics are extremely rare, as in Entebbe on the northern edge of Lake Victoria in East Africa (Table 3–3). To the southeast near Nairobi, models are much rarer than their mimics, and in this area over 30 percent of all mimics are imperfect in design. These data (Ford, 1964) suggest that natural selection is not sufficiently strong to maintain the proper modifiers for the main mimetic alleles in an area where models are rare. Other evidence for the

Table 3–3 The frequency of imperfect mimics relative to models in female *Papilio dardanus* in two East African areas.

	MODELS	*P. dardanus* (♀) MIMICS	
LOCALITY	(TOTALS)	TOTALS	PERCENT IMPERFECT
Entebbe	1949	111	4
Nairobi	32	133	32

SOURCE: Ford (1964).

evolution of dominance in *Papilio dardanus* comes from wide geographic crosses in which imperfect mimetic expression of the main gene controlling color pattern results when it is crossed into an unusual (that is, different) genotypic background of modifiers, similar to the situation in the nonmimetic polymorphism of the Mexican platyfish.

UNITY OF THE GENOTYPE

A unity or cohesion of the genotype has been recently emphasized by Ernst Mayr and other authors as having the central role in the evolution of populations. In pre-Mendelian interpretations, species were taken as inseparable wholes. Post-Mendelian workers in some cases have gone to the other extreme, down to an atomotic approach; that is, the change in one gene or a change in gene frequency is defined as evolution. At this point, authors have lost sight of the importance of interactions of genes in pleiotropy and polygeny. Learner (1954) began the recent trend toward emphasizing the genetic homeostasis of a gene complex.

The evidence for unity in the genotype is varied and perhaps largely circumstantial. When one maximizes a trait with intensive artificial selection, undesirable side effects almost always result. If this selection is relaxed, then the character moves back most if not all of the way to its previous coadapted genetic state. Other evidence for the cohesion of the genotype lies in the fact that structural genes may account for only 50,000 of an average 5 million cistrons in an organism's DNA. The remainder, as regulator genes, may be more important in the functioning of the organism. Epistatic effects, for instance, may be largely caused by regulator genes. The narrowness of hybrid zones where two species meet is a commonplace observation, yet this hybrid zone ought to widen rapidly after meeting of two species owing to intergression of genes in backcrosses to the parental species. For example, in Europe the hooded crow and carrion crow joined together 8000 years ago, yet today the hybrid zone is only 100 miles wide (Mayr, 1963). Extensive study on linkage of genes suggests that an ultimate genetic strategy is to prevent recombination of favorable groups by putting genes together into one chromosome. This can actually be an evolutionarily efficient mechanism by tying together valuable genes into supergenes in unalterable chromosome strands. Speciation in such cases must involve the major reconstitution of such linkage groups.

The consequences of such cohesion are of vital importance to evolution on the population level. Because the fitness of an allele depends in part on the genetic background, the entire collection of genes that affects a particular factor is important. The process of selection does not involve single genes. Any given selective pressure simultaneously affects whole packages of genes. Ontogeny, that is, the developmental sequence of the organism, is the most important stage as regards effects of genes, though the outcome is usually read in the form of variation among the offspring when they themselves reach reproductive age. Genetic changes in the form of clines of characters, establishment of ecotypes, and so forth, reflect the adaptive nature of most geographic variation. Yet geographic variation is not as common as previously thought. Many species do not vary phenotypically across whole continents. Mayr (1963 and in earlier publications) was the leading advocate of the interpretation that this invariability in certain species was due to extensive gene flow. By 1973,[1] Mayr was saying that gene flow is still important but is not the main explanation for the cohesion of the phenotype of a species. Techniques of gel electrophoresis showed that often there did not even exist cryptic geographic variation in allelic forms of enzymes. Instead, there was commonly a constant set of enzymatic alleles in such phenotypically nonvariable species populations. So a cohesion or unity of the genotype across wide geographic ranges seems to be reflected in these gells. The uniformity of frequency of rare alleles, through time and space as in the Mexican platyfish, which we looked at earlier, show how strong this cohesion of the genotype can be.

The classic theory of geographic speciation said that geographic barriers usually separated a widely distributed species into two groups of populations, each of which would then diverge on its own in response to different selection pressure. Now it appears, according to Mayr (1973[1]), that this is the exception rather than the rule. Instead, it seems that a founder population on the periphery of a species range often supplies the material that will expose a cohesive genotype to strong and new selective forces that will allow conditions for speciation. Major chromosomal reorganization will be required to break up the cohesion of the genotype.

This emphasis on the cohesion and indeed conservative nature of

[1] *First International Congress of Systematic and Evolutionary Biology, Boulder, Colo., August 9, 1973.*

the genotype explains some outstanding problems in evolutionary biology. Persistent rudimentary organs, such as gill arches in vertebrate embryos, reflect ancestral conditions in ontogeny and presumably remain, under this explanation, because of the genes being locked into a cohesive genotype chracteristic of these animals. Cohesion of the genotype may also explain the extreme evolutionary inertia in such long stagnant evolutionary lines as the maidenhair tree, *Ginkgo*. The breakout from a cohesive genotype may explain the sudden explosive expansions in the fossil record of long stagnant lines in other groups. It may also explain such problems as why all terrestrial vertebrates develop as quadrupeds and all insects develop as hexapods. About 24 recognized animal phyla exist today. These are the modern representatives of a limited group of early founder organisms that solved particular problems; their basic characters have remained inherited for hundreds of millions of years down to today. Thus, the genes are not the units of evolution nor are they the targets of natural selection, but it is the inherited phenotypic expression of the *genotype* itself that becomes the unit for natural selection. In Mayr's words: "The most difficult problem in evolution has been to break out of the cohesion of the genotype."

Likewise, H. L. Carson (1973[1], 1975) recently emphasized a theory involving the unity of the genotype as an explanation for the ways in which organic evolution forms new species. Carson distinguishes the formation of adaptations from the formation of species. Formation of adaptations is well documented by evolutionists and involves microevolutionary events, which do not seem to result in new species, whereas the mode of origin of species is still a major problem in evolutionary biology. Carson proposes that there are two kinds of genetic systems in an organism: (1) an open variability system, where the genetic components are freely open to recombination and selection by natural selection; and (2) a closed variability system, where the elements of the system cannot be separated by selective pressure and still result in a viable organism under normal circumstances. In this latter case, blocks of genes are thought to be associated with a single homologue of a chromosome pair and perhaps captured in the form of inversions. Inversions could conceal a great deal of genetic variation in the closed genetic system, whereas in an open system, with recombination, great variability among progeny could result with selection removing the less viable offspring. The basic closed genetic system for a species seems to be obligatory and spread throughout that species. Gene flow would not be expected to affect this closed genetic system, which is probably the core of what constitutes a species. What might happen to change this closed genetic system? Most theories of allopatric or sympatric speciation deal with gradualism. Carson suggests saltational moves with rather catastrophic change (i.e., forced reorganization) in the closed genetic system. Natural selection would be forced to make something out of this disorganized collection of blocks of genes. Hence, the founder of the new species would be a rather unusual organism compared to its fellows, because of disorganization of its original genome. Carson feels that such disorganization could occur if natural selection was temporarily removed from a population since strong natural selection was necessary to maintain the closed system.

In Carson's view, natural selection of the hard type (*K* selection) is removed from some generations, but it would be hard to conceive of all *r* selection being eliminated for such periods. These two types of selection will be discussed in more detail in Chapter 5, but may be briefly defined here. In *r* selection (referring to the instantaneous rate of increase *r* in the population growth equation), a high rate of increase is favored in an unstable environment. Operating in populations below the carrying capacity of the environment, the favored organism is the generalist that utilizes resources rapidly and disperses into new habitats before any competitors arrive. In *K* selection, of principal importance in populations at or near the saturation level or carrying capacity *K* of a stable environment, the favored organism is the specialist who has good competitive abilities and a rather low rate of increase *r*.

When selection is absent, and during the resultant flush of variation, genetic recombination becomes very great. A severe population crash follows, with perhaps a single founder surviving, while significant selection is still absent. The founder individual survives not because of his genotype but by chance. He survives *in situ* despite the environmental disaster, is blown to another area, or somehow survives accidentally. The new closed variability system might be genetically incompatible with the old one, due to disruptive effects on the original closed variability system. These ideas, then, would explain the narrowness of hybrid zones and other phenomena, such as the fact that subspecies form through gradual development with selection on the open variability system, and do not result in the development of new species. Carson's definition of a species, then, becomes "a reproductive community of populations which share a common closed variability system." No hybridization occurs unless the selection is relaxed, and then variability may occur as well as speciation. A number of evolutionists are now attempting to test these hypotheses on cohesion of the genotype against the other explanations of evolutionary change that have been advanced.

ISOLATING MECHANISMS

Isolating mechanisms prevent the cohesive genotype of a species from being subjected to dilution or genetic interbreeding with other populations. The term was coined by Theodosius Dobzhansky in the first edition (1937) of his classic work *Genetics and the Origin of Species.* These mechanisms are defined as biological devices that prevent interbreeding of two sympatric or potentially sympatric species. In this sense, then, geographic isolation represents an extrinsic isolating mechanism because it operates purely through spatial separation. The principal isolating mechanisms may be classified as follows [after Mayr (1963) and Littlejohn (1969)].

PREMATING MECHANISMS

The isolating factors operate before mating can successfully occur.

1. *Habitat isolation*
   Individuals in the respective populations occur in different habitats in the same general region.
2. *Seasonal isolation*
   Mating or flowering periods occur at different seasons.
3. *Behavioral (ethological) isolation*
   Mutual response between males and females of different species does not occur to a sufficient degree to achieve mating.
4. *Mechanical isolation* (unsuccessful copulation)
   Mating or pollination is unsuccessful due to physical noncorrespondence of the genitalia or floral parts.

POSTMATING MECHANISMS

The isolating factors operate after mating but before the ensuing offspring can successfully reproduce themselves.

1. *Gametic mortality* (no fertilization)
   Sperm or pollen tubes of one species are not attracted to the eggs or ovules or do not survive in the reproductive ducts of another species.
2. *Zygotic mortality* (death of fertilized egg)
   The hybrid zygotes fail to develop because of genic incompatibility or lack of certain genes early in ontogeny.
3. *Hybrid inviability* (F_1 hybrid weak or dies before reproduction)
   The hybrid is insufficiently viable to reproduce.
4. *Hybrid sterility* (no, or deficient, functional gametes from F_1; F_2 affected).
   The hybrid F_2 or backcross offspring fail to develop or are sterile upon reaching the age for reproduction (normally because of chromosomal problems).

THE ROLE OF HYBRIDIZATION

Hybridization or the interbreeding of two unlike individuals is a result of the breakdown of isolating mechanisms. Mayr (1963) defines hybridization as the crossing of individuals belonging to two unlike populations that have secondarily come in contact. In our brief review here, we shall consider a few general principles concerning hybridization.

When individuals of two unlike populations cross, whether these populations be varieties of the same species or two different species, segregation in the F_2 and later generations from such hybridization produces an incredibly large number of recombinations and types. This result derives from Mendelian segregation of the genetic factors responsible for the intervarietal or interspecific differences (Stebbins, 1950). From the viewpoint of our preceding discussion on the cohesion of the genotype, it is especially interesting that correlations between groups of parental characteristics are always evident among hybrids. The very large number of recombination types is by no means a random sample of the total possible array of possible recombinations of the phenotypic characters of the parents. Many different recombinations are of similar types, near the intermediate line between the two parental species or varieties. Thus, many of these characters act as if they were linked. There are at least two explanations for this phenomenon. First, in flower characters such as floral tube shape and length, there is evidence that these differences are associated with genically controlled differences in hormone activity. Hence, a given gene present in the hybrid without a matching allele produces an intermediate effect with half a dose of the normally available amount of gene product. Second, genetic linkage will be more strongly evident in F_2 and F_3 populations of offspring if the character differences are governed by multiple factors than if they are controlled by single factors. For instance, in a hybrid tobacco plant with a low haploid number of chromosomes ($N = 9$), some linkage is almost certain to occur between any two characters which happen to be governed by nine or more genes. Such linkages will eventually be broken up in later generations, but they will have a great effect on the distribution of variants in the immediate progeny of a cross between subspecies or species.

It should also be noted that the partially sterile offspring of natural interspecific hybrids in the wild have a much greater chance of producing offspring from crossing with the abundant viable pollen or sperm and eggs of the numerous plants or animals of the parental species which surround them, than from the scant, poorly viable pollen or sperm and eggs produced by the few F_1 individuals that may be present. Thus, offspring of most natural interspecific hybrids are far more likely to represent backcross types to the parents than true F_2 segregates. This situation will be accentuated in later generations by natural selection. A combination of characters represented by the parental species have been tried over many hundreds or thousands of generations and have shown themselves to be adaptive. The chances that any new combination will prove equally adaptive are relatively small. Hence, the nearer a backcross segregate approaches one or the other of the parental species in characters of adaptive value, the greater its chances of survival. The expected normal result of interspecific hybridization in nature, then, would be the reversion of hybrid offspring toward one or the other of the parental species.

In its ultimate evolutionary importance, hybridization depends directly on the environment in which it takes place (Stebbins, 1950). This generalization is a direct outcome of the previous statements. Hybridization between well-established and well-adapted species in a stable environment will have no significant outcome, or will be detrimental to the species population—hence, the importance of isolating mechanisms. But if the crossing occurs under rapidly changing conditions or in a region that offers new habitats to the segregating offspring, many of these segregates may survive and contribute in a greater or lesser degree to the evolutionary progress of the group concerned.

GEOGRAPHIC SPECIATION

The primary cause of speciation has long been held to be geographic isolation of sets of populations, which allows genetic divergence to occur unimpeded by gene flow (Mayr, 1963). For intermediate stages in geographic speciation, one would expect to find geographic variation in species characters. Abundant geographic variation in morphological characters, behavior, habitat occupation, and even degree of development of isolating mechanisms, occurs in virtually all known groups of organisms. In many groups that have been analyzed throughout their areas of distribution, one finds populations in every stage of apparent divergence, from simple polymorphisms and local genetic strains or races through subspecies to semispecies to superspecies to distantly related species and members of different species groups or subgenera.

Sometimes this geographic variation will include a complicated mixture of species attributes. In the Florida mosquito *Aedes taeniorhynchus,* populations are polymorphic for diets required for ovarian development in the females (O'Meara and Evans, 1973). Mosquito populations inhabiting mangrove swamps have females which mostly possess the capacity to produce eggs on a blood-free diet. But where grassy salt marshes are the mosquito's principal habitat, most females lack this capacity. The ability to produce eggs without a blood meal increases from northern to southern populations (Figure 3–7).

Much evidence advanced in support of the concept of geographic speciation was consistently interpreted in the light of interruption of gene flow. In the view of supporters of the synthetic theory of evolu-

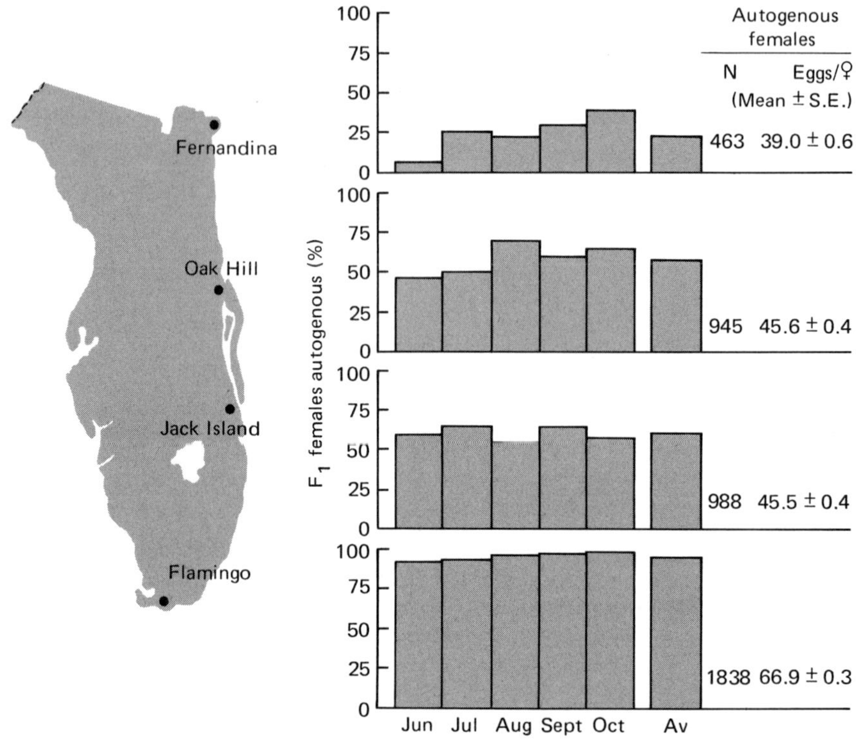

Figure 3–7 Geographic and seasonal variation in frequency and expression of autogeny (lack of requirement for a blood meal to develop and lay an initial egg hatch) in F_1 *Aedes taeniorhynchus* female mosquitoes in four coastal Florida sites. Monthly field collections of biting mosquitoes were made at each site from June through October 1971. Most mosquitoes in the southern mangrove swamps can produce eggs on a blood-free diet, but most females lack this autogenous capacity in the northern grassy salt marsh. (Source: G. F. O'Meara and D. G. Evans, 1973, *Science, 180:* 1291–1293.)

tion popular from the late 1930s to about 1972 or 1973, geographic speciation is an extension of the process of microevolutionary change, and this is implicit in the definition of a species. Thus, Mayr (1963) has defined species as "groups of actually or potentially interbreeding natural populations which are reproductively isolated from other such groups." This definition implies three corollaries, which Mayr expands extensively: (1) within a species the flow of genes is unimpeded, (2) the flow of genes within the species maintains the species' genetic and phenotypic integrity, and (3) differentiation is not possible unless gene flow is interrupted. Mayr (1963) further states that geographic barriers provide the commonest and only effective interruption of gene flow. Thus, he defines geographic speciation as "the genetic reconstruction of a population during a period of geographic isolation." Note that this statement can be interpreted in at least two major ways. Geographic isolation is necessary *per se* to interrupt gene flow. Also, selective forces will be different or be the same in each area and cause or prevent differentiation. The views of many biologists (for example, Ehrlich and Raven, 1969, and now even Mayr, 1973[1]) currently emphasize the significance of the type of selection present rather than the presence or absence of gene flow. In other words, gene flow may or may not be going on, but what is important is the amount and type of selection pressure present. We examine these views in more detail below, but first let us consider alternate interpretations for a classic situation involving evolutionary divergence.

On the various islands of the Galapagos one finds series of 15 different races of giant tortoises (Figure 3–8). The structural variation in shell type was pointed out to Charles Darwin in 1835 by the Spanish governor of the islands. The traditional interpretation of this distribution of differing shell types among the tortoises would be that geographic isolation had occurred as each new population was founded by a single individual drifting over from already populated islands. With gene flow interrupted because of the intervening water, the new population was free to evolve in any direction that local selective conditions warranted or demanded. The key question is whether this genetic reconstitution is accomplished *because* of the interruption of gene flow or if it is caused simply by strong selection. Previously, it was thought that gene flow prevented divergence but as we shall see in the following section, genetic reconstitution can be maintained in the face of terrific amounts of gene flow as long as there is strong selection. Recently, as we have seen above, evolutionists believing in the cohesion of the genotype have said that in a new situation such as a peripheral population on an isolated island, the founder population may be exposed to new selection pressures (or the lack of them) and the cohesive species genotype may break up and allow adaptive radiation. Hence, this would explain the diversity of tortoise types on different islands of the Galapagos Archipelago. Speciating populations pass through a genetic revolution in each case, with a highly unstable period involving a great loss of genetic variability. Breakdown of genetic homeostasis at this time brings about reconstitution or loss of previously existing balanced gene arrangements.

GENE FLOW AND DIFFERENTIATION OF POPULATIONS

Gene flow is said to be relatively common in nature, and the traditional view of its role is twofold: it prevents divergence of populations, and it promotes the cohesiveness of species. Let us look at several well-studied organisms to examine critically the importance of gene flow in maintaining population integrity and preventing or allowing variation to arise.

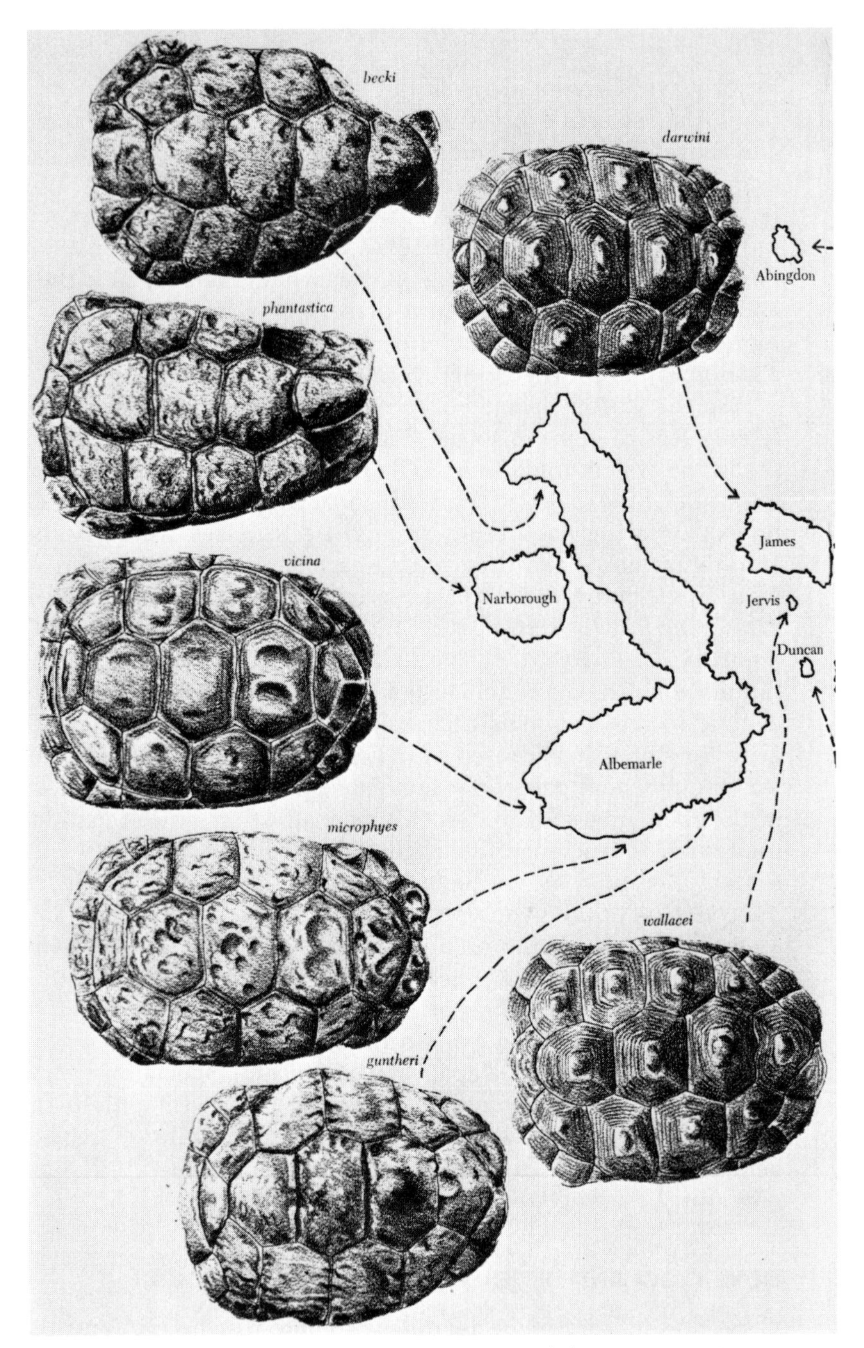

Figure 3–8 The carapaces of 13 of the distinctive races of the giant tortoise, *Geochelone elephantopus,* in the Galapagos Islands. Saddle-backed races tend to be present on the more arid islands, whereas dome-shaped forms are prevalent on the wetter central and southern islands. (Source: S. Carlquist, 1965, *Island Life,* Natural History Press, New York, pp. 370–372.)

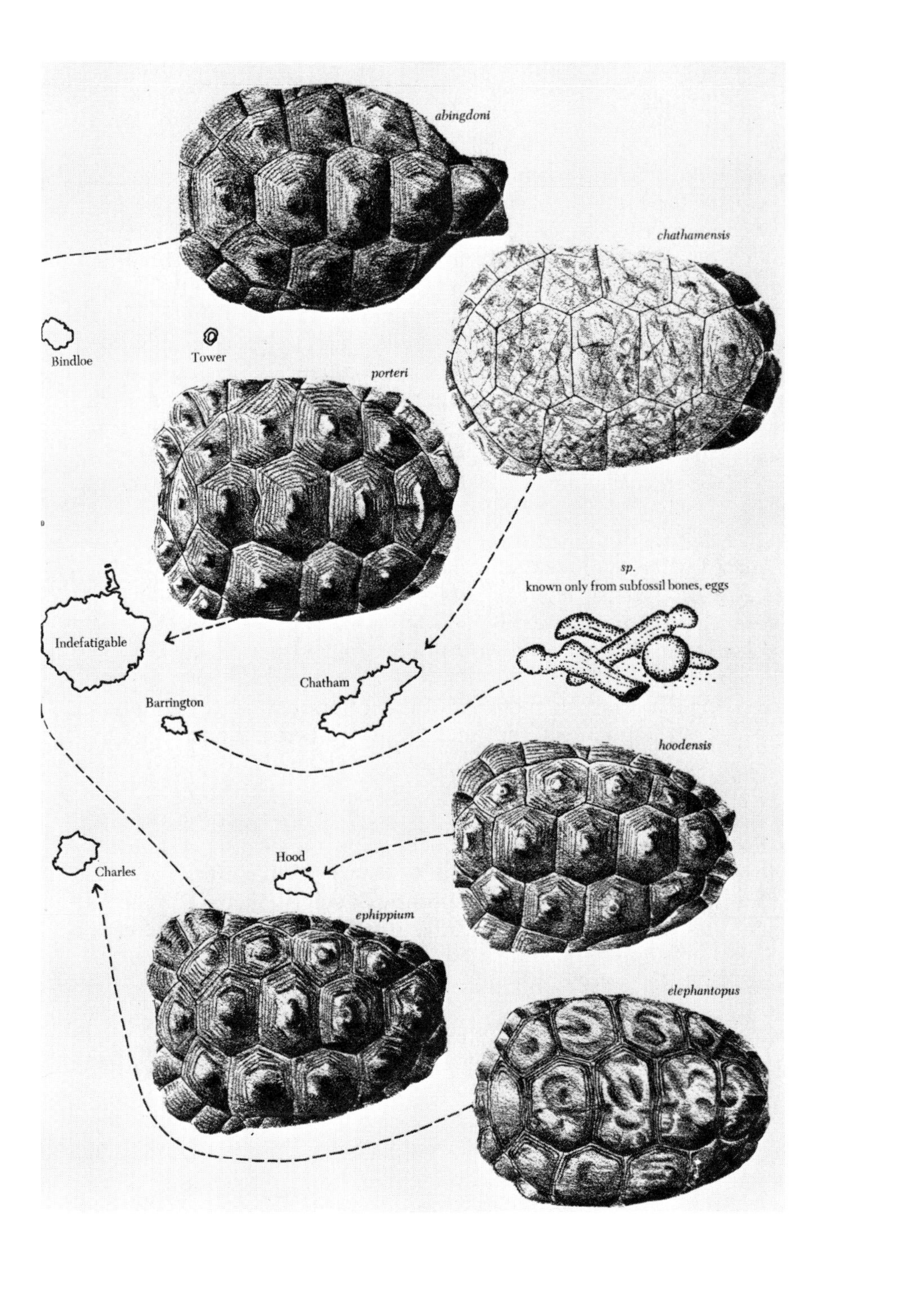
abingdoni
chathamensis
Bindloe
Tower
porteri
sp.
known only from subfossil bones, eggs
Indefatigable
Chatham
Barrington
hoodensis
Hood
Charles
ephippium
elephantopus

In a series of papers (1952 to present, and reviewed in Ford, 1965) on the satyrine butterfly *Maniola jurtina*, Dowdeswell and Ford and their co-workers have shown that the distribution of spotting on the underside of the hindwings is a sensitive index of variability and microevolutionary divergence. For instance, across southern England through a 230-mile range of cold and dry to warm and humid climates, we find a characteristic stabilization of spotting pattern with the males unimodal at two spots, and the females unimodal at zero spots on the hindwings (Figure 3–9). In other places, such as in the isles of Scilly off the coast in the English Channel, these frequencies vary quite a bit from island to island population. Much more radical variations were found in female spot distributions so work was concentrated on that sex. However, there is a highly significant correlation between the spot average in males and the spot average in females of the same population.

In the vicinity of the Devon–Cornwall border in southern England, the typical English stabilization of spotting pattern changed to a different Cornish pattern (Figure 3–9) in a matter of a few yards and in the absence of any detectable ecological barrier. Female spotting changes from unimodal at zero spots to becoming bimodal at zero and two spots. During the first nine years after the discovery of this southern English–Cornish spotting difference borderline, the situation remained much the same except that the zone of transition moved by somewhat erratic steps 22 miles eastward (Dowdeswell and McWhirter, 1967). Extraordinarily, at this boundary the *Maniola* on both sides seemed to be flying back and forth across the spotting zone of transition, yet despite this extensive apparent gene flow of females carrying eggs and males flying across to mate with the other females, this spotting distribution and the spotting averages remained sharply distinct. This was presumably because of selection for different gene complexes on either side of this boundary; although the exact causes remain unknown to present, they are probably associated with coadapted gene complexes.

Dowdeswell and McWhirter (1967) surveyed *Maniola jurtina* population samples throughout Europe and Northern Africa, and they find a series of geographically separated stabilizations, each clearly defined in terms of spot distribution. From museum samples, these stabilizations seem to have remained at the same level for at least the last 60 years. They also find parallelisms in spotting frequencies within other Palearctic species of *Maniola,* and point out that these general patterns of spotting on the forewings and hindwings are a universal characteristic of the family Satyridae, the family to which *Maniola* belongs. Spotting patterns of this type appear in several other butterfly families. Interestingly, they suggest that the gene systems involved are probably similar in nature, or in other words, "The genes controlling spotting appear . . . to be trans-specific, trans-generic, trans-familial

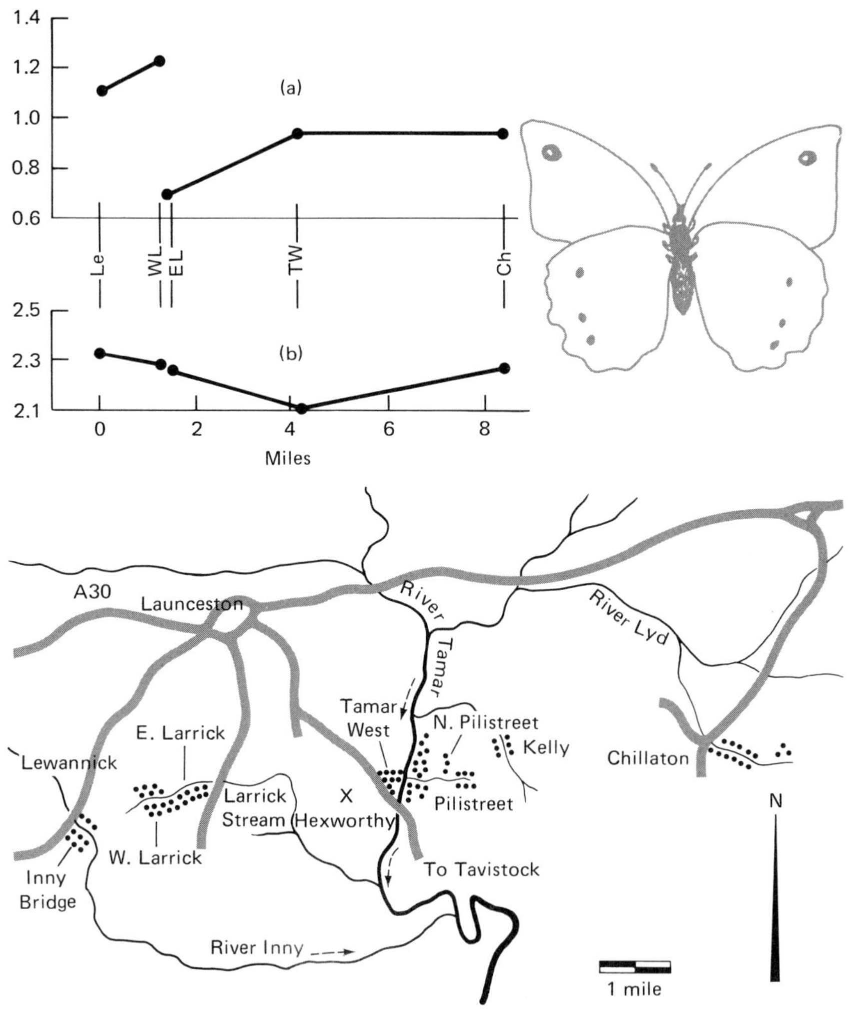

Figure 3–9 Spotting variation in *Maniola jurtina*. The bottom map shows the central transect studied by Ford and Dowdeswell across the Devon–Cornwall border where the "East Cornish" and "Southern English" hindwing spotting stabilizations of the butterfly *Maniola jurtina* met in 1956. (The river Tamar is the Devon–Cornwall county boundary here.) The localities where the samples were obtained are indicated by dots. In the diagram above, the average spot numbers of (a) the females and (b) the males are plotted against the distance along the transects (Le, Lewannick; WL, West Larrick; EL, East Larrick; TW, Tamar West; Ch, Chillaton). (Source: E. R. Creed, W. H. Dowdeswell, E. B. Ford, and K. G. McWhirter, 1959, *Heredity, 13:* 377, top; 375, bottom. Reprinted in E. B. Ford, 1971, *Ecological Genetics,* Methuen, London, pp. 82, 83.)

and therefore of great antiquity." A logical deduction from this conclusion is that the gene systems influencing spotting must have been established for longer than the genera and species now existing have been around. Many other interacting genes in the satyrine gene complex must have been under a long process of selection and have evolved against the background provided by the genes controlling spotting. Thus, they suggest that very ancient genes (palaeogenes) tend to maintain themselves in freely breeding populations at certain frequencies. These frequencies do not change even with extensive gene flow but are maintained by selection. When prolonged violent alterations in selection take place, these gene frequencies may change, and they do so in relatively large steps which are detectable through the characters controlled by the palaeogenes.

Variability in the snails *Cepaea nemoralis* and *C. hortensis* (Figure

Figure 3–10 The European land snail *Cepaea hortensis* (upper figure) shows variation in banding and coloration similar to that found in *C. nemoralis*. Visual predators such as the song thrush (*Turdus ericetorum*) eat the most conspicuous snails by taking them to nearby stones and cracking their shells. A thrush anvil stone with shells of *C. nemoralis* and *C. hortensis* is shown in the lower figure. (Photos by W. H. Dowdeswell.)

3–10) has been well studied from the viewpoint of bird selection and the maintenance of balanced polymorphism. Day and Dowdeswell (1968) measured dispersal and natural selection in these two species at a place called Portland Bill in England. This Portland Bill area is known as "The Strips" because of remnants of an extremely ancient form of cultivation, present since Saxon times. Instead of having their fields of long strips separated by hedges, the Saxons left, and farmers still leave today, green banks of unplowed turf. In this locality, *Cepaea* occurs on these green banks, which are 2 to 6 ft high and run parallel with each other for about 400 yards in length. Day and Dowdeswell discovered a distinct and fairly steep cline of banding morph frequencies along two of these strips, from one end with less than 10 percent of the population banded to the other end where 65 percent of the population was banded, over a distance of 300 to 400 yards. During four years of study, the pattern of polymorphism in the strips remained substantially unchanged, indicating a stabilization between the inherent variation of *C. hortensis* and the selective forces acting upon it. Yet from the marking experiments, the snails seem to wander in all directions, both along the banks and between them across seven yards of plowed ground. The general picture that emerges is of a fluid population moving in every direction, unimpeded by ecological barriers. Uninterrupted gene flow along the length of the strip certainly occurs, but it takes place rather slowly. Day and Dowdeswell found that the highest average annual rate of speed was 94 yards per year, which is probably a maximum for *Cepaea,* as other studies have shown average rates of 3.3 to 7.4 yards per year. So again we get a picture of reasonably free gene flow, but selection overriding this gene flow and maintaining different frequencies of genes in even closely adjacent areas.

Our final example of gene flow in a cline with selection overriding its effects is in Kettlewell and Berry's (1969) study of a cline in the Caradrinid moth *Amathes glareosa* in the Shetland group of islands just northeast of the main British Isles. This moth has a distinct melanic form (*edda*) in Shetland, which is controlled by a single gene (black having near complete dominance). This melanic form decreases in frequency from 97 percent in the north of the 70-mile long group of islands to about one percent in the south (Figure 3–11). There is an apparent barrier to gene flow in the center of this cline, where the phenotypic frequency decreases by 50 percent (from 50 to 60 percent to less than 5 percent melanic) over a distance of about 15 miles. One would expect to find in this area a barrier to gene flow separating populations living under different ecological conditions. When Kettlewell and Berry looked in this area, they found the Tingwall Valley crossing the middle of this region which consists largely of agricultural habitat unfavorable to the moth. They set up a large-scale capture–mark–re-

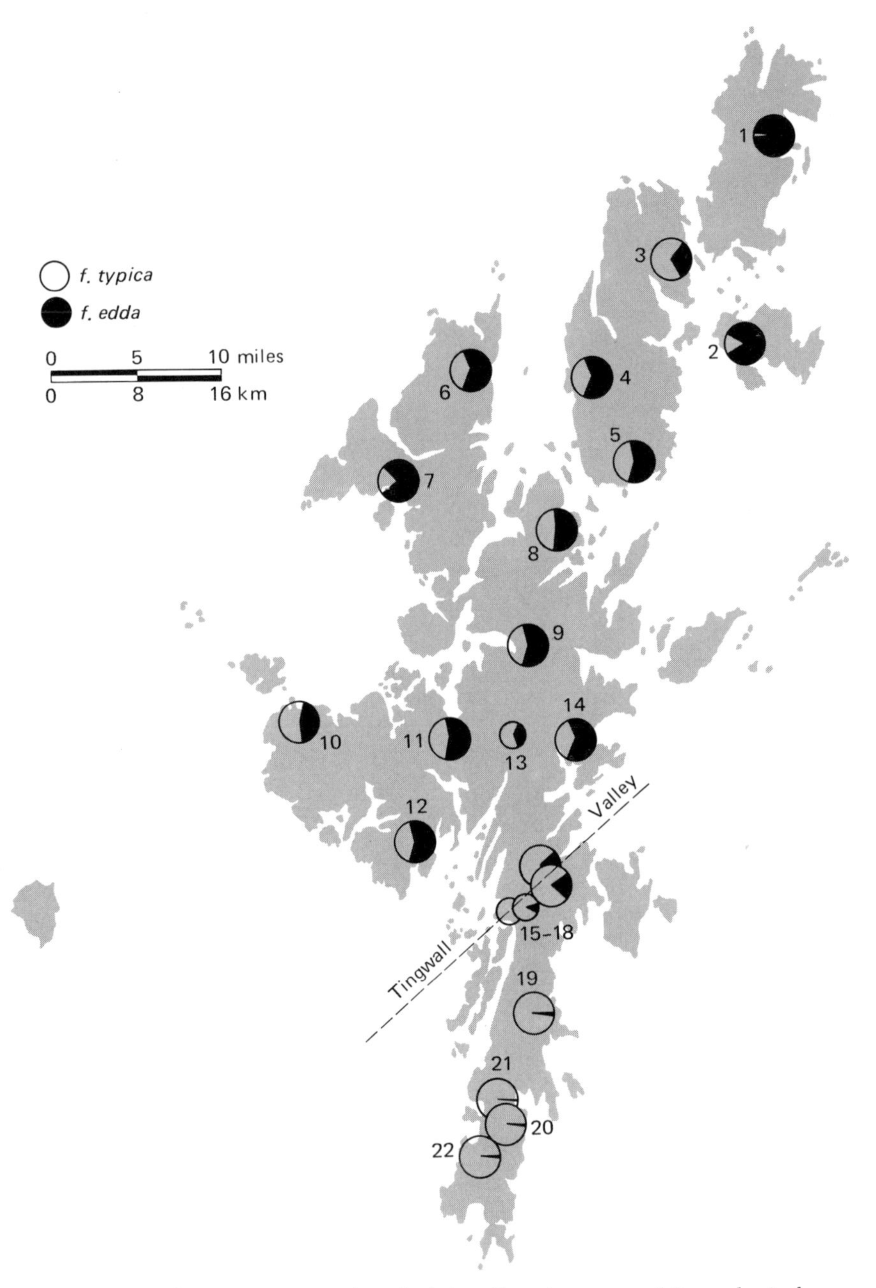

Figure 3–11 A frequency map of typical *Amathes glareosa* and its melanic form *edda*, comprising 20,000 records from 22 localities in Shetland. The dramatic change in frequency at the Tingwall Valley is unrelated to noticeable habitat changes. (Source: B. Kettlewell, 1973, *The Evolution of Melanism*, Oxford Univ. Press [Clarendon], New York, p. 188.)

lease–recapture experiment and marked over 1600 moths on both sides of the valley. Of the 65 recaptured, only one had crossed the valley. This would seem to confirm the presence of a barrier to gene flow, but it turned out that the frequencies of the melanic form *edda* were identical on both sides of the valley. The apparent barrier had nothing to do with the frequency change observed in this general region, for it was in the 2½ miles immediately north of the valley where the major frequency change occurred and more upsetting, this region was perfectly uniform habitat. Additionally surprising data involved the fact that the moths could fly up to half this distance (the mean net flight distance of recaptured individuals was half a mile). So with all this gene flow going on, there was still a considerable frequency change over 2½ miles of uniform habitat. Kettlewell and Berry postulated a major change in selective forces in this area similar to that at the Devon–Cornwall boundary of *Maniola jurtina* spotting types, because these gene frequencies throughout the cline are maintained at a constant level from year to year.

It therefore became apparent by the late 1960s that gene flow between populations seems to have little to do with holding species together as similar phenotypic units, because generally it is not a very common phenomenon and even where it does occur, selection on the local population level determines whether or not this gene flow will have any effect on the observed patterns of differentiation, that is, whether selection acts as a cohesive force or a disruptive force. Ehrlich and Raven (1969) and Endler (1973) present compelling evidence that differentiation along environmental gradients and in isolated or adjacent populations may be frequently independent of gene flow. If populations are subjected to different selective forces, they will differentiate but if they are not, they will tend to remain similar.

The factual evidence presented by Ehrlich and Raven (1969) for this view may be summarized as follows: gene flow is usually very restricted in nature and is far less significant in population structure than commonly thought; long isolated populations often show little differentiation; and populations freely exchanging genes but under different selective regimes may show marked differentiation. From the evidence that has accumulated, then, it seems unwise to generally view species of sexually reproducing organisms as groups simply sharing a common gene pool, though it may be true in particular instances. This conclusion is of general importance because the common-gene-pool notion has promoted the idea in evolutionary biology that "a species is an evolutionary unit, and that gene flow among its populations makes it such a unit."

What processes might be critical to the stability and multiplication of species, then? Many isolated populations having no gene flow show a remarkable lack of differentiation. Because gene flow in many cases is of little or no importance in maintaining the phenotypically similar units we call species, populations in these groups are presumably kept similar by existing under similar selective regimes. Yet in other instances, there may be great differentiation in populations having considerable gene flow between them. Differentiation readily occurs in areas with empty niches and depauperate faunas, such as the Galapagos Islands. Instead of mere isolation between islands as a key factor causing geographic speciation, evolution in the archipelago may have been more based on selection to fill various niches. Isolating mechanisms in this view may be considered as a common but not universal *result* of the process of speciation. They are certainly not a *cause* of speciation. Under different selective regimes, especially influenced by competition (Chapter 11), the gene pool of each of several populations changes, and the genetic differences eventually lead to incompatibility.

Ehrlich and Raven (1969) suggest that the true evolutionary units are local interbreeding populations, not species. There is substantial evidence for extremely local patterns of differentiation in both plant and animal populations, suggesting very close adjustment to local changes in selection. Along with other evolutionists, these authors suggest that the most basic forces involved in the differentiation of populations may be antagonistic selective strategies, one for close tracking of the environment, providing short-term adjustment to selective changes, and one for maintaining coadaptive genetic combinations, that is, combinations in individuals that have high average fitness in environments that are inevitably variable through time. This concept fits in remarkably well with the views of Mayr and Carson on the importance of the cohesion of the genotype and coadapted gene complexes in maintaining the integrity of species and the factors that are necessary to break up this coadaptive gene complex in order to promote speciation. Gene flow itself, then, may play a rather insignificant role in evolution as a whole.

1 2 3 **4** 5 6 7 8 9 10 11

Population Size: Growth and Dynamics

A population is composed of a number of genetically similar individuals living in a limited framework of time and space. More specific definitions often involve reference to growth processes. Lamont Cole (1957) defines a population as "a biological unit at the level of ecological integration where it is meaningful to speak of a birth rate, a death rate, a sex ratio and an age structure in describing the properties of the unit." In and of themselves, these rates and ratios are merely statistical descriptions of the present status, whereas the notion of dynamics indicates what happens over time. Hence, Elton (1933) introduced the term "population dynamics" to emphasize the place of active change in the study of populations.

Population dynamics is concerned with rates of increase, fluctuations in numbers, and the relationship of numerical changes to environmental factors that influence the population. Basic to the investigation of such phenomena is the estimation of the absolute number of organisms that compose the population, usually at periodic intervals in time. These measurements are usually far more difficult in zoological than in botanical studies.

POPULATION SIZE AND CENSUSING

Where the number of individuals in a particular area has been assessed, we know the *density* of a population. In a broad sense, there exist two ways of measuring the density of a population: (1) determine the *absolute density* by counting or estimating all the plants or animals of that species in the area, or (2) determine the *relative density* by counting a sample portion of the population and thus establishing an *index of relative abundance.*

Methods of estimating *relative density* are primarily used to assess trends in population size. Samples must therefore represent a constant (though unknown) proportion of the total population (Andrewartha, 1961). The implicit assumptions are that the sequential sampling takes place in uniform habitat and weather conditions, that there is random distribution of individuals in the habitat, and that the organisms are uniformly conspicuous to the observer. Different methods of estimating relative densities range from fecal pellet counts to listening for vocalizations to snap-trap kills of small mammals per night of trapping. In applied ecology, an economic entomologist may be studying the fluctuations in a population of scale insects in a Florida orange orchard. Every week, he takes a sample of 300 leaves of orange trees in that orchard and counts the number of scale insects on them. If the first week's sample yields 625 scale insects and if our sampling procedure meets the above assumptions, the following week's sample will faithfully reflect a true change in the scale-insect population size for the orchard. The entomologist could estimate the *absolute density* of the orchard scale-insect population if he knew the total number of leaves in the grove, but this would be impracticable to count, perilously inaccurate to estimate, and also unnecessary since relative densities are sufficient to indicate the extent of infestation and the point at which spraying may be required. There are so many variables involved in correcting relative indices to absolute abundance that it is more satisfactory for the population biologist to determine the actual population size in the first place if that is at all practical.

Measurement of *absolute density* provides base-line data for estimating rates of population change, and if data are collected on age and sex of the individuals composing the population, estimates of birth rates and death rates will allow construction of a life table (Chapter 7) and determination of some specific characteristics of population growth, as we shall shortly see. Effective methods of determining absolute population density include the five following standard procedures, the first being a direct total count and the others being dependent upon density per unit area sampling or census methods.

Counting the whole population directly

An accurate count of every individual living in an area is often possible with populations inhabiting small areas or composed of organisms with large bulk living in relatively open habitat. Thus, desert mountain populations of the evening primose *Oenothera organensis* could be counted directly in New Mexico because only several hundred localized individuals compromise the species (Wright, 1964). Direct counts are most easily done with vertebrates, especially birds or large herbivores on open plains or arctic tundra rather than with insects or other motile invertebrates. Whooping cranes, African wildebeest, caribou and deer herds, and polar bears may be censused by aerial photographs. Whales of some species are reduced enough in numbers that observers on whaling vessels and scouting planes provide a reasonably accurate estimate of their population numbers.

Strip censuses

A strip census is used for rough counts in wildlife management work with species or areas where it is impractical to count the entire population. The number of birds or large mammals is counted along a transect line of travel, within a strip area of ready visibility. Inaccurate counts will result if individuals are not randomly scattered or if there are differences in the conspicuousness of different individuals, some flushing while others remain hidden. If there is significantly less chance of sighting individuals that are further away from the line of travel, or as one moves into new types of habitat, the strip census will omit individuals thereby giving an inaccurate representation of population levels.

Quadrats

The quadrat sampling method in its numerous modifications has been usefully employed with plant populations and relatively immobile species of invertebrates, fish, reptiles, and small mammals. It is also commonly utilized with breeding and nesting bird and amphibian populations. The simplest procedure involves choosing small areas (normally squares), or *quadrats,* at random throughout a larger geographic area which contains the whole population. The size of the quadrat is chosen such that its population may be precisely counted, and of course the total area of the population must be known to allow extrapolation from the absolute population size of the quadrat to the estimated population size of the total area. The number of quadrats required to give a statistically reliable representation of the population density must be determined.

Most problems encountered in accurately determining population density by the quadrat method result from populations being distributed nonrandomly in the space which they could occupy (Chapter 6). Hence, foliage-inhabiting insects may be clumped in distribution on particularly favored host plants, such that randomly taken sweep-net samples need to be abundant enough (600–1000 sweeps of the net) and distributed over a sufficiently large volume of foliage space to reflect in unbiased fashion the distribution of those insects. Predatory insects such as assassin bugs (Reduviidae) may exhibit a negatively contagious distribution with very regular spacing throughout the foliage. Randomly distributed or regularly distributed species are safely sampled by smaller quadrats than species with clumped distribution. To determine the size of the sample quadrats, one looks for nonrandom distribution in small-sized plots. If population distribution is clumped, many of these plots will contain no individuals, whereas others will contain an excess of individuals. This situation represents a deviation from the typical Poisson distribution expected with random distribution (Snedecor, 1956), wherein the majority of the quadrats should contain intermediate numbers of individuals.

In a Poisson series showing random distribution, the mean number of individuals should equal the variance of that number, according to the formula

$$\frac{\Sigma(x-\bar{x})^2}{(n-1)}=\bar{x}$$

or dividing through by $\bar{x}$,

$$\frac{\Sigma(x-\bar{x})^2}{\bar{x}(n-1)}=1$$

The letter x represents the number of individuals in each quadrat, x is the mean number of individuals in all quadrats, and n equals the number of quadrats (a range of 20 to 40 replicate quadrats is ample; Snedecor, 1956).

If the value obtained in the above formula is about one, then the individuals in the n quadrats are randomly distributed. If the value obtained is significantly *greater* than one, the individuals are showing a clumped or contagious distribution and there is considerable variation of the value of each x out from the population mean $\bar{x}$ because of this patchiness. If the value is significantly *less* than one, the organisms are distributed in a regular or negatively contagious fashion and the number of individuals in each quadrat is very close to the mean number ($\bar{x}$) per quadrat for the entire population.

Capture per unit effort

During certain seasons of the year, the population biologist may find a closed or stabilized population of adult individuals that are not reproducing or showing mortality and where individuals are not emigrating

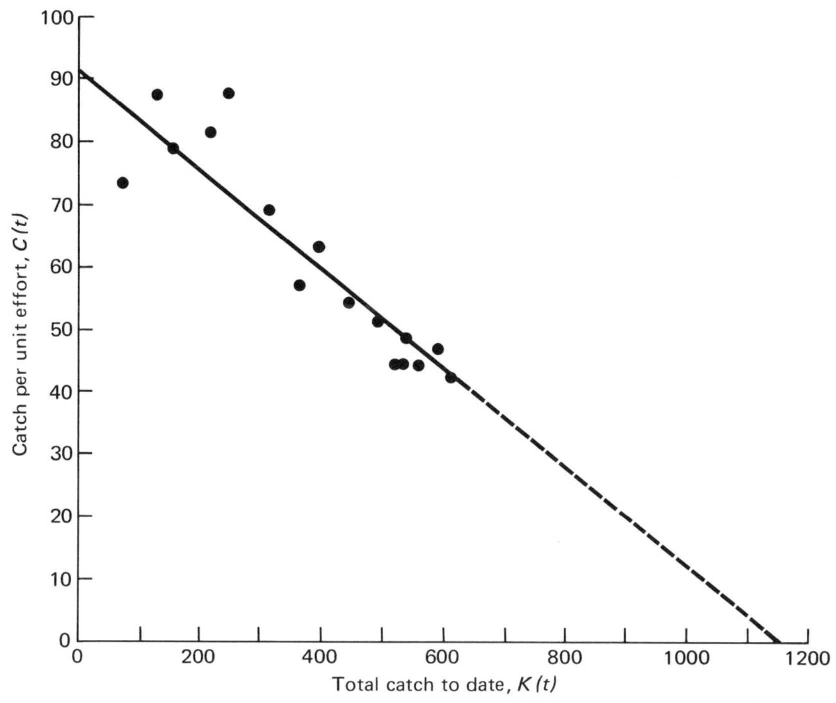

Figure 4–1 Application of the capture per unit-effort method for population estimation. Here the total population ($K = 1170$) is calculated by extension of a straight line through circles showing successive catches per unit effort $C(t)$ plotted against the accumulating total catch $K(t)$. (From S. C. Kendeigh, 1961, *Animal Ecology*, Prentice-Hall, Englewood Cliffs, N.J., p. 35. Source: D. B. DeLury, 1947, *Biometrika, 3:* 147–167.)

or being increased by immigration. Under these special circumstances, and provided that environmental conditions remain constant, it is possible to mark the population daily and have the number of new individuals captured or discovered and censused with each subsequent visit become less and less, eventually reaching zero. DeLury (1947) showed that one could determine the total population of animals in an area by extrapolation from a few catches. The number of new individuals captured by unit of effort is plotted against the cumulative number of animals captured, and the resulting straight line is extended to zero (Figure 4–1). The increasing percentage of marked animals in the total number captured in the population on successive dates can also be extended to 100 percent in the same manner to extrapolate the total population size (Kendeigh, 1961).

Capture–recapture methods

Many populations of mobile animals can be best censused using capture–recapture methods. These procedures also yield data on population movements, survivorship, and age and sex structure, which are highly important parameters to establish in investigating the population biology of a species. The Lincoln Index (or Petersen estimator) is the simplest such method and is the basis of more complex mathematical models. This intuitively easy-to-understand method was discovered independently and applied to natural populations by three biologists working on vastly different organisms. C. G. L. Petersen of the Danish Biological Station used the relationship with fish populations in 1896. F. C. Lincoln of the U.S. Fish and Wildlife Service developed the method in 1930 to estimate the number of ducks on the North American continent. C. H. N. Jackson originated the formula in 1933 to determine the density of African tsetse flies. R. A. Fisher and E. B. Ford (1930) mentioned this method as a theoretical possibility for population estimations and applied it in later work (Fisher and Ford, 1947).

The method basically depends on capturing a sample in a population area, marking them in a distinctive and permanent manner, releasing them for rediffusion throughout the population, and capturing a new sample after a short interval. The ratio of recaptured marked individuals to the total number marked originally should in theory be the same ratio as the total sample of marked and unmarked animals captured on that second trapping is to the total original population. In standard format, the Lincoln Index is written as

$$P = \frac{M_1 N_2}{R_2}$$

where P is the population estimate at time 1 or time 2 or both (see assumptions below), M_1 is the number marked and released into the population at time 1, N_2 is the total number captured at time 2, and R_2 is the number of marked individuals recaptured at time 2. When R_2 is large (i.e., more than about 20 animals) the estimate of P is reasonably unbiased, but if there are fewer recaptures, Bailey (1951, 1952) has shown that the estimate tends to be too large and that it is less biased if 1 is added to N_2 and R_2:

$$P = \frac{M_1(N_2 + 1)}{(R_2 + 1)}$$

Bailey's correction factor can also be used in the modified Lincoln Index incorporated in multiple recapture models. Generally, the greater the percentage of the population which is marked and subsequently recaptured, the greater is the accuracy of the calculated population size.

The accurate use of the Lincoln Index requires certain assumptions.

1. Marking does not affect the mortality or behavior of the animals,

the marks are not lost, and marked individuals are readily recognized.
2. The marked animals mix randomly in the unmarked population following their release.
3. The population is sampled each date at random. The age, sex, or mark of an animal does not influence its chance of initial capture or subsequent recapture.
4. Either there is no birth or immigration (dilution) or there is no death or emigration (loss). If there is loss but no dilution, the estimate is valid for the population on the day of release (time 1). On the other hand, if there is dilution but no death loss, then the estimate is valid for the day of recapture (time 2) (Sheppard and Bishop, 1974). With no dilution and no loss, the estimate is valid on both days.
5. Sampling must be performed at regular time intervals and the time spent in sampling must be negligibly small in relation to the intersampling period.

In attempts to meet the requirements of assumption 1, investigators have employed a wide variety of marking and tagging methods. Mammals are marked with dye patches, plastic neck collars (even with radio transmitters), tatooing (especially on the external ear), ear notching, or toe clipping. Birds may be dyed or banded with aluminum or plastic bands placed around their legs. With reptiles, the toes of lizards are generally clipped (not more than one per foot), whereas on snakes, scales are removed from conspicuous locations on the body. Turtles are marked with metal tags on the edge of the carapace or are riveted through the front flipper (marine species), numbered with paint, coded with carapace notching via a file, or with toe clipping. In amphibians, marking is commonly accomplished by toe clipping, tags, or punctures in the webbing between the toes. Fish are generally marked by fin clipping or by attaching numbered tags to the jaw, gill cover, or fin. Insects are marked with paint or dyes, and even radioactive tracers. Dustan (1965), for instance, tagged the Oriental Fruit Moth, *Grapholitha molesta,* with radioactive phosphorus (P^{32}) for flight and dispersal studies. This moth will drink liquid readily from moist surfaces, so Dustan fed newly emerged adults with a solution of P^{32}, which effectively labeled them for detection by a Geiger-Muller counter for at least a week (the half-life of P^{32} is 14.2 days). Marking by such methods insures no damage or externally visible mark to attract predators; hence, it may prove useful with mosquitoes and other delicate insects.

The limitations imposed by assumptions 2 and 3 may create considerable errors if violated. Knowledge of the normal behavior of the organisms in the population is necessary to verify that a standard one-day intersampling period is sufficient to allow random mixing of

Table 4–1 Summary of properties of some popular capture–recapture models. An essential component of each model is that the animals involved in multiple-recapture studies be given a date-specific mark at every capture or (preferably) an individual identification mark.

MODEL	NO. OF SAMPLES REQUIRED	RATE OF LOSS	RATE OF GAIN	PARAMETERS EST.	BEST WORKED EXAMPLE	COMPUTER PROGRAM	REMARKS
Lincoln Index	2	0 Constant 0	0 0 Constant	$P_1 = P_2$ P_1 P_2	—	—	To be used when only two samples are available.
Fisher and Ford (1947)	3 or more	Constant	Variable	(a) P for every sample time except first (b) Single constant survival rate (see remarks) (c) Gains and losses	Reanalysis of data of Dowdeswell, Fisher and Ford (1940) by R. J. White. Available on request from J. A. Bishop at University of Liverpool.	White and Sheppard in Bishop and Sheppard (1973)	Model calculates a single survival rate which is assumed to remain constant for duration of experiment. Requires less data (recaptures) than Jolly's and Manly and Parr's models (Bishop and Sheppard 1973)

Jolly (1965)	3 or more	Variable	Variable	(a) P for every sample time except first and last (b) Survival rate for each intersample period except last two (c) Gains and losses	Jolly (1965) and Southwood (1966)	Davies (1971)	Calculates separate survival rates for inter-sampling periods. These assume that every animal has the same chance of surviving until next sample (i.e. mortality is independent of animal's age)
Manly and Parr (1968)	3 or more	Variable	Variable	As by Jolly's model	Manly and Parr (1968)	Calculations simple but program by J. S. Bradley & R. J. White available from J. A. Bishop.	Calculates a separate survival rate for inter-sampling periods. These do not assume that mortality is independent of age

SOURCE: Sheppard and Bishop (1974).

marked animals. If the catchability of different individuals is variable due to sex, age, size, or other character, the situation may demand separate analyses of subpopulations (for example, males and females may have to be treated independently).

Most of the multiple-recapture models derived from the Lincoln Index try to remove assumption 4 to make allowance for gain (causing dilution) and loss. Hence, they attempt to represent more precisely the natural situation. The most popular approaches are listed in Table 4–1. Generally, as one proceeds down this summary table, one is dealing with capture–recapture methods that need increasing amounts of data in the form of recaptures and repeated days of sampling. However, as Sheppard and Bishop (1974) point out, these models are increasingly realistic and give a more complete series of estimates of parameters of the population. The choice of the type of analysis for a capture–recapture study will depend on the number of data that may be obtained. When sampling time is limited, only the Lincoln Index may be applicable. Where comparatively few recaptures are made in three or more samples, Fisher and Ford's (1947) method may be the most appropriate, but where more data are available, Jolly's (1965) formulation or even Manly and Parr's (1968) method may be used.

Kelker (1940) and other authors (for example, see Chapman, 1954) have proposed a "capture–recapture" method of population estimation based on change of composition through selective removal. Kelker's method makes use of the naturally distinguishing marks of a species such as sexual dimorphism or any other recognizable distinction such as those of different ages or different morphs of a polymorphic species. One determines the initial proportion of the different forms, removes a known number of one form, and then samples the population again to determine the new ratio. From the change in the ratio, the size of the total population can be calculated as follows:

$$P = K_\alpha \div \left[D_{\alpha 1} - \frac{(D_{\beta 1} D_{\alpha 2})}{D_{\beta 2}} \right]$$

where α and β are the two sexes (or other distinct forms) in a population, α is the form in which part is removed during the interval t_1 to t_2, and β is the form which is not affected. Then K_α is the number of component α killed, D is the proportion of the population represented by a component at t_1 or t_2, D_α is the proportion (as a decimal) of α in the population before any were removed, and D_β is the proportion (as a decimal) of the population represented by β after K individuals of α were removed. This procedure finds its main utility with game animals, among which individuals of a certain sex or age class are killed. It would not be useful with insects or many other organisms where the selective removal of one type of animal would seriously prejudice the study of population characteristics other than size. As an example of its use in game species, consider a mule deer population in which α = males, β = females, $D_\alpha = 0.50$ and $D_\beta = 0.50$, $K_\alpha = 100$ bucks killed by hunters, and $D_\alpha = 0.40$ with $D_\beta = 0.60$. Then $P = 100/0.17$, or 588 deer.

Another intriguing adaptation of the capture–recapture method is based upon frequency of recapture (Craig, 1953). Craig's method makes some assumptions about the frequency of recapturing marked individuals 1×, 2×, 3×, or more. If the animal is mobile but stays within the general area, and if individuals are collected at random, marked, and immediately released again, a certain number will be recaptured once. If they are marked again (or recorded as recaptures) and released, a certain proportion will be recaptured a second time, and so forth. Then the number of animals marked once (f_1), twice (f_2), three (f_3), . . . (f_x) times are part of a frequency series, the other term of which represents those animals that have not been caught or marked at all. Craig (1953) assumed that the frequency of recapture could be described by a modified Poisson distribution function. Of his three proposed methods based on this model, we can consider the following relation:

$$P = (\Sigma x f_x)^2 \div (\Sigma x^2 f_x - \Sigma x f_x)$$

where P is population size, x is the number of times an individual had been marked, f_x is the frequency with which individuals marked x times had been caught, and $\Sigma x f_x$ is the total number of recapture times for all animals that were marked and recaptured [viz. (1 × the number caught once) + (2 × the number caught twice) + (3 × the number caught thrice) + etc.].

Valid application of the model requires that the animals be very mobile so as to randomize their chances of recapture almost immediately after release, yet they cannot leave the habitat. The method has been used successfully with butterflies (Hanson and Hovanitz, 1968) and could be applied to other large conspicuous vagile animals under limited circumstances (Southwood, 1967).

POPULATION GROWTH AND DYNAMICS

Given the means to ascertain population size and trends in population numbers, how do we measure population growth and change from a given base number? Population growth is a function of three factors: *birth rate*, *death rate*, and *movements* of individuals in and out of the population. If one eliminates immigration and emigration or if they balance, the rate of population growth becomes a function of the birth and death rates, and the difference between the two.

The so-called crude birth rate is merely the number of births per unit of population. The specific birth rate is the birth rate for a particular cohort in the population, such as two-year-old animals. The crude and specific death rates are similarly calculated. Then the rate of natural increase of a population becomes

$$R_i = R_B - R_D$$

or

$$= \frac{P_2 - P_1}{P_1}$$

where P_1 and P_2 are the population sizes at time 1 and time 2, respectively. The true rate of natural increase has been defined as the rate at which the fertile cohort of the population is increasing (Dublin and Lotka, 1925). The true rate may equal the rate of natural increase but probably will not. Sharpe and Lotka (1911) have shown that if age-specific mortality and reproductive ratio are constant, a stable age ratio will develop in that population and with this the rate of population increase becomes constant. Then all age groups are increasing at the same constant rate, and the rate of increase will equal the true rate of increase.

The growth curve

If a small population is presented with optimal environmental conditions, it will tend to expand at a geometric rate. Should each female leave two females on the average in the next generation, and each male be replaced by two male offspring, the population will double in size at every generation. In this case, the net replacement rate per generation is two individuals, or $R_0 = 2$. To generalize, if the population is increasing a constant rate, then the population size N at some time t will be equal to the initial population size N_0 multiplied by the rate of increase R per generation compounded by the amount of elapsed time t, or

$$N_t = N_0 R^t$$

This finite rate of increase is comparable to the growth rate for compound interest on bank deposits, which if compounded quarterly or at other regular intervals would assume this form. However, if our compound interest rate was to be compounded continuously instead of several times a year, or if our population had breeding going on all the time rather than seasonally in nonoverlapping generations, then $R^t = e^{rt}$, where e is the constant 2.7183 (the base of natural logarithms) and r is a constant called the intrinsic rate of increase. Thus in a time interval of one unit, of $R = 2$, then $e^r = e^{.69315}$ and $r = 0.69315$, the intrinsic (or instantaneous) rate of increase for that population.

The intrinsic rate of increase r merely equals the individual birth rate b_0 (the number offspring one individual will have on the average per unit of time) minus the individual death rate d_0 (the average number of deaths in the population per individual per unit of time). Under ideal conditions with optimal environmental space, food, and absence of competitors and predators, r would assume its maximum possible value which is termed r_{max}, the *maximum intrinsic rate of increase,* or the *biotic potential* of the species. This inherently possible growth rate interacts with the environmental resistance, i.e., all the limiting factors in the environment acting upon that particular population. As a result of this interaction, populations tend to have a characteristic pattern of increase or *population growth form* that represents a modification of the expression of the maximum possible logarithmic population growth,

$$N_t = N_0 e^{rt}$$

or

$$\frac{dN}{dt} = rN \quad \text{(in derivative form)}$$

Exponential growth curves

In special environmental situations where unlimited growth may occur, population size increases rapidly, in exponential or compound interest fashion as described in the above equation (Figure 4–2). The new individuals are continuously added to the principal balance and thus continuously provide an expanded base of reproductive individuals.

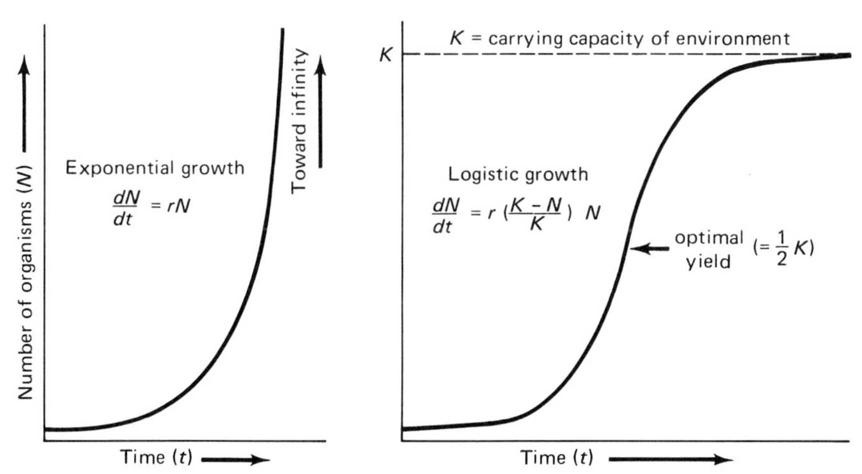

Figure 4–2 Two basic forms of population increase: the exponential growth curve (left) and the logistic growth curve (right).

Exponential growth is abruptly terminated when the carrying capacity of the environment is surpassed and environmental resistance becomes effective more or less suddenly. The population then crashes abruptly. Such growth curves are characteristic of rapidly reproducing and maturing annual plants, seasonal insect flushes, and man's population growth in recent years, but in general they are very short-lived phenomena for obvious reasons.

Logistic growth curves

When considered over long periods of time, all populations of organisms, with one or two exceptions (for example, man), tend to show a zero rate of increase; in other words, the interaction of biotic potential and environmental resistance results in a more or less constant population size. Hence, the young growing population experiences detrimental effects of increased density, and growth is progressively inhibited until the population reaches some asymptotic level, or *carrying capacity*, which represents the maximum number of individuals that can be supported in a given habitat. The pattern of population growth which results is the S-shaped or sigmoid growth form, called the *logistic growth curve*, in which growth starts slowly, accelerates rapidly in exponential form, and then decelerates and continues thereafter at a more or less constant level (Figure 4–2).

The logistic growth curve tends to be characteristic of a larger organisms with longer life cycles and lower biotic potentials, although yeasts and *Drosophila* exhibit ideal sigmoid growth as readily as mice (Figure 4–3). Sigmoid growth of a population may be seen in newly founded colonies or in populations recovering from very low levels, for example, fur seals abruptly protected from the hunting activities of sealers (Figure 4–4). Most of the time, an established population will be existing at or near its carrying capacity (Chapter 5).

Logistic growth may be expressed in a differential equation, called the *logistic equation*, which states the rate of growth with several variables inserted to describe the effects of density. The equation was derived by the French mathematician Verhulst in 1838 as a model of population growth in a limited environment, and Pearl applied it in 1920 to the growth of the population of the United States. The Verhulst–Pearl equation takes the form

$$\frac{dN}{dt} = rN\left(\frac{K-N}{K}\right)$$

which is merely the exponential equation

$$\frac{dN}{dt} = rN$$

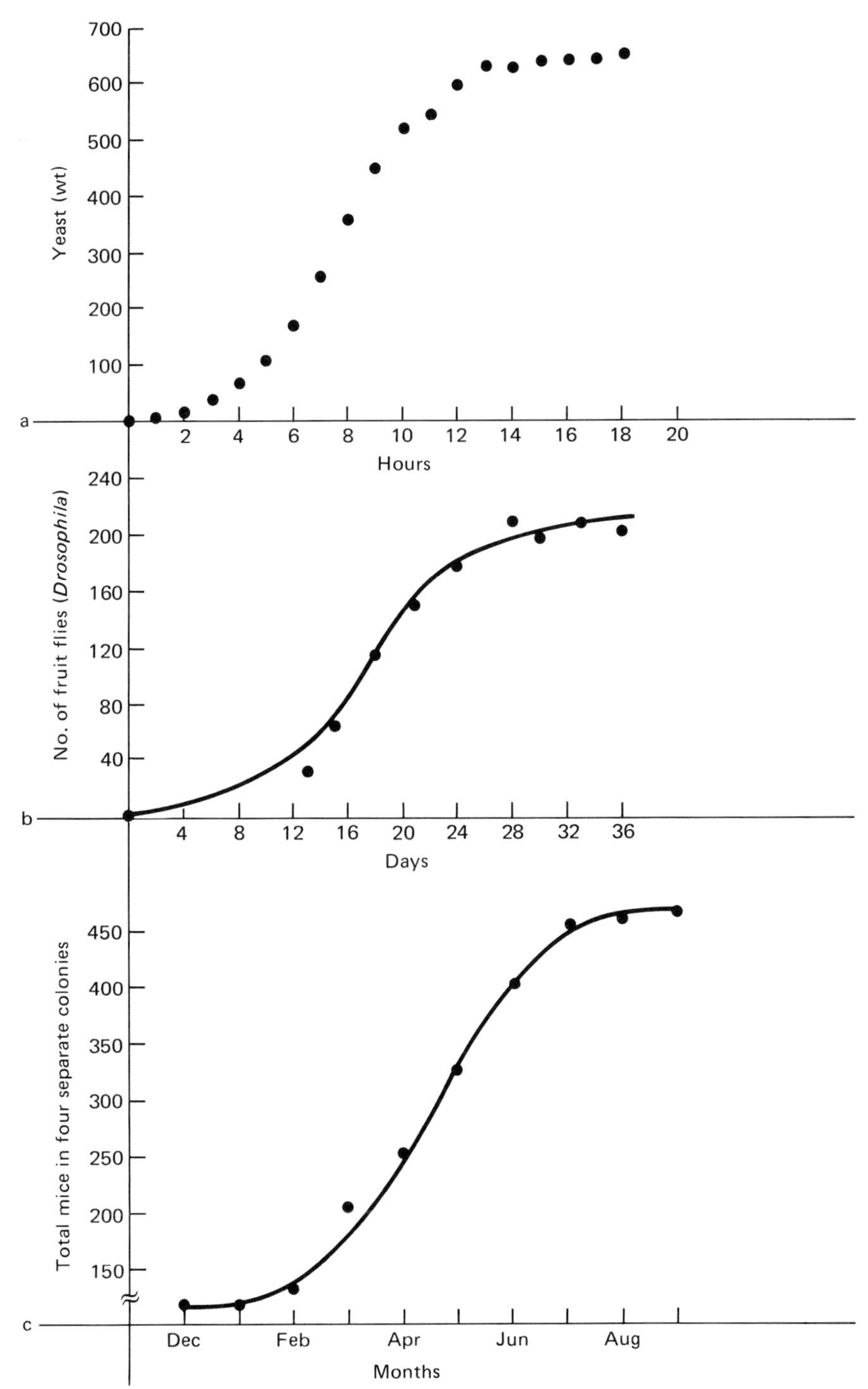

Figure 4–3 Three sigmoid growth curves for (a) yeast cells, (b) fruit flies (*Drosophila*), and (c) mice colonies. The top two curves closely approach the mathematical logistic curve. (Source: J. J. Dinsmore.)

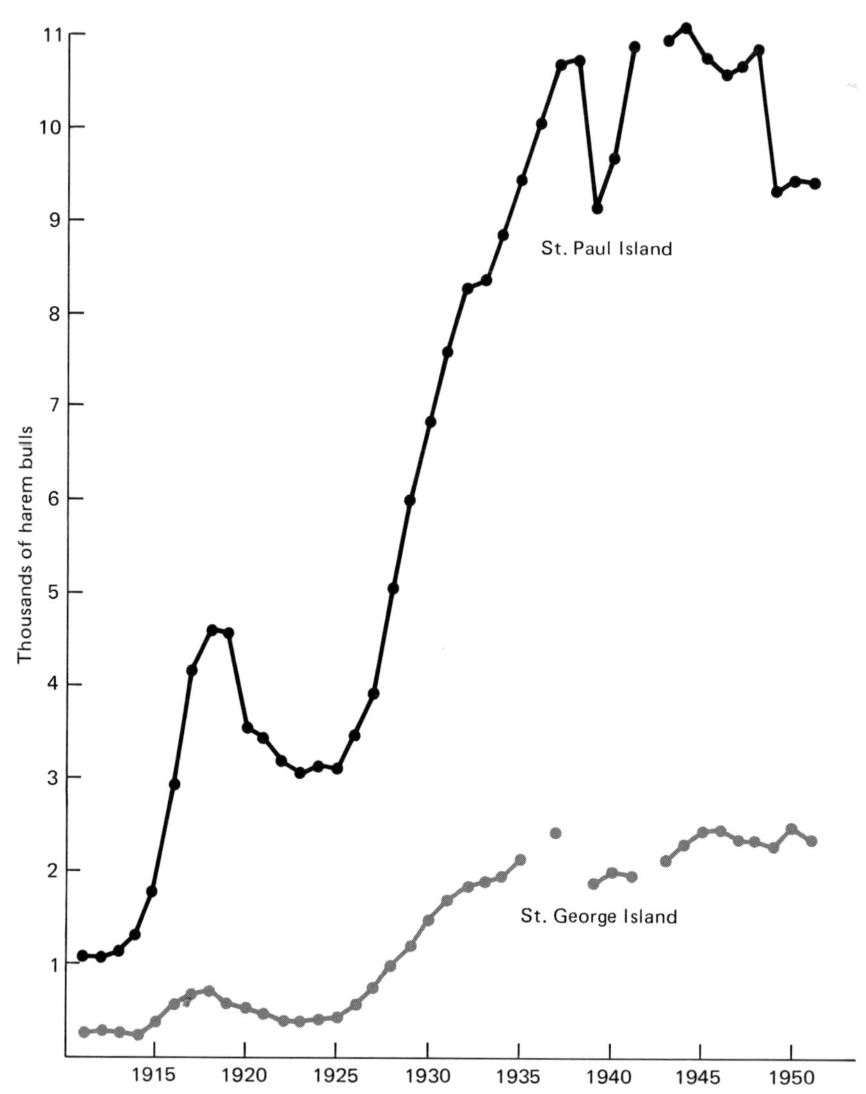

Figure 4–4 The logistic growth in numbers of harem bulls in fur seal herds on two of the Pribilof Islands following cessation of pelagic seating in 1911. (Data from K. W. Kenyon and V. B. Schaffer, 1954, *U.S. Fish and Wildlife Service, Spec. Sci. Rep.* No. 12.)

multiplied by the term $(K - N)/K$, an expression that indicates that as N increases, dN/dt decreases. When $N = K$, the upper limit or carrying capacity of the environment, and the curve flattens out, the term equals zero, and $dN/dt = 0$ (see Figure 4–2). When N is still close to zero and the population is just beginning to expand into the available environment, dN/dt is very close to equaling rN; that is, the growth is almost exponential. As density increases, the maximum rate of growth slows, marked by the inflection point in the curve.

The characteristics that make the logistic curve useful to population biologists are (1) it is a convenient and intuitively satisfactory mathematical description of a frequently observed pattern of the population growth; (2) it gives an insight into factors that affect population growth; (3) it has been and still remains a useful demographic tool if not overextrapolated from insufficient data; and (4) it allows biologists to relate population density to rates of population growth. Slobodkin (1961) has criticized the logistic equation as unrealistic for the oversimplifications that it entails. The equation assumes that all individuals (N) regardless of age or sex have identical ecological properties and that these individuals respond instantaneously to environmental change. It also assumes that there is a constant upper limit to population size, and that the rate of gain of the population is directly related to the difference remaining between this upper limit and the population level at any one time. The environmental characteristics assumed to be present are a limited amount of space and a constant food supply. These problems entail consideration of the basic concept of an

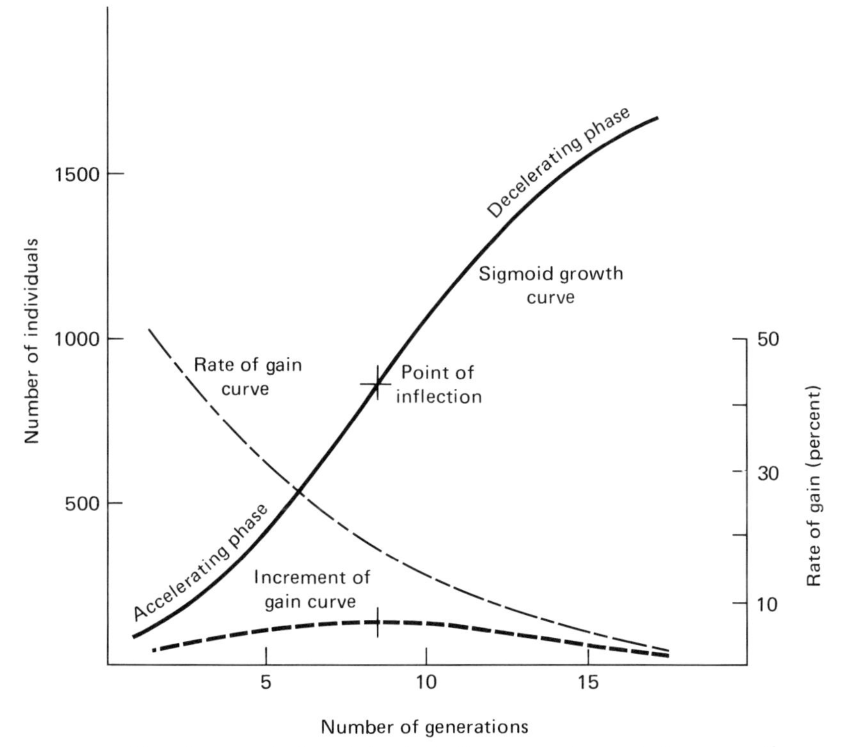

Figure 4–5 The characteristics of the logistic curve of population growth. (Source: J. J. Dinsmore.)

equilibrium level for the population and the factors affecting population regulation, many of which act in a density-dependent fashion (see Chapters 5 and 7 for further discussion).

Before considering the regulation of population size in the next chapter, let us close the present section by examining briefly the properties of a population that lead to logistic population growth. The logistic curve has basically three parts (Figure 4–5), aside from its initial starting point and ending asymptote at the carrying capacity. The lower, near-exponential part of the curve is the accelerating phase. Here the increment of growth per time is increasing but the rate of growth itself may be increasing, decreasing, or constant. The increment depends on the growth rate and size of the population. At the point of inflection, the increment of growth is at its maximum and switches from increase to decrease. In the decelerating, upper phase

Table 4–2 Population growth in three hypothetical populations with different rates of increase and increments of increase.

R_i	INITIAL POPULATION	INCREMENT OF INCREASE	FINAL POPULATION
		POPULATION A	
1.00	100	100	200
0.90	200	180	380
0.80	380	304	684
0.70	684	479	1163
0.60	1163	698	1861
0.50	1861	931	2792
0.40	2792	1117	3909
0.30	3909	1173	5082
0.20	5082	1016	6098
0.10	6098	610	6708
0.00	6708	0	6708
		POPULATION B	
0.50	100	50	150
0.44	150	66	216
0.39	216	84	300
0.34	300	102	402
0.29	402	117	519
0.24	519	125	644
0.21	644	135	779
0.18	779	140	912
0.15	912	136	1048
0.12	1048	126	1174
0.10	1174	117	1291
0.08	1291	103	1394
0.06	1394	84	1478
0.05	1478	74	1552
0.04	1552	62	1614
0.03	1614	48	1662
0.02	1662	33	1695

POPULATION C

R_i	INITIAL POPULATION	INCREMENT OF INCREASE	POP. SIZE (AD. + YG.)	MINUS CONSTANT 20% LOSS	FINAL POPULATION	r'
1.00	100	100	200	40	160	0.60
0.93	160	149	309	62	247	0.54
0.86	247	212	459	92	367	0.49
0.80	367	294	661	132	529	0.44
0.74	529	391	920	184	736	0.39
0.69	736	507	1243	249	994	0.35
0.63	994	626	1620	324	1296	0.30
0.58	1296	752	2048	410	1638	0.26
0.53	1638	868	2506	501	2005	0.22
0.48	2005	962	2967	593	2374	0.18
0.44	2374	1045	3419	684	2735	0.15
0.41	2735	1121	3856	771	3085	0.13
0.37	3085	1141	4226	845	3381	0.10
0.34	3381	1150	4531	906	3625	0.07
0.32	3625	1160	4785	957	3828	0.06
0.31	3828	1187	5015	1003	4012	0.05
0.29	4012	1163	5175	1035	4140	0.03
0.28	4140	1159	5299	1060	4239	0.02
0.27	4239	1145	5384	1077	4307	0.02
0.26	4307	1120	5427	1085	4342	0.01

SOURCE: Data provided by James J. Dinsmore.

of the curve, both rate and increment of growth are declining, as shown in the bottom two curves.

Some appreciation for the effects of the rate of gain and the increment of growth on logistic curves may be obtained by reference to Figure 4–6, plotting data on growth in several hypothetical populations with different increments (Table 4–2). In Population *A*, the sigmoid shape is clearly a function of the increment of growth; the point of inflection is at a high population level where increment of increase is maximum. During the accelerating phase, the rate of growth could be doing any of three things: increasing, decreasing or staying constant, through relative mortality and movement rates of individuals. In Population B, the rate of increase R_i is one-half that in Population A, and at some population level over 1000, the R_i values are depressed to almost zero at an equilibrium level. In Population C, the rate of increase R_i is identical to that of Population A, but a constant 20 percent loss of individuals is taken from the population by predators or other factors. This depresses the rate of growth accordingly and the population reaches a lower carrying capacity. For Population D, the rate of increase values are depressed at a more rapid rate than in Population A, and the asymptotic carrying capacity is achieved much sooner.

These curves suggest that in situations in which it is desirable to promote maximum population growth, as with many game popula-

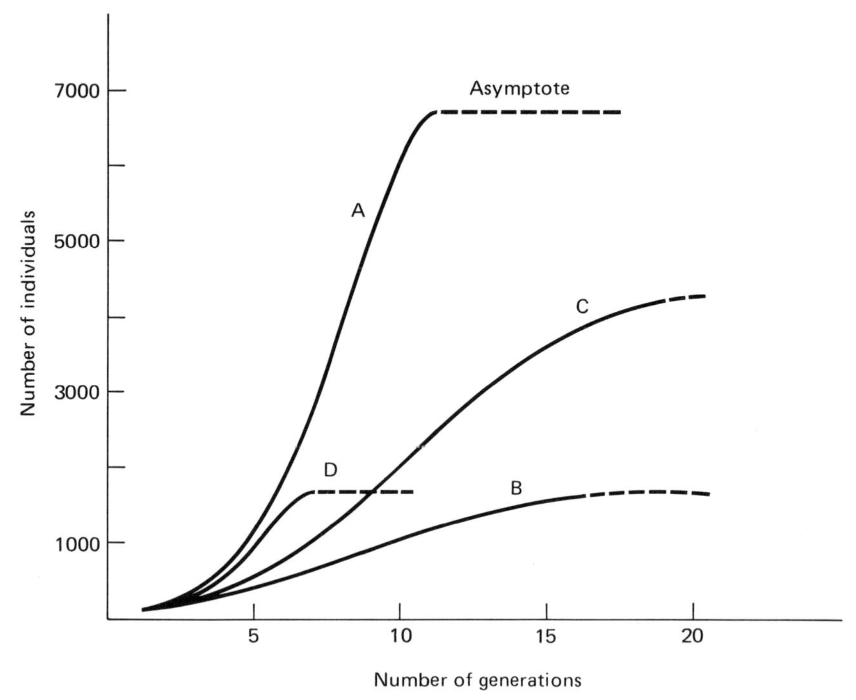

Figure 4–6 Population growth in four hypothetical populations with different rates of increase and increments of increase (see Table 4–2 and text for explanation). (Source: J. J. Dinsmore.)

tions, the harvest would be best controlled at the point of inflection where the increment of growth is at its maximum. The asymptote is determined by density and is reached when the rate of gain equals zero. However, the time-specific life table (Chapter 7) offers a finer analysis of the dynamics of population growth and the influence of sex and age structure in natality, mortality, and rates of increase. Before considering those factors in Chapter 7, we shall look at fluctuations in population size and dispersion, and the regulation of population growth.

1 2 3 4 5 6 7 8 9 10 11

Population Fluctuations and Regulatory Systems

Population growth does not continue indefinitely; it levels off or declines once the carrying capacity of the environment has been reached or exceeded. If conditions allow nearly exponential growth of the population before that point, population increase can be truly spectacular. Introduced species on oceanic islands often show such explosive growth if no competitors are present. In 1957 an Ecuadorian fisherman released three goats on Pinta Island in the Galapagos Archipelago where there were no other herbivorous mammals. During 1971–1972 the National Park Service shot some 30,000 feral goats on that island, hardly making a significant decrease in the estimated 100,000-plus population. Other introductions or founder populations do not fare as well, of course, most becoming extinct soon after initiation. In 1909, 48 greater birds of paradise (*Paradisaea apoda apoda*) were captured on the Aru Islands south of New Guinea in the Pacific and were brought to Little Tobago Island located in the southern West Indies, 1 mile off the coast of Tobago. The founding population was augmented by three more birds in 1912, and several died shortly. By 1913, between 16 and 30 survived; in 1955, the population numbered 11 birds. In 1965–1966, Dinsmore (1970) found only seven birds. The presence of many competing frugivorous (fruit-eating) birds, as well as perhaps lack of proper habitat and possibly inbreeding, may be contributing to the decline of the *apoda* population.

POPULATION FLUCTUATIONS

Fluctuations in population number seem to be found in all species, whether introduced or resident. The relative stability of a biological community often can be measured by the effect that fluctuation in one member species of the community has on the population levels of other members of the community. Community stability, then, tends to be a function of the number of species that occur at each trophic level (hence, the food-chain position of a species becomes important in determining its effect on stability) and of the number of trophic levels. Hence, biotic communities tend to be most complex in the equatorial tropics and least complex in desert and arctic areas. The environmental extremes and relatively recent nature of deserts and the arctic have not fostered the species diversity found in other temperate or tropical communities. Whereas the number of species is high in the tropics, the number of individuals per species is highest in the arctic. Thus, in the arctic and in the desert, we find simple food webs with a low number of species. Adding trees and warmer temperatures, even in the boreal forest adjoining the arctic, increases the number of species in food webs and hence should increase community stability through establishing alternate food chains (Figure 5–1). However, the seemingly complex food web in the boreal forest actually has great instability because of its prey base. The three species of grouse (ruffed, spruce, and sharp-tailed), snowshoe hare, mice, voles, red squirrel and muskrat all show some degree of population fluctuation. The waterfowl and songbirds are migratory, and the ground squirrels hibernate. Thus, the principal stable prey populations are the ungulates, but of the listed predators only the wolves feed on them. The result is great instability in the boreal forest, with extremes in population numbers ranging from great abundance of both prey and predator species to nearly simultaneous crashes of both groups.

Within a particular biome such as the temperate deciduous forest of the eastern United States, stability may increase with habitat complexity. Thus, comparing stability of songbird populations in five types of woods, Brewer (1963) found that the greater the complexity of the stand, the greater the number of species inhabiting it and the lower the variation in songbird population levels (Table 5–1). MacArthur and MacArthur (1961) examined the physical properties of eastern forests and in general found that complexity of the stand increases the number of resident bird species; that is, the height profile of foliage density was directly related to species diversity in the bird community.

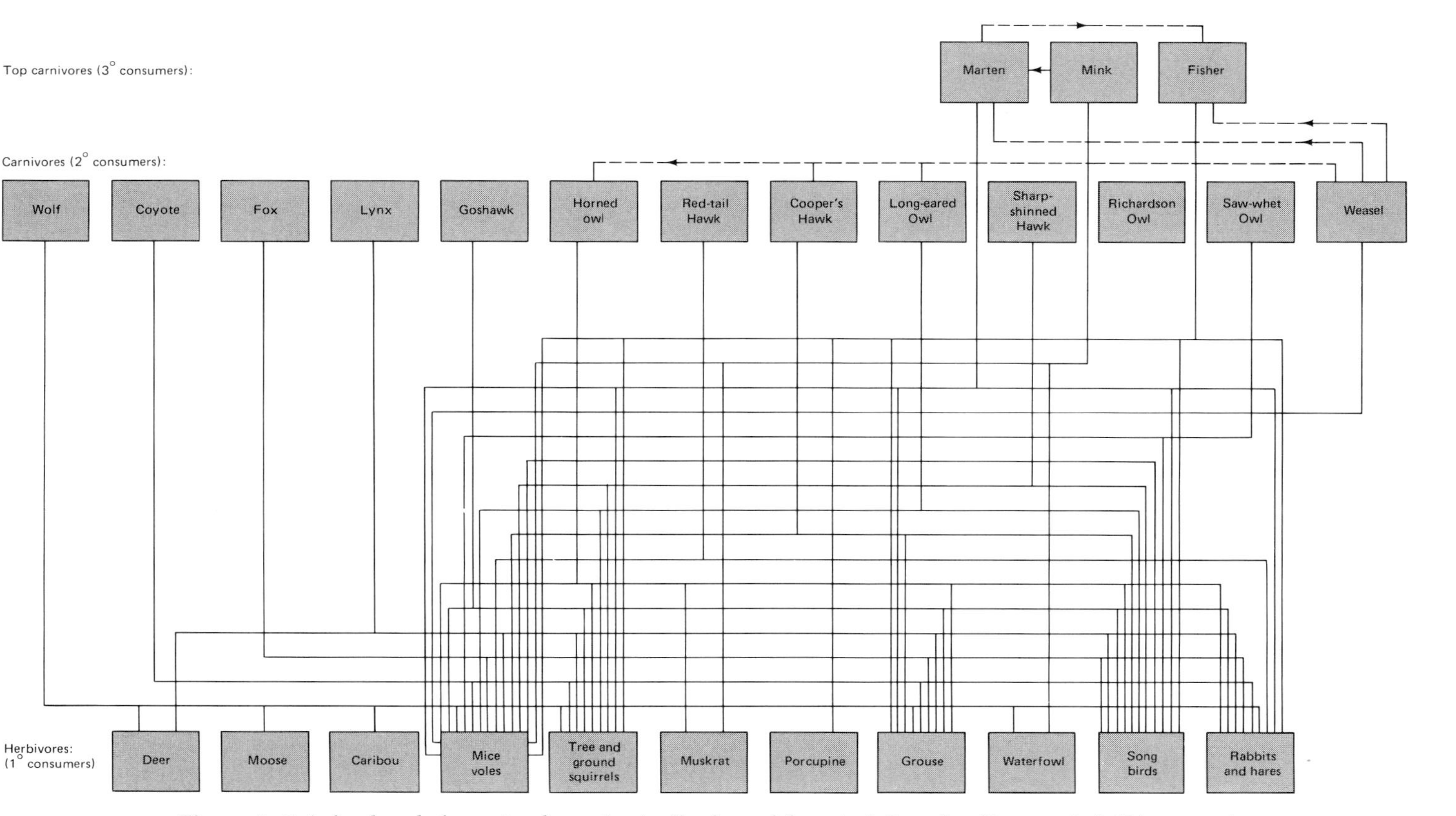

Figure 5–1 A food web for animal species in the boreal forest of Canada. (Source: J. J. Dinsmore.)

Table 5–1 The number of breeding songbird species and number of pairs per study stand in five types of deciduous woods.

	UPLAND OAK	MAPLE–ELM–OAK	BEECH–MAPLE	LOWLAND BEECH–MAPLE	FLOOD PLAIN FOREST
Mean number of breeding species	15.6	23.3	24.7	25.8	27.6
Coefficient of variation for mean number of pairs	18.0	15.3	14.2	8.1	8.2

SOURCE: Brewer (1963).

In the supposedly most stable biotic communities of the equatorial region, population structure and size remain relatively stable in some species while varying greatly in others. Thus, the paleotropical scincoid lizard *Emoia atrocostata* maintained a population size of 400–600 individuals in a 3.6-ha mangrove forest area in the Philippines during a three-year period (Alcala and Brown, 1967), and the neotropical teiid lizard *Ameiva quadrilineata* was observed to maintain fairly stable population densities of 5.1 to 10.0 juveniles per acre and 6.9 to 15.8 adults per acre over three years on Tortuguero beach in Costa Rica (Hirth, 1963). In a 1960–1971 banding study of ant-following antbirds of three species in the tropical rain forest of Barro Colorado Island, Willis (1974) showed that the small species, the spotted antbird (*Hylophylas naevioides*), remained stable in density at about 20 pairs/km^2. A medium-sized species, the bicolored antbird (*Gymnopithys bicolor*), decreased from about three pairs to 1.5 pairs/km^2. A large species, the ocellated antbird (*Phaenostietus mcleannani*), declined from 1.5 pairs/km^2 to near extinction, with only one female remaining in early 1971. Two of three other species that regularly follow army ants showed relatively stable populations, but a third large species, the barred woodcreeper (*Dendrocolaptes certhia*) declined from two pairs to local extinction. Prior to 1960, a very large ground cuckoo that follows ants had already become extinct on the island. Thus, the three largest of the seven original species that were regular ant-followers were extinct or nearly gone by 1970. No clear reasons for the declines were found by Willis (1974), except that annual mortalities of adults were high in ocellated antbirds (about 30 percent) compared to spotted antbirds (15–17 percent) and nest losses were perhaps higher in the former (96 percent compared to 94 percent). In bicolored antbirds, nest mortalities were slightly lower (88 percent) and adult mortalities intermediate (about 25 percent). These antbird species renest repeatedly during long nesting seasons, up to 14 times per year for ocellated antbirds. However, to replace females of this species under the present

Barro Colorado conditions, 19 nestings per year would be needed. Concurrent surveys of all birds of the island showed that 45 species of breeding birds, equaling 22 percent of the avifauna present when the island was made a reserve five decades earlier, had disappeared by 1970. No new bird species replaced them. The growth of the forest had crowded out 32 second-growth and forest-edge species, whereas 13 of the lost species were forest birds that apparently were lost in part because of the small size of the reserve and its isolation from contiguous tropical forest.

Tropical butterflies also reflect variable responses to a seemingly stable environment, although density fluctuations appear to depend in large part upon length of adult life and associated attributes of the species' life history. Thus, the life expectancies of two distasteful species of *Parides* swallowtails (*P. anchises* and *P. neophilus*) on the neotropical island of Trinidad are between 5 and 10 days, and their numbers increase notably at times a month apart during the wet season, perhaps reflecting the high points in density for the first and second generations of the season, respectively (Cook et al., 1971). On the other hand, the gregariously roosting distasteful heliconiine butterfly *Heliconius erato* has a life-span of 50 to 90 days in Trinidad (Turner, 1971) and relatively stable population sizes through the year, at least in Costa Rica (Benson, 1972). Another Trinidad heliconiine species, *Heliconius ethilla,* lives up to 180 days in the wild. Ehrlich and Gilbert (1973) found that the size of a *Heliconius ethilla* colony occupying a ridge in northern Trinidad remained remarkably stable over a 27-month period, which covered about 27 overlapping generations. The recruitment rate was about six butterflies per day (three of each sex) and the total yearly production of individuals in the colony was approximately 2200 (relatively great compared to other tropical heliconiine populations). The constancy of population size (around 130 to 190 individuals) appears to result from a combination of high preadult mortality (over 99 percent), long adult life, and limited adult food resources, which in turn limit total egg production.

The nondistasteful nymphaline butterfly *Marpesia berania,* which roosts gregariously at night in the rain forest of Costa Rica (Figure 5–2), has been shown to have life spans of over five months and maintain long-term stable population numbers in their roosts during the January to March dry season and the first month (April) of the wet season (Benson and Emmel, 1973). However, later in the wet season, recruitment decreases to near zero and the adult population drops in August to half or less of the dry season levels (Table 5–2). In this study, the empirically determined rates of recruitment mortality (Figures 5–3 and 5–4) accurately predicted the initial absolute population size at the time of marking, indicating that the adult population was under equilibrium conditions during the dry season and the first month of the wet season.

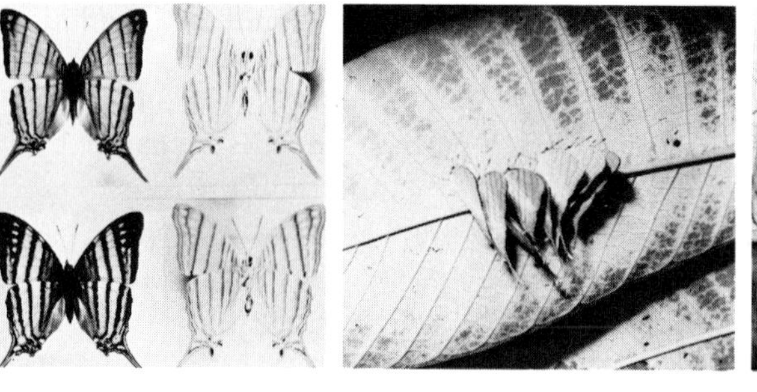

a b c

Figure 5–2 (a) Males (top) and females (bottom) of the daggerwing butterfly *Marpesia berania* (Hewitson) (Nymphalidae: Nymphalinae) giving dorsal and ventral views, respectively. (b) Top and (c) frontal view of butterflies on a nocturnal roost in Costa Rica, as seen from below with eight individuals on the underside of a roost leaf prior to marking. Note outward orientation and close arrangement of the individuals. (Source: W. W. Benson and T. C. Emmel, 1973, *Ecology, 54* (2): 326–335. Photos by T. C. Emmel.)

Table 5–2 Census data for a roosting population of *Marpesia berania* butterflies along a stream bed at the Osa Peninsula, Costa Rica. See text for explanation.

DAYS FROM MARKING	MARKED INDIVIDUALS IDENTIFIED: MALES	?	FEMALES	UNMARKED INDIVIDUALS IDENTIFIED	TOTAL OF KNOWN STATUS	TOTAL PRESENT
0						
1		32		1	33	33
14	11		6	5	22	47
16	15		10	14	39	48
	11		7	5	23	48
17	9		7	9	25	45
	4		5	5	14	45
19	8		5	7	20	48
	4		3	6	13	48
20	5		4	8	17	44
	4		5	3	12	44
24	14		8	14	36	45
27	11		8	13	32	54
31	13		8	15	36	47
35	9		6	15	30	(32)
37	7		6	17	30	(38)
42	4		4	15	23	(40)
51	6		8	27	41	50
62	5	1	5	22	33	57
74	9		7	24	40	52
86	7		11	39	57	57
94	10		7	33	50	53
157	1		2	15	18	18
158	1		1	16	18	20

SOURCE: Benson and Emmel (1973).

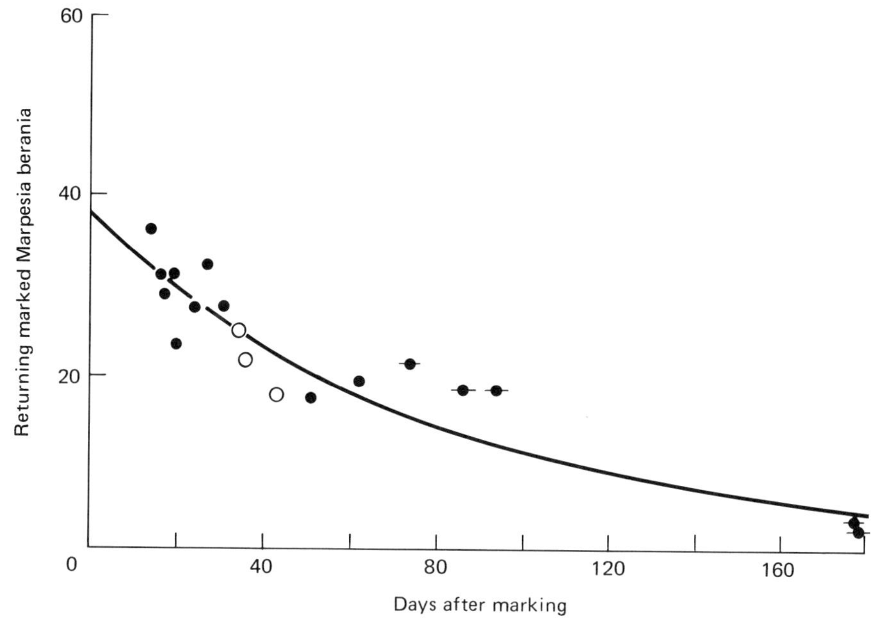

Figure 5–3 Graph showing the decrease in marked returns (survivors) of *Marpesia berania* on the Creek Roost at Osa Peninsula, Costa Rica. The closed circles (●) represent datum points used to calculate the regression line, indicated by the solid line. The slashed closed circles (-●-) represent datum points omitted from the regression analysis because of possible bias by age-dependent mortality and decreasing sample sizes. The open circles (○) are biased because of incomplete sampling and are therefore omitted; the points were estimated using an assumed population size of 50 individuals at the times of censusing. The equation describing the regression line is $S_t = 37.93\ e^{-0.01255t}$ (Source: W. W. Benson and T. C. Emmel, 1973, *Ecology*, *54*(2): 326–335.)

Fluctuations in population size may occur at different time levels. Season-to-season trends are commonly observed and will be discussed in detail in Chapter 10. These are usually caused by seasonal fluctuations in availability of relevant environmental resources such as food or by seasonal variation in appropriate physical limiting factors for growth, such as temperature and rainfall. Year-to-year trends may involve equilibrium levels, in which population size shows almost no fluctuation, or considerable fluctuation, in which the fluctuations may be cyclic or irregular. Peaks in the cyclic trends may be spaced some years apart, as we shall see shortly. These cycles normally result from annual fluctuations in biotic and physical limiting factors. This equilibrium versus fluctuation classification of population-density status is rather arbitrary. At what point does random variation about an equilibrium level become an irregular or cyclic fluctuation?

One statistical measure of population stability is the coefficient of

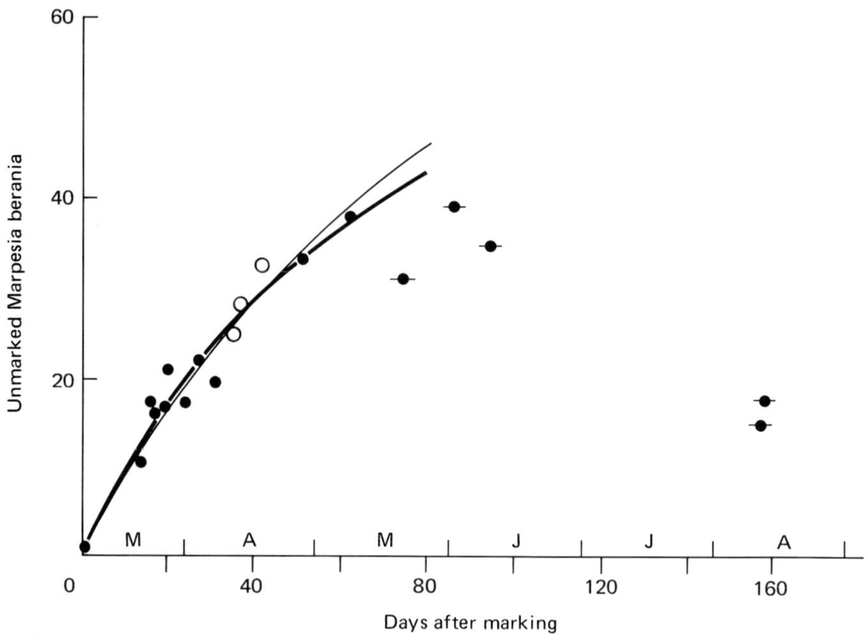

Figure 5–4 Graph showing the increase in unmarked surviving recruits of *Marpesia berania* on the Creek Roost. The closed circles (•) represent the datum points used to calculate the regression equations. The lighter upper line represents a nonlinear least-squares estimate given $m = 0.01255$ with a rate of recruitment R being estimated, $N_t = (0.9070/0.01255)\ (1 - 3^{-0.01255t})$. The lower, heavier line is the regression ,line where the same technique is used to estimate both m and R, $N_t = (1.0336/0.01986)\ (1 - e^{-0.01986t})$. The slashed closed circles (-•-) and the open circles (○) are datum points that were omitted from the calculations because of consistent departure from the theoretical relationship and incomplete sampling, respectively. (Source: W. W. Benson and T. C. Emmel, 1973, *Ecology*, *54*(2): 326–335.)

variation, which shows the variation of a series of population levels (for example, yearly values) around their mean value. Most populations have a coefficient of variation below 25 percent, and this figure is arbitrarily designated as the threshold below which the population is in equilibrium. Thus in a 15-year survey of birds on 65 acres of climax maple–beach forest in Ohio, Williams (1947) found the following mean number of breeding pairs of three species of songbirds (means and coefficients of variability of the number of pairs were calculated from Williams' annual data):

	Mean no. pairs	*Coefficient of variation*
Red-eyed vireo	25.4	24.2
Ovenbird	17.9	55.7
Wood thrush	14.3	39.7

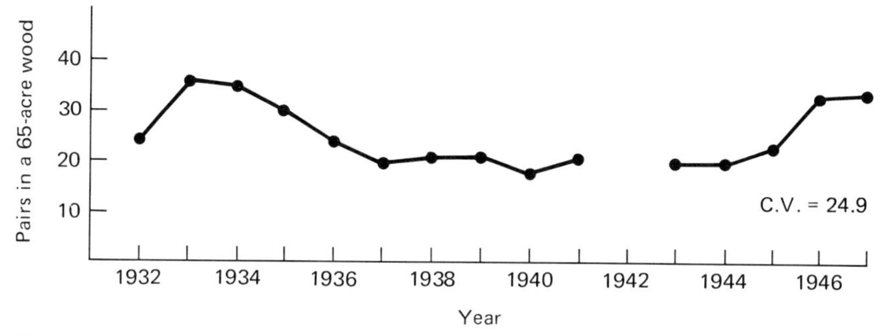

Figure 5–5 The number of breeding pairs per year and the coefficient of variation as an index of population stability in the red-eyed vireo in a 65-acre plot of climax beech–maple forest in Ohio. (Source: A. B. Williams, 1947, *Audubon Field Notes, 2:* 231.)

On the basis of comparing the values for the coefficient of variation, the red-eyed vireo population appears to be in equilibrium (Figure 5–5) while the ovenbird and the wood thrush show considerable annual fluctuations in population size.

Even for populations of the same species, however, direct comparisons of this statistic as an index to population stability may often prove invalid. First, of course, the size of area and hence sample size involved in the population index will affect the statistical value. Small areas tend to have greater population fluctuations than large areas where many population counts are pooled (for example, for all of Canada); the coefficient drops when used with pooled data. Second, the time span over which population data are gathered will influence conclusions. The longer the sampling time span, the greater the probability of some marked deviation occurring after even long stability. Despite these drawbacks, wildlife ecologists often find the coefficient of variation (C.V.) to be a useful means of differentiating populations in equilibrium (C.V. < 25 percent) and populations with fluctuating size levels (C.V. > 25 percent). There is far less agreement over the division between irregular and cyclic fluctuations in population size.

CYCLIC CHANGES IN POPULATION SIZE

No population stays at a constant size, but if it exceeds the "normal" range of variation about a mean level, how can we differentiate population cycles from irregular or random oscillations? Most ecologists agree that cycles involve some degree of regularity or repetition of peaks at particular intervals, but the problem has been to define the degree of regularity required to distinguish cyclic and random oscillations. Davis (1957) suggested that population cycles would entail regularity that was greater than would be found in chance occurrences; that is, cyclic phenomena might recur at intervals somewhat variable in length, but the variability would be less than one would expect by chance. Davis also suggested that not only does fluctuation have to be nonrandom, but there has to be an element of predictability in the occurrence of population peaks. The distinction between cyclic and random fluctuations becomes more obvious with long series of population data, and with long-term cycles more than with short-term cycles.

The two most common cyclic intervals are the short-term oscillations of three to four years, as in the several species of arctic lemmings in Europe and North America, and the long-term oscillations of nine to 10 years, as in the Canadian lynx and the snowshoe hare. The 10-year cycle has been particularly well documented for the lynx and hare across Canada through fur returns recorded by the Hudson Bay Company since 1845 (Figure 5–6), and even longer for lynx in specific areas of Canada (to 1735). These fur returns are assumed to reflect actual population peaks and depressions, although the amplitude of the charted cycle may have been affected at times by the price of furs, variance in delivery of furs such as lost shipments, and the number of active trappers. In recent years, the fur market has fluctuated greatly and largely invalidated the demographic usefulness of fur records for particular species. From 1945 to 1960, for instance, the price of long-haired pelts dropped so low that trapping effectively was halted. Foxes experienced a great population peak in the early 1950s that went unrecorded by trapping. Another factor to be considered in analyzing fur returns is that peak years in fur returns may actually come later than the timing of peaks in wild populations. Furbearers are easier to trap

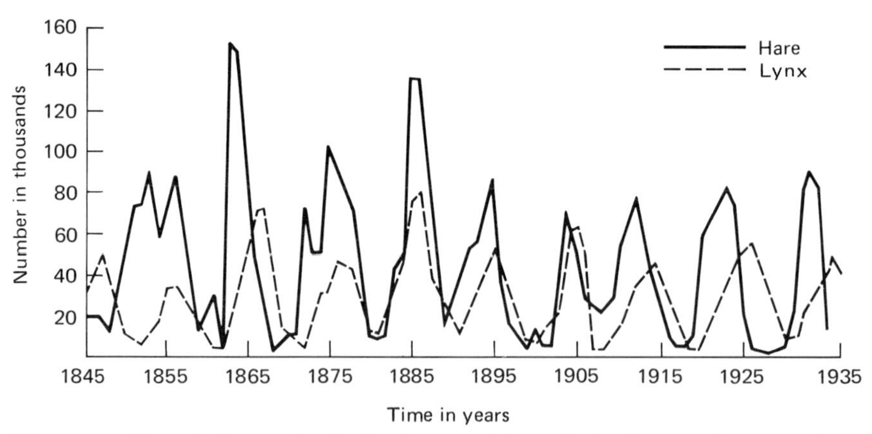

Figure 5–6 Population cycles in the lynx and its principal prey, the snowshoe hare, in Canada. The vertical axis gives the number of pelts received by the Hudson Bay Company. (After E. P. Odum, 1959, *Fundamentals of Ecology*, 2nd ed., Saunders, Philadelphia, p. 193. Redrawn from D. A. MacLulich, 1937, *Univ. Toronto Studies, Biology Series, No. 43*.)

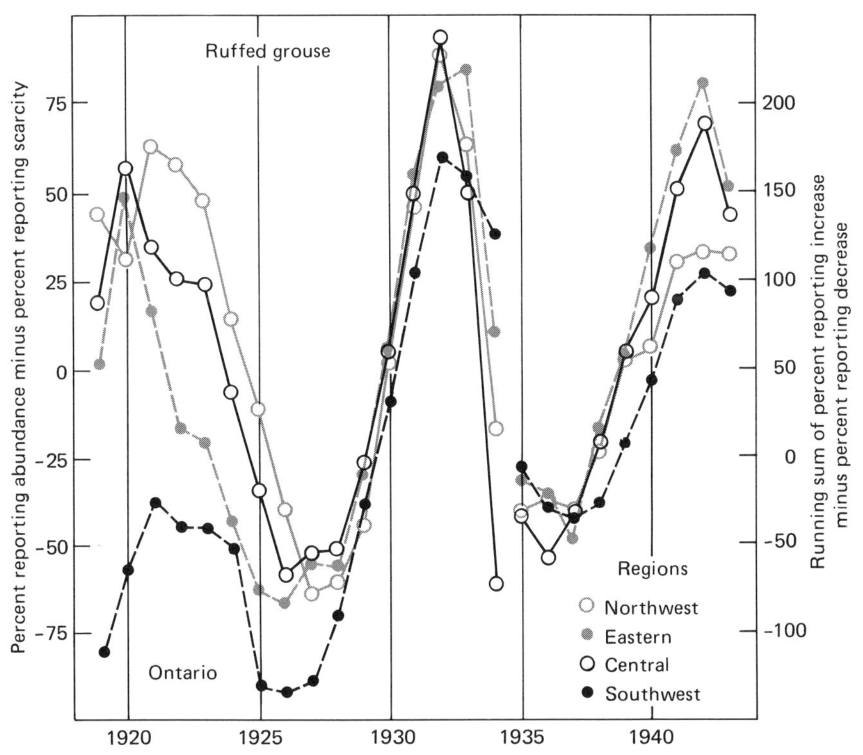

Figure 5–7 Synchrony of population cycles in ruffed grouse from four regions of Ontario, Canada, from 1918 to 1944, as determined by questionnaire data. (After L. B. Keith, 1963, *Wildlife's Ten-year Cycle,* Univ. of Wisconsin Press, Madison, p. 33.)

in winter (and their coats are in more attractive condition); hence, the winter trapping peak will follow the true spring and summer population peak and probably be somewhat less than that true peak because of mortality in the intervening period. This situation may be remedied in part by plotting data as the population index for the *biological* year preceding the actual winter year trapping date.

In addition to fur records, questionnaires are used in querying competent observers and hunters about non-furbearing animals such as grouse and hares. Hunting kill data may have even more biases than fur returns as estimates of comparative population sizes over a period of years. However, one can compare local and regional population levels of the same species and even different species to establish the synchrony of nonrandomness over an area, which may suggest regular cycles (Figure 5–7). Keith (1963) has utilized this general approach to demonstrate that a number of boreal and arctic species have 10-year cycles (Figure 5–8). The causes of this cycling may be considered as part of the general topic of population regulation.

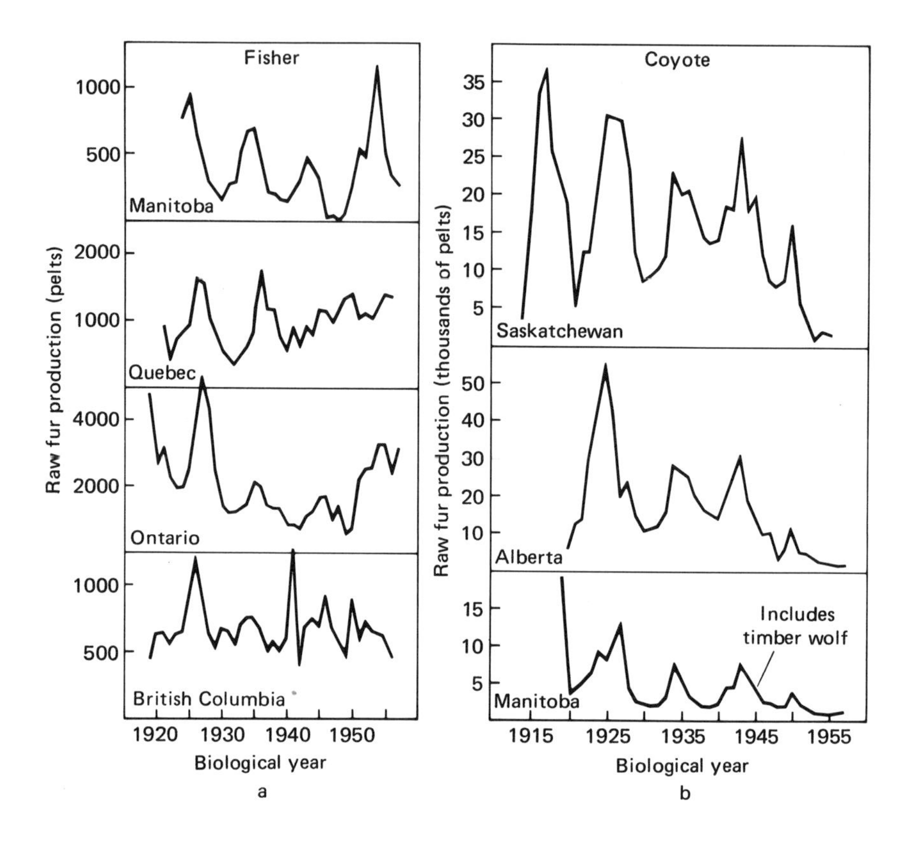

REGULATION OF POPULATION SIZE

Probably no area of population biology has generated more polemic argument and debate than has the regulation or lack of regulation of population size in nature. As noted in discussing the two major forms of the growth curve (Chapter 4), limiting factors of some sort, which we have loosely grouped under the term of "environmental resistance," always slow or stop population growth. Often the population slides into a sharp decline in numbers that may continue to extinction, but is more often reversed as biotic or physical conditions again improve for the individuals in the population. This series of growth spurts and declines form the basis of the commonly observed population fluctuations we have examined in the preceding section.

There seems little disagreement among ecologists about the concept of carrying capacity, at least providing that it is defined at a specific point in time. Even among anthropologists and cultural geographers, the calculation of the carrying capacity of the environment of some particular primitive human society and the improvisation of formulas for the determination of carrying capacities have been fairly common objectives in recent years (Street, 1968). A population size of

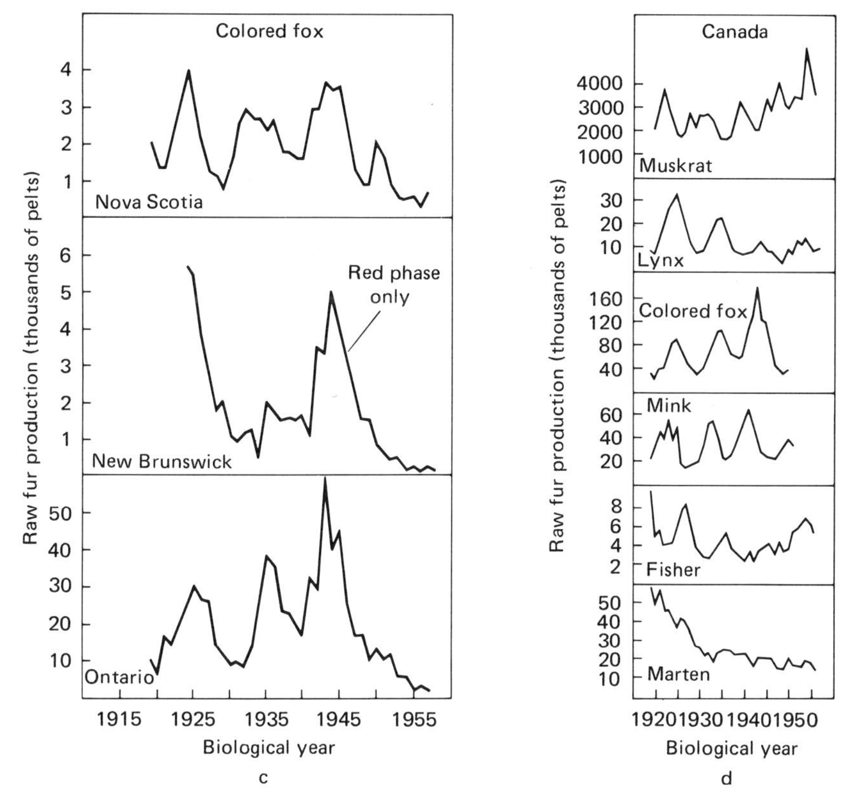

Figure 5–8 Ten-year cycles in populations of arctic and boreal North American mammals: (a) Fisher fur returns from four Canadian provinces; (b) coyote fur returns from each of the three prairie provinces; (c) colored fox fur returns from three eastern provinces of Canada; (d) Canadian fur returns for six alleged cyclic species from all provinces. (Source: L. B. Keith, 1963, *Wildlife's Ten-year Cycle,* Univ. of Wisconsin Press, Madison.)

a particular maximum level can be reached for a limited area, given specific and limited amounts of food, shelter, living space, and other resources. The fact that this carrying capacity will vary from month to month or even from day to day with the seasons and other environmental circumstances leads one to be cautious about advocating a static carrying capacity or a stable equilibrium point for a particular population, and in fact this fact has led some ecologists to surmise that the notion of a "balance of nature" with respect to population size is demonstrably false (Ehrlich and Birch, 1967).

Considerably greater controversy arises over the assignment of relative influence of various forces in affecting population numbers below, at, or above this carrying capacity. Two basic schools of thought are represented among population ecologists. One school argues that extrinsic environmental forces outside the population itself influence population numbers. These factors are termed *density-independent* because the action or effect is constant regardless of the number of individuals in the population. The other school argues that intrinsic forces generated within the population regulate population size below the carrying capacity of the environment. These factors are said to be *density-dependent* because the effect on the population varies with the density.

Density-independent factors

Density-independent influences exert the same effect on the population regardless of the numbers of individuals present. Climatic factors such as seasonal changes in temperature, rainfall, and daily photoperiod length produce density-independent effects. Heavy snow during severe winters may bury browse or seeds so deeply that deer and other animals such as quail are unable to obtain sufficient food and the result is an exceptionally high mortality, especially among recent young, which have less stored fat reserves than do mature adults. Generally, density-independent factors are most frequently observed to govern the population growth of small organisms such as insects or plankton with short life cycles and high biotic potentials. Weather and other physical environmental factors such as soil or water temperature are quite important in determining the length of favorable periods of annual growth for such organisms. If multiple-brooded, they generally lack stability in their population size, which changes radically through the seasons.

The chief supporters of the role of density-independent factors have been invertebrate-oriented ecologists who have worked mainly with insect populations (Andrewartha and Birch, 1954; Ehrlich and Birch, 1967). They feel that density-independent factors affecting the rate of increase cannot regulate population growth to any one level, because "regulation" in itself suggests a homeostatic feedback mechanism that functions with density. Hence, they argue that there is no equilibrium point for a population because the environment is never constant. The population persists in nature because environmental forces such as weather act upon the rate of increase during the short reproductive period, causing either a rapid increase or decline, and when the population drops to a low level, the remaining individuals are able to find favorable habitat and survive until the next reproductive period occurs. To build a model that deals with the ebb and flow of numbers in populations, they emphasize that these propositions must be included (Ehrlich and Birch, 1967):

1. All populations are constantly changing in size; there is no known case of exact replacement (for example, one female now by one female alive a generation from now) in a natural population.

2. The environments of all organisms are constantly changing, with changes on different time scales (diurnal, seasonal, long-term, etc.) going on simultaneously.
3. Local populations, within which there is relatively free movement of individuals, must be recognized and their structure investigated if changes in population size are to be understood.
4. The influence on population size of various components of environment will vary with population density, among species, among local populations, and through time.

Andrewartha and Birch (1954, p. 648) further argue that by their very nature, population limits are theoretical quantities which must be deduced and that they cannot be expected to be observed by counting the numbers of animals in natural populations. Thus, they relegate to the category of "allegorical" such expressions as "density-dependent factors," "balance," "steady-density," "control," and "regulate."

The yearly peaks and depressions in abundance of the small sap-sucking insect *Thrips imaginis* in Australia is one of the best-studied examples of density-independent effects on population density (Figure 5–9). Davidson and Andrewartha (1948a,b) found that the thrips populations on garden rose flowers reach a minor peak of abundance in the southern winter (August), increase dramatically as late spring approaches (September to December) and then become very scarce until the following August. This pattern of population growth results from the fact that in southern Australia flowers are abundant in spring but not during the long dry summer or in the vegetative growing period of the cool winter. With a great increase in the number of rose blooms, the abundant food and living space provide an unlimited environment for expansion of the thrips populations. They never have a chance to become excessively abundant on individual flowers because the long dry summer arrives first and the thrips must emigrate from the dying blooms, usually expiring themselves in the search for new flowers. Therefore, there is not an absolute shortage of food (which would become a density-dependent influence), only a seasonal and density-independent change in the accessibility of food. Hence, the initiation and decline of the rather regular seasonal peaks in the abundance of thrips each year (Figure 5–9) depend on the influence of the climate, whereas the amplitude of that particular annual peak depends on the overwintering level of the thrip population at the start of the principal spring growing season.

As might be expected from their mode of action, density-independent factors should affect the dynamics of taxonomically diverse populations in the same area if they are of wide significance in limiting population numbers. Relevant data from natural populations is seriously lacking. Ehrlich et al. (1972), however, observed dramatic effects on the biota of a subalpine area around Gothic, Colorado, when

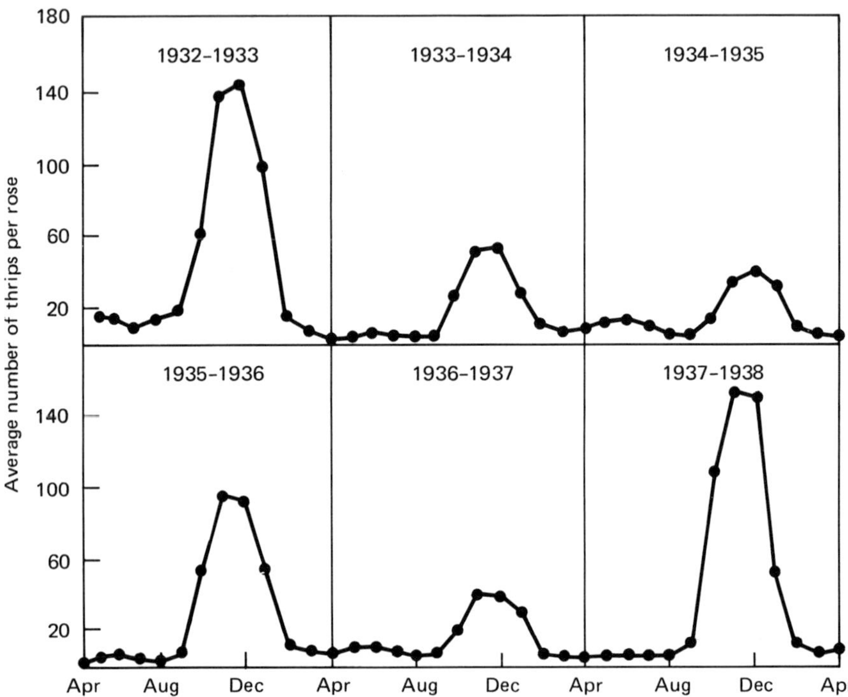

Figure 5–9 Seasonal fluctuations in Australian thrip populations living on rose blooms, over six successive years. Maximum reproduction and population growth always occurs in the Southern Hemisphere spring and summer (October through January). (Source: R. H. MacArthur and J. H. Connell, 1966, *The Biology of Populations*, Wiley, New York, p. 133. From H. G. Andrewartha and L. C. Birch, 1954, *The Distribution and Abundance of Animals*, Univ. of Chicago Press.)

unusual spring weather was climaxed by a late June snowstorm. Damage to the 683 species of herbaceous perennial plants in the region was extensive; 50 to 90 percent of the flowers produced by these plants failed to reach maturity due to the intervention of four days of snow and a heavy freeze. Because more than 80 percent of the 806 species of vascular plants in the Gothic area are herbaceous perennials, this serious damage had tremendous effects on the total biota, in contrast to the probable result in more mesic environments at lower elevations where the herbaceous flora does not constitute such a prominent component of the vegetation. The size of insect and many small mammal populations was notably depressed, and the storm caused the extinction of at least one *Glaucopsyche lygdamus* butterfly population that had been under study for several years. Ehrlich et al. (1972) suggest that following the dynamics of a taxonomically and geographically diverse sample of populations through time will offer the firmest evidence for choosing between models emphasizing density-dependent feedback controls on population size and those emphasizing apparently random fluctuations in size, dependent on complex fluctuations in the environment that must be considered essentially random.

Density-dependent factors

Density-dependent factors exert effects that vary in proportion to the size of the population, hence serve to regulate the population in a homeostatic or feedback process. These factors are responsible for the steady-state population sizes seen near the carrying capacity level of the logistic growth curve. They operate throughout population growth, however, not solely at one point in time (although their initial effects cause the inflection point observed in the sigmoid growth curve). Thus, they normally begin to act well below the carrying capacity and their effects are intensified as the upper population limit is approached. There, at the theoretical equilibrium density, the production of offspring ought to exactly balance the loss of adults by death or emigration. In actuality, as emphasized by Ehrlich and Birch (1967), a natural population is never truly at numerical equilibrium but constantly experiences fluctuations. Should the population size exceed the equilibrium density, the density-dependent factors exert even stronger effect and cause a greater rate of loss. This brake operates until the population size approximates the equilibrium density or carrying capacity again. Likewise, if the population size drops below the equilibrium level, the effects of the density-dependent factors will decrease sufficiently to permit the population to increase. Because growth and reproduction involve time, it is obvious that there will frequently be a significant time lag between cause and effect. Hence the population size will oscillate about the hypothetical but rarely attained equilibrium density.

The classical density-dependent view, then, is that as a governor mechanism regulates the speed of an engine, so density-dependent factors regulate population growth, preventing overpopulation in most plant and animal species. Thus, they (rather than density-independent influences) are said to be the factors primarily responsible for the achievement of a steady-rate or mildly fluctuating density level at the carrying capacity of the environment.

Density-dependent factors, being intrinsic to the population, are usually clearly biotic in character and involve interactions with other organisms. *Predators, parasites,* and disease *pathogens* such as bacteria, viruses, and protozoans, act as density-dependent agents upon a population. Predators are able to locate and secure prey more easily in populations of high density and provided that they are not swamped by extremely high numbers of potential prey, they operate as a very efficient density-dependent factor. Parasites and pathogens spread

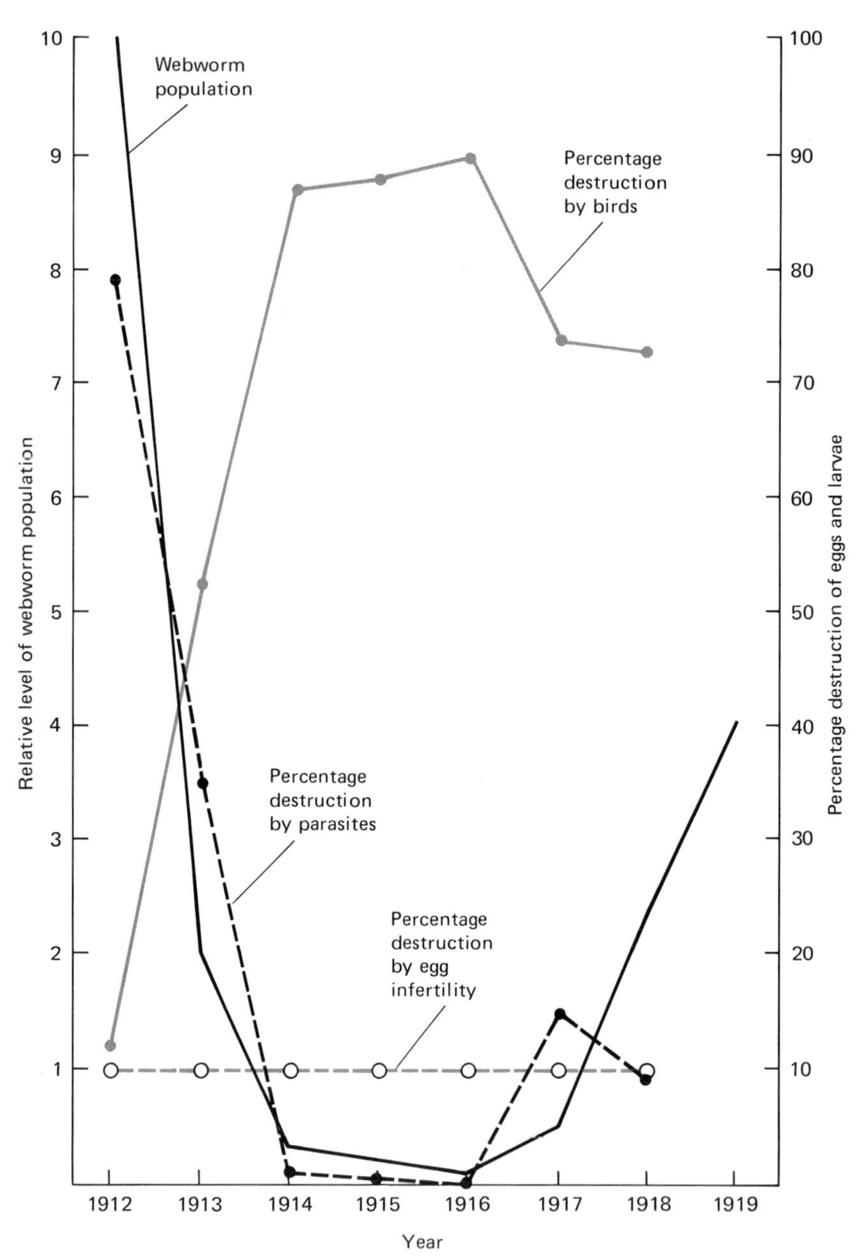

Figure 5–10 Changes in population levels, parasitism, and predation in a fall webworm population in New Brunswick. (Data from J. D. Tothill, 1922, *Bull. Can. Dept. Agric., 3:* 1–107.)

more easily from organism to organism in high-density populations of either plants or animals; therefore, they tend to reduce the density of the hosts through increased mortality. The early biotic theories dealing with natural control of animal populations emphasized the role of parasites and predators—the forerunners of modern density-dependence theory (see Solomon, 1949 for discussion of the history of natural control theory). Hence, Howard and Fick (1911) concluded in the initial biotic control proposal that populations of two economically important moth species were kept under control by a complex of parasites, which caused increasing host destruction as the population of hosts increased in density. This facultative or density-dependent control was aided by disease only when the parasites failed to limit population numbers; hence, microbial diseases were thought to have survival value since they prevented the host moths from completely decimating the available food supply and preventing future reproduction altogether. In studying changes in populations of the fall webworm moth in New Brunswick, Tothill (1922) found striking changes in the amount of parasitism and predation (Figure 5–10). Weak-flying *Apanteles* parasitic wasps accounted for 18 percent of webworm loss in 1912, but in 1913 it caused neglible losses. The strong-flying *Carpoplex* parasites caused 41 percent mortality in 1912 and 34 percent in 1934, hence holding its success in the face of a major decline in the host population. When the webworm population was exceptionally depressed, parasitism was almost zero. After the webworms began increasing in density, *Apanteles* renewed its attacks (having survived on alternate hosts) but *Carpoplex* had dispersed and was extinct in the host population area. Figure 5–11 shows the relative effects of bird predation and wasp parasitism for different population densities of the webworm. It is not all unusual to find reverse trends in two types of density-dependent control agents. In this case, birds do not distinguish between parasitized and nonparasitized insects, but take them in a nondiscriminatory way. Thus at low webworm densities when birds are effectively eating most of the available webworms, they are interfering with (indeed superseding) the controlling operations of the parasites. The percentage of predation by European tits on pine-eating moth and sawfly larvae in Holland has been shown to decline as the host population increases (Figure 5–12) (Tinbergen, 1949; Lack, 1954), and predation by tits on winter moth caterpillars in an oak woodland shows a similar relationship (Figure 5–13).

Intraspecific *competition* is commonly an important density-limiting process in natural populations. Individuals of the same species compete for a resource in limited supply; under such conditions the population can accommodate only a certain maximum density without becoming unstable. Food and space are normally the environmental resources in short supply, although because an energy source is essential for the maintenance of a population, the food supply is invariably involved either directly or indirectly (through other limited resources) in the regulation of population numbers. When the rate of reproduction is excessive and the equilibrium density is surpassed by new animals added to the population, com-

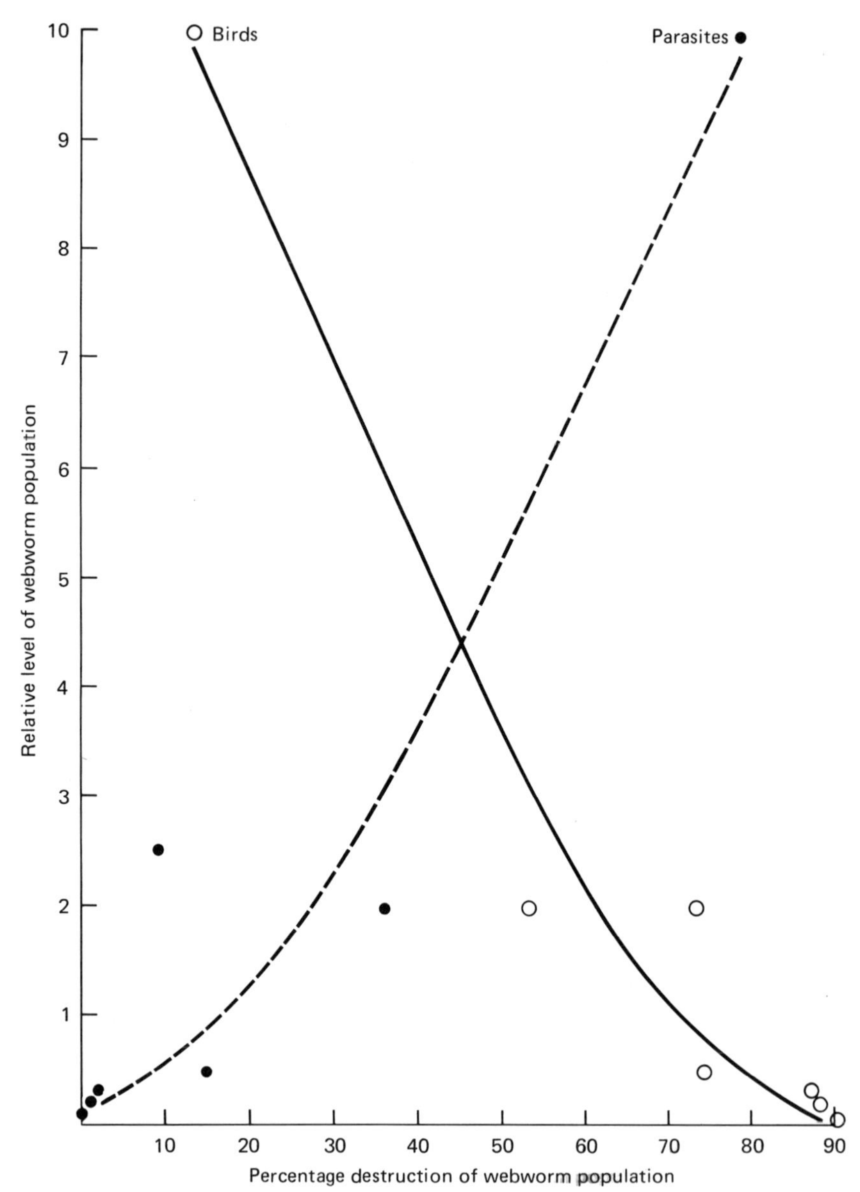

Figure 5–11 The density-dependent action of birds and parasites on a fall webworm moth population. (Data from J. D. Tothill, 1922, *Bull. Can. Dept. Agric., 3:* 1–107.)

petition results in an increasing mortality rate which continues to accelerate as the population increases. Often mortality exerts its strongest effects on egg production and survival of immature stages and hence the addition of new reproductive animals to the population is

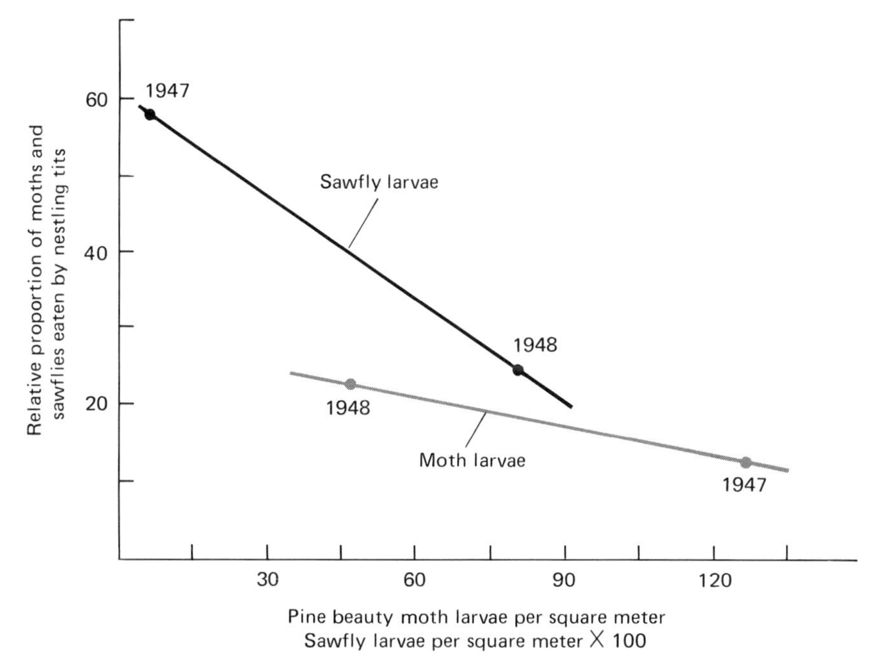

Figure 5–12 Predation by tits on pine-eating larvae in Holland. (Data from D. Lack, 1954, *The Natural Regulation of Animal Numbers,* Oxford Univ. Press. From L. Tinbergen, 1949, *Nederl. Bosch. Tijdschr., 4:* 91–105.)

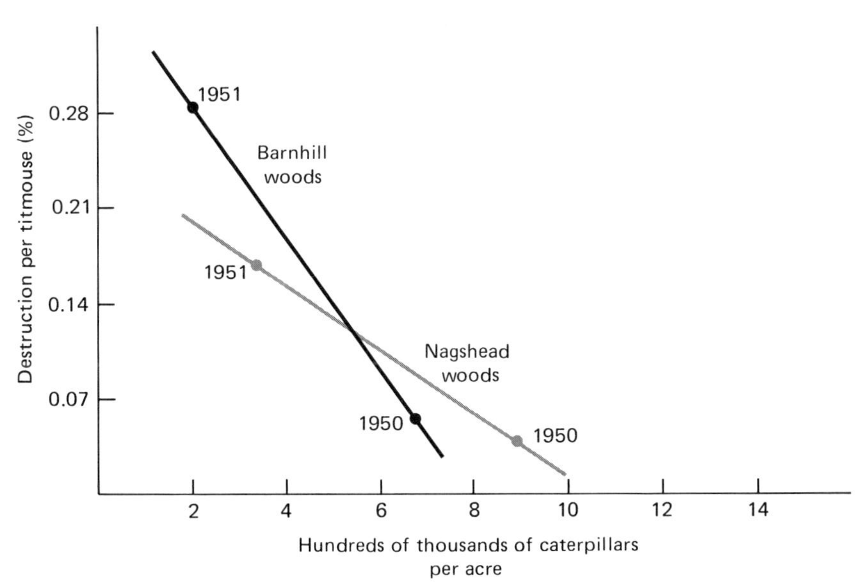

Figure 5–13 Predation by tits on winter moth caterpillars in an oak woodland. (Data from M. M. Betts, 1955, *J. Animal Ecol., 24:* 282–323.)

curtailed, eventually dropping the birth rate back to a level that yields an approximate equilibrium density.

One of the best documented examples of the influence of intraspecific competition in a population is the work of A. J. Nicholson (1954, 1957) with the Australian sheep blowfly, *Lucilia cuprina*. Nicholson introduced small numbers of *L. cuprina* into laboratory population cages and then maintained predetermined conditions constant, including the supply of such depletable resources as food and water at a constant rate. A careful day-to-day record was kept of the numbers of the adults and the various developmental stages; otherwise, the population was left to its own devices. Environmental conditions were manipulated in subsequent experiments. The experiments with a constant supply of food (e.g., 50 grams of liver daily for the larvae and water and dry sugar for the adults) revealed that the number of adults in each cage varied with great violence and with an evident periodicity (Figure 5–14). The cause of this fluctuation was obvious after the larvae were examined. When the numbers of adults were high, the vast numbers of eggs that were laid caused all of the provided food to be consumed while the larvae were still too small to pupate; consequently, no adults offspring resulted from eggs laid during such periods (see vertical

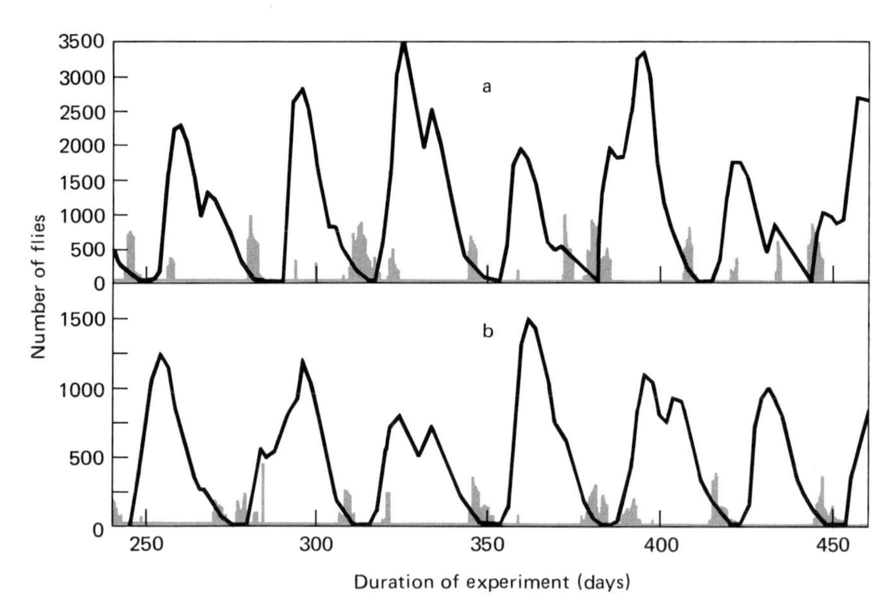

Figure 5–14 Population cage experiments with competition in Australian sheep blowflies. Heavy line indicates number of adult *Lucilia cuprina* flies in cage; vertical lines indicate numbers of adults eventually emerging from eggs laid on dates of plots. Both cultures were subject to identical conditions except that the daily quota of larval food was 50 g in (a) and 25 g in (b). (Source: A. J. Nicholson, 1957, *Cold Spring Harbor Symp. Quant. Zool., 22:* 153–173.)

lines in Figure 5–14). The adult numbers therefore dwindled progressively, until a point was reached at which the intensity of larval competition became so reduced that some of the larvae attained a sufficient size to pupate and metamorphose into egg-laying adults about two weeks later. During this period, the adult population continued to drop, thereby further reducing the intensity of larval competition, and permitting increasing numbers of the larvae to pupate and produce later adults. A new rise in the adult population then ensued, and when it reached a high level at which too many eggs were laid to permit larval survival, a new cycle began. With the two-week lag between the initiation of corrective reaction and its operation upon the population, the population-density change inducing the reaction would continue unabated for this period thereby causing alternating excessive over- and undershooting (to zero for the adults) of the equilibrium level. The average numerical level of a population controlled by food supply varied directly with the quantity of this supply; contrast Population A with Population B in Figure 5–14, where A had twice as much food as B (note that for ease of comparison, the scale of B is double that of A). These two graphs show clearly that the average population of B approximates 50 percent that of A, and that the violence and other characteristics of the oscillation are unaltered by the difference in quantity of food supply.

After demonstrating that larval competition was unquestionably the dominant governing factor producing the population oscillations he observed, Nicholson (1957) set up further experiments to test the role of adult competition for food and water. He found that several density-dependent factors would act in complementary fashion to exercise a profound influence upon the density, age structure and pattern of numerical change of the population. The culture illustrated in Figure 5–15 had limited larval food, unlimited water and ground liver for the adults, but contained little sugar. The effect was to shorten the life of the adults when numbers were high, causing sharp population drops, but to allow normal life-spans when numbers were low. The conditions for the culture illustrated in Figure 5–15 were similar, but the adults had unlimited sugar and limited water. The average adult life-span was shortened considerably, particularly when their numbers were highest. In consequence this population laid very few eggs, which reduced larval competition and therefore increased the average number of flies emerging each day. This factor raised the mean adult population to a higher level than for similar populations provided with adequate supplies of water. Here the availability of water was clearly a complementary regulating factor, for it reacted to density and its effect was associated with that of larval competition for food. The mean population size rose and the entire pattern of popula-

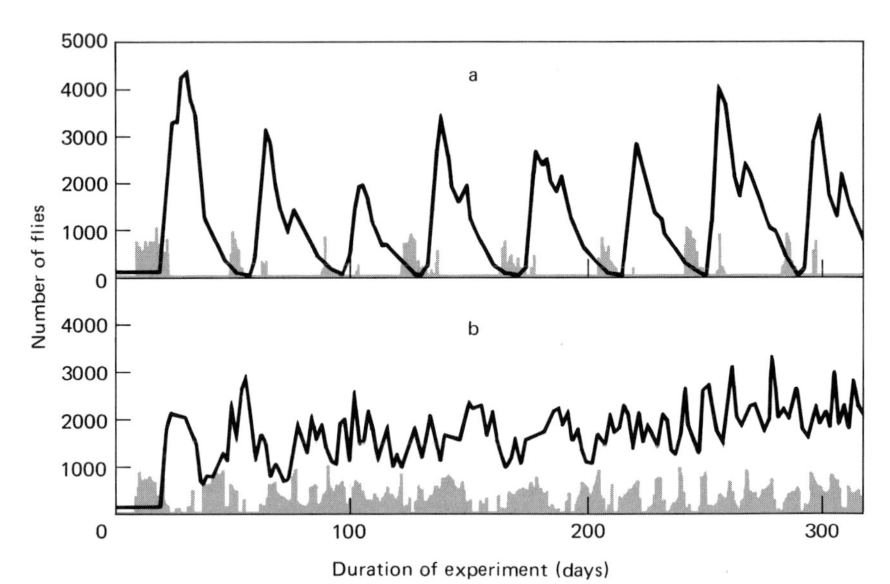

Figure 5–15 Effects produced by complementary governing factors on sheep blowfly populations otherwise governed by competition of larvae for 50 g of meat. (Source: A. J. Nicholson, 1957, *Cold Spring Harbor Symp. Quant. Zool., 22:* 153–173.)

tion fluctuations was changed to more closely approximate an equilibrium level.

Emigration from the population also acts as a density-dependent process. At high population levels, emigration of a portion of the adults or young can regulate the population density of the original area. In the black-tailed prairie dog (*Cynomys ludovicianus*) towns of South Dakota, King (1955) has shown that emigration acts as a population-regulating mechanism. Each year, adults older than two years, mostly females, will leave their home territory in a town and emigrate to new areas on the outskirts to initiate new coteries (family social units). The younger inexperienced animals are able to remain behind in the center of the town in their natal territory while the mature, experienced prairiedogs establish residence on the edges of the town where exposure to predators is greatest. Over a period of years, the population density will remain more or less constant. The periodic mass wanderings of Norwegian lemmings (*Lemmus lemmus*) in the mountain regions of Scandinavia occur during peak fluctuations in population size and serve to relieve population density (Clough, 1965). Regardless of population density, lemmings make two annual trips in normal years, usually short journeys from the drier winter habitats where a heavy protective snow cover occurs, to the lower wet peat lands during the summer, where grass and moss are most abundant. These relatively short trips occur during April and May, and then again in late summer and fall before the snow comes, and often involve vertical movements up and down the mountain slopes (Kalela, 1961).

Other processes operating in a density-dependent fashion include *physiological* and *behavioral* control mechanisms. The social stresses caused by crowding have been shown to act on individual rodents through the endocrine system (Christian and Davis, 1964). In vertebrates, these endocrine-induced changes are particularly associated with the pituitary and adrenal glands. Increasing population density leads to inhibition of sexual maturation, decreased sexual activity, and inadequate milk production in lactating females. Stress caused by crowding also curtails the number of offspring produced per litter, through spontaneous abortion or absorption of early embryos into the uterine wall of the pregnant female. *Genetic* differences may cause polymorphic behavior or viability in a population, inducing one genotype to migrate while another genotype remains resident or allowing one type to be better at raising its young under crowded conditions (Chitty, 1957). Complex components of *habitat suitability* and *food supply* may fluctuate in availability in time and space without reference to competition (Singer, 1972).

Generally, species obviously controlled by density-dependent factors tend to be larger organisms such as birds, mammals, shrubs, and trees. They tend to have longer life cycles, usually extending beyond a single year. Their fecundity, that is, their rate of production of offspring is lower and hence they have lower biotic potentials and more stable population sizes than populations controlled principally by density-independent factors. Let us now look at these highly important density-dependent factors in more rigorous detail.

Predation

Observations of predator and prey cycles in the abundance of arctic mammals and birds prompted population ecologists very early to consider predation as a density-dependent regulator of population growth (Elton, 1924). Today it is commonly believed to serve as a chief source of density dependent influences in the regulation of population numbers in a large percentage of plant and animal species (Wilson and Bossert, 1967), and an abundance of theoretical models have been derived to describe its effects.

The classic Lotka–Volterra equations of predators–prey interaction were independently proposed by Alfred J. Lotka (1925) and Vito Volterra (1926), both of whom expressed the rate of growth of predator and prey populations by simple differential equations. Consider the prey as species 1 and the predator as species 2. Then in its simplest form the rate of growth in numbers of the *prey* population, dN_1/dt, will have two components. One represents the possible reproductive rate of the prey population in the absence of predators, (r_1N_1), where r_1 is the birth rate of an individual, and N_1 represents the number of individuals alive in the prey population. The individual birth rate of the prey is not directly dependent on the abundance of the predator. The second component represents the removal of prey from the population by predators, that is, the death rate, which is directly proportional to the abundance of the predator and may be most simply described as the product of the prey and the predator populations (N_1N_2, which is proportional to the probability of a chance encounter between predator and prey), multiplied by a coefficient of predation P. Thus, the overall *prey population gorwth* may be expressed as

$$\frac{dN_1}{dt} = r_1N_1 - PN_1N_2$$

The growth rate of the *predator* population also has two components. It is proportional in the first case to the number of prey individuals that the predator population succeeds in capturing (aPN_1N_2), which simply recognizes that the individual birth rate of the predator depends upon the amount of food available, which in turn depends upon the density of the prey population and the efficiency a with which predators convert food into offspring. From this term, we must subtract the death rate of the predator population (which does not depend on the prey as much, and can be expressed as a constant rate of death d_2 times the number of predators or d_2N_2). The resulting equation for *predator population growth* becomes

$$\frac{dN_2}{dt} = aPN_1N_2 - d_2N_2$$

As Ricklefs (1973) points out, this Lotka–Volterra model includes several simplifying assumptions: the prey population is experiencing no complicating interaction with its own food supply; the relationship between predator numbers and prey abundance is linear; and mortality of the predator population is independent of density.

The most useful prediction of the Lotka–Volterra model is that predator and prey populations will oscillate. The upper diagram of Figure 5–16 shows that the number of predators increases as one goes up the right side of the vertical line (which represents the state where both prey and predator populations are in equilibrium and $N_1 = d_2/aP$, a constant value), and hence the predator population is growing (+). It increases because when there are more than d_2/aP prey, the birth rate of the predator population exceeds its death rate. To the left of this line the predator population declines (−). Below the horizontal line (where predator and prey populations are in equilibrium and $N_2 = r_1/p$, a constant value), the prey population is growing. With the

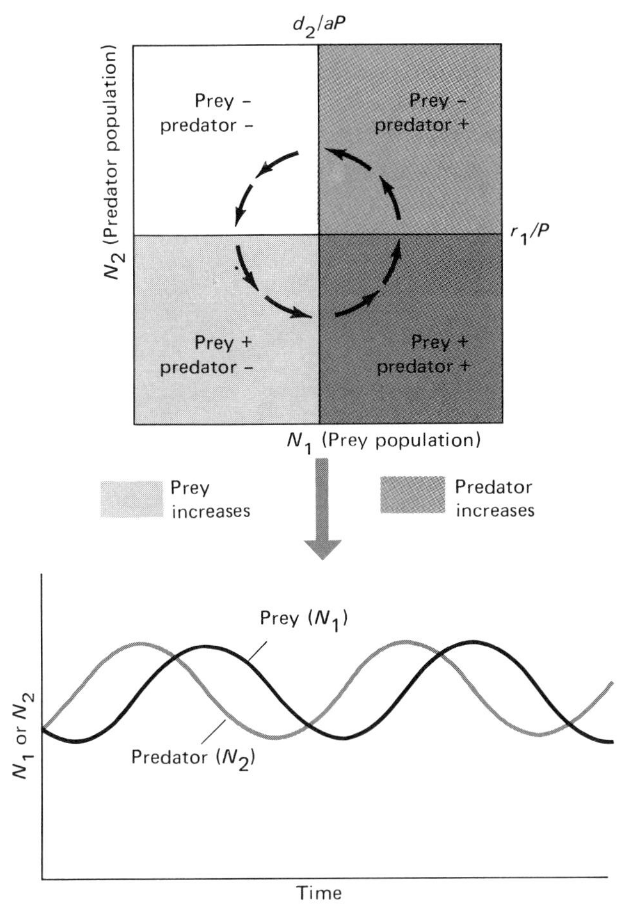

Figure 5–16 Predator–prey interaction as predicted by the Lotka-Volterra equations. The upper graph shows the joint abundance of the two interacting populations. The lower graph shows the result when the abundances of the two species are plotted as a function of time. (Adapted from E. O. Wilson and W. H. Bossert, 1971, *A Primer of Population Biology*, Sinauer, Stamford, Conn., p. 132.)

decrease in the number of predators (less than r_1/p of them), the death rate of the prey population falls below its birth rate.

The arrow trajectories of predator and prey populations in the upper graph of Figure 5–16 represent the joint population sizes of the predator and prey as they change through time. When displaced from the equilibrium point at the center, they will continue to fluctuate, following a circular counterclockwise path (because a buildup or decrease in the prey population must precede an increase or decrease in the predator population). The resulting oscillations of the predator and prey populations through time are shown in the lower graph of Figure

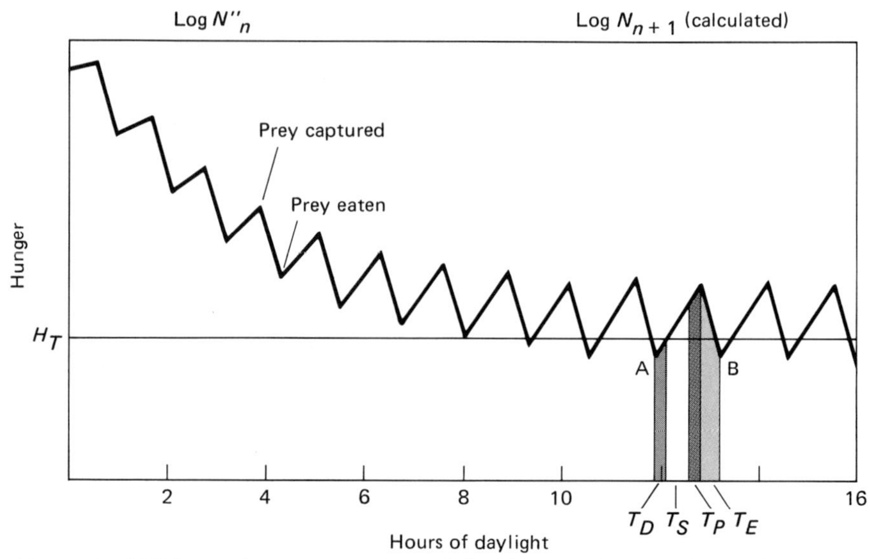

Figure 5–17 Schematic representation of the changes in a predator's hunger during 16 hr of daylight. H_T = hunger level that just triggers attack; T_D = time spent in a digestive pause; T_S = time spent searching for prey; T_P = time spent in pursuit of prey; T_E = time spent in eating prey. The predator in this system was the Carolina praying mantis. (Source: C. S. Holling, 1963, *Mem. Entomol. Soc. Can., 48:* 1–86.)

5–16. Such oscillations could form natural population cycles of the type seen in Canadian lynx and snowshoe hare populations (Figure 5–6). Three other agents besides predation, however, have also been shown to cause population cycles: mass emigration (for example, in the Norwegian lemming), physiological stress resulting from overcrowding, and genetic changes in the population.

Increasing efforts have been made to refine the hierarchy of predation components that regulate a population so that more precise predation regulatory models may be stated. Holling (1965, 1966) experimentally examined 10 components of predation by praying mantids, including the rate of successful search, time of exposure to prey, handling time, hunger, learning behavior by the predator, inhibition by prey, exploitation, interference between predators, social facilitations, and avoidance learning by prey. Each of these components in turn (for example, hunger) could be broken down into *their* components (Figure 5–17) and the hierarchical array involved, including the interplay of the rapidly multiplying numbers of components, bogs down all known methods of modeling. Holling (1966, p. 48) expressed the problem as follows:

> It was first thought that all the effects [of components] described here could be incorporated into a differential equation which could be integrated to yield an expression for the number of prey attacked at different

> prey densities. It quickly became evident, however, that the inclusion of each fragment became progressively more difficult as the model was expanded. In fact, the point was quickly reached where the partially synthesized equation became intractable.

By using a system of sequential cyclic equations and handling the cumbersome calculations with a computer program, the difficulty of analyzing just ten components on the first hierarchical level was met. Although the methodology of using this approach of the systems hierarchy or experimental component analysis is formidable, it appears to offer the greatest promise for attacking the complexities of population control by predation, competition, and other major and minor factors.

Attempts to analyze and model selected features of the predator–prey interaction continue apace with the holistic modeling, and reference may be made to the recent review of some of these models by Pianka (1974). As an example of the restricted (but productive) approach, Salt (1974) experimentally investigated the roles of predator density, prey density, and their interaction in the control of the capture rate by the predator. The protozoan *Didinium nasutum* served as the predator and *Paramecium aurelia* as the prey. The results showed that both predator density and prey density were potential controls and that prey density was a stronger regulator than predator density (Figure 5–18). Salt postulates that the two controls are alternates, that the capture rate is set by whichever control is the more restrictive, and that the prey density control is probably a fixed internal control while the character of the predator density control could not be determined. From these laboratory experiments, Salt calculated that if one introduced a single *Didinium* predator into a reproducing *Paramecium* population with a density of 52 animals per 100 microliter of water, within 24 hr the densities of predator and prey would be such that the regulation of the capture rate would have been transferred from prey density to predator density control. Hence, natural selection has acted on this predator to maximize the profitability of its hunting through developing a negative feedback control on hunting rate. In other words, the variable control of capture rate provides a mechanism for prolonging the contact between the *Didinum* and their prey in order to maximize reproduction and produce a large number of new predator individuals. The benefits to the prey species are a by-product of the self-interest-promoting mechanism of the predator. Royama (1971), in analyzing data from many predator systems, concluded that natural selection has operated on all predators to maximize the profitability of their hunting and that variable negative controls of noninterference type have evolved only in the higher vertebrates, namely birds and mammals. Salt's (1974) findings suggest that such apparently sophisticated controls may be present in lower organisms and be the product of natural selection.

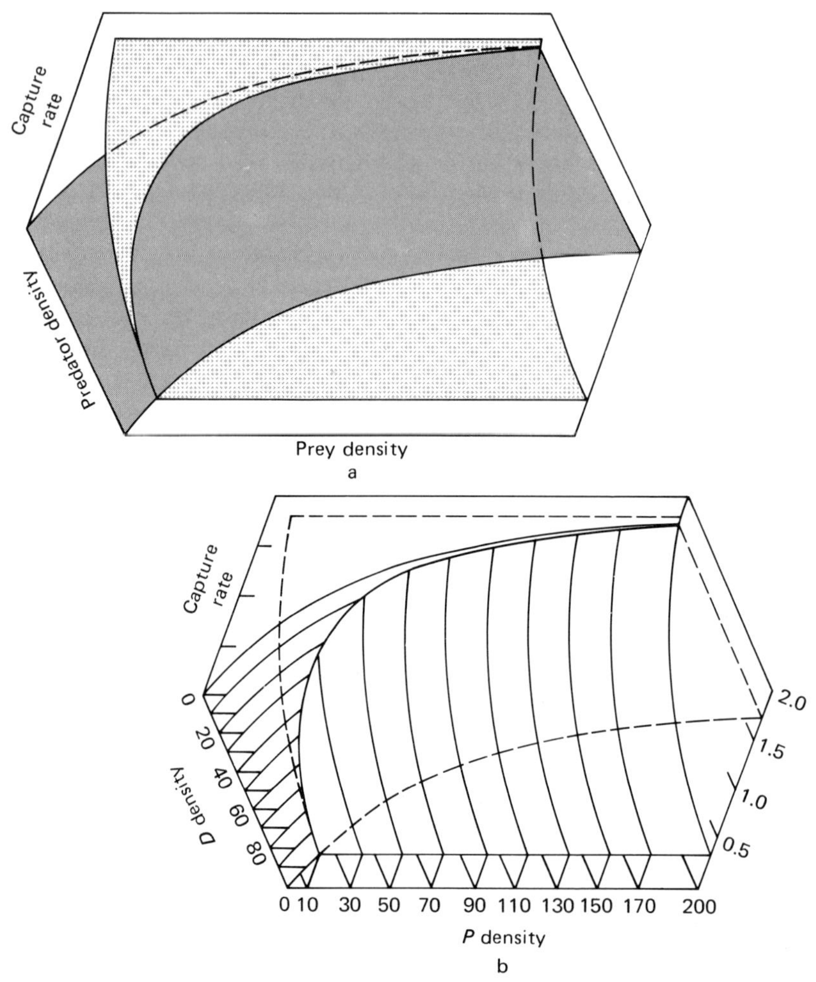

Figure 5–18 (a) Diagram of the simultaneous action of predator and prey density as controls of the capture rate. Capture rates set by prey density at all combinations of predator and prey density lie on the light-colored curved plane. Capture rates set by predator density lie on the darker curved plane. (b) Diagram of the realized capture rates at all combinations of predator and prey density when the more restrictive of the two controls determines the capture rate. Densities are given in number of animals per 100 μl. Capture rate is given as the number of *Paramecium* captured per *Didinium* per hour. (Source: G. W. Salt, 1974, *Ecology, 55:* 434–439.)

Competition

Competition for resources between individuals of the same population was another of the earliest biotic influences to be championed as a chief regulating factor in population ecology. The foundations of competition theory were laid by A. J. Nicholson (1933), who has been the leading advocate of competition as the single key factor in population regulation. The principal arguments in his thesis are as follows.

Densities of a given species tend to differ consistently from place to place, and such differences in average population level can commonly be attributed to obvious differences in climate or other environmental factors. Therefore, a stabilizing or balancing mechanism must exist that tends to maintain different average population levels between areas. Nicholson's view, contrasting sharply with that of Ehrlich and Birch (1957), is that "without balance the population densities of animals would be indeterminate, and so could not bear a relation to anything." Observed associations between weather and population changes do not indicate that densities are determined or caused by weather, but only that weather may vary the densities.

The factors responsible for balance are almost entirely some form of competition. Balance is defined as "the state of a system capable of effective compensatory reaction to the disturbing forces which operate upon it, such reaction maintaining the system in being." Competition is "the state of reciprocal inference which occurs when animals having similar needs live together and which influences their success [in survival and reproduction]."

The terms control and regulation refer to factors capable of producing balance. Because only those factors operating more severely against high populations than against low populations (that is, dependent upon density) are capable of producing balance, then by definition only *density-dependent factors* are capable of controlling or regulating populations.

Competition takes two forms: (1) chemical and physical, and (2) biological. It occurs in alteration of the chemical and physical nature of the environment by the activities of individuals (such as found among populations of plants, microorganisms, and some grazing animals), and it may be expressed in difficulty in *finding* the requirements of life, or difficulty of natural enemies in *finding* a particular host or prey species (commonly seen among insects).

The basic problem of competition can be analyzed by studying the characteristics of random searching among animals. Although searching by individuals themselves is probably not random, but strongly oriented as a result of innate behavioral patterns and experience, the fact that searching by one individual does not influence the area searched by another means that searching *within* favorable environments by the population as a whole tends to be random. The *area traversed* represents the total amount of searching carried out by an individual, (i.e., the swath it cuts through the area) while the *area covered* represents the increment of area not previously traversed by other animals (that is, the area of successful searching). Thus, it is possible to construct a "competition curve" (Figure 5–19) showing the relationship between

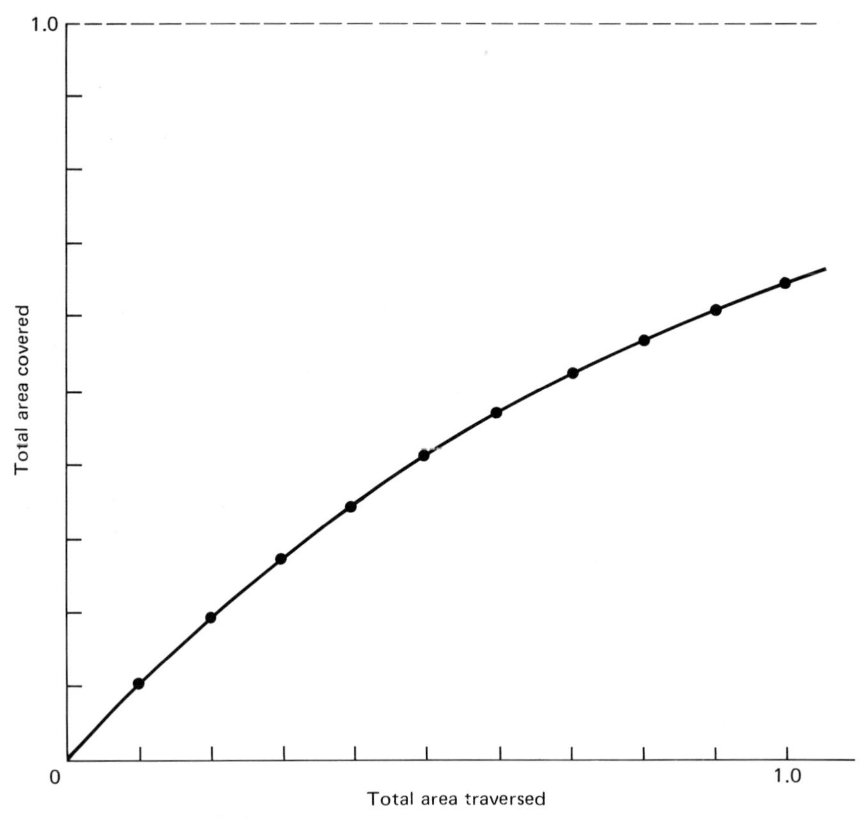

Figure 5–19 A Nicholson competition curve. (Data plotted from Table 5–3.)

population searching (total area traversed) and successful searching (total area covered), from these data (Table 5–3). An analogy to the effects of competition/theory may be drawn between the progressive diminution in the rate at which an area is covered, and the decreasing success among competing individuals as populations increase in density.

Competition results in surplus animals being destroyed, or in failure to produce a surplus, thereby preventing further population increases. Increased competition causes animals to become less successful and thus to survive at a lower rate or produce fewer progeny. Under completely stable environmental conditions, populations would increase until the density-dependent regulatory action of competition fixed population size at a particular level. This population condition was called the "steady state" population by Nicholson (1933). Populations are always tending toward the steady state, but due to the continuous changes in environmental conditions, the theoretical steady state density is also continuously changing. Hence populations are constantly striving for stability around a continually varying theoretical level.

Table 5–3 An example of the type of data which Nicholson used to construct a competition curve.[a]

AREA TRAVERSED	TOTAL AREA NOT PREVIOUSLY COVERED	AREA COVERED	TOTAL AREA COVERED	TOTAL AREA TRAVERSED
0.1	1.0	0.1	0.1	0.1
0.1	0.9	0.09	0.19	0.2
0.1	0.81	0.081	0.271	0.3
0.1	0.73	0.073	0.344	0.4
0.1	0.66	0.066	0.410	0.5
0.1	0.59	0.059	0.469	0.6
0.1	0.53	0.053	0.522	0.7
0.1	0.48	0.048	0.570	0.8
0.1	0.43	0.043	0.613	0.9
0.1	0.39	0.039	0.652	1.0

[a] See Figure 5–19.

The experimental evidence for Nicholson's competition theory has been considerable, primarily from his work and that of his students with laboratory populations of the Australian sheep blowfly *Lucilia* (e.g., Nicholson, 1955, 1957). These experiments show that in complex environments, competition for critical factors lowers the population size before food or another resource becomes so low that the population goes to extinction. Among wild populations of blowflies, Nicholson believes that competition *per se* is capable of controlling population density in any situation. Thus, blowfly populations can be said to increase because of an increase in environmental resources (e.g., sheep carcasses) or because competition is decreased.

Criticism of the competition ideas of Nicholson has come on several grounds. Thompson (e.g., 1939) feels that the concept of balance, and that only density-dependent factors can cause this balance, is not true in most natural populations. Whereas Nicholson believed density-dependent factors to prevent the population from becoming extinct, Andrewartha and Birch (1954) believe the population can exist in the absence of density-dependent effects. They emphasize the importance of habitat heterogeneity, with spatial patchiness and fluctuations occuring in physical components of the environment. The chances for extinction in any one part of the range in any year may be high in the absence of density-dependent factors. However, chances for extinction by density-independent factors in all parts of the range are quite remote. Milne (1957, 1958, 1961, 1962) attacks the mathematical basis of the competition curve and other Nicholsonian proposals, negating the implied randomness of the searching (random searching being one of the basic ideas of the competition theory). Milne's criticisms from an ecological viewpoint include, as in the case of parasitic wasps, the clear existence of protective niches where prey can hide from its enemies. If the number of hosts in protected niches exceeds the number necessary to just maintain the host population, then the parasite population has adequate food from the excess. The concept of host niche protection thwarts the idea of parasite population control by competition, as control depends on the number of physical hiding places. Of course, competition among the hosts for adequate hiding places must play a role in the parasites controlling that population. Milne also points out that in parasitic insects, fluctuating physical factors (especially weather) affect parasites and hosts differently and may cause synchrony or asynchrony in the life cycles of the two species, controlling or not controlling the host and parasite populations, without reference to the biotic effects of competition. Nicholson counters Milne's concept of imperfect density-dependent influences by asserting that the alleged interference of density-independent factors merely reduces the effectiveness of predators or parasites and causes them to regulate their prey or host at higher population levels. Lack (1954) supports Nicholson's ideas in reviewing natural regulation in bird populations, whereas other authors use the same data to support density-independent theories. The controversy remains unresolved today.

Most recent ecologists have focused their attention and modeling efforts upon *interspecific competition,* a subject to be discussed in detail in Chapter 11. As with predation theory, Lotka and Volterra described the effects of competition between two species on the regulation of population growth. The logistic growth equation for species 1 is modified to incorporate the effect of the species 2 competitor, and similarly species 2 is affected by species 1 (see Chapter 11 for the two competition equations and discussion). Usually, one or the other species becomes extinct, or both may die out. Under certain circumstances, it is possible for two species competing for the same limited resources to stably coexist. In one well-documented experiment (Ayala, 1971), the relative fitnesses of two species of *Drosophila* under competition in laboratory populations were shown to be inversely related to the relative frequencies of these species (Figure 5–20). This frequency-dependent fitness leads to a stable coexistence of the two species in spite of their competition for limited resources.

Modeling treatment of intraspecific interactions have been less widely developed, but interest in this area is increasing. Wiegert (1974) developed realistic, general equations of population growth, separately representing scarcity of renewed material resources and scarcity of fixed resources related to space. These equations show the differential impact of these two factors on competitive coexistence. Wiegert proposes that the equations suggest a general model of competition covering any number or kind of resources, consumers, and intraspecific as well as interspecific interactions. Simulation on digital or analog computers is required for assessment of coexistence and sta-

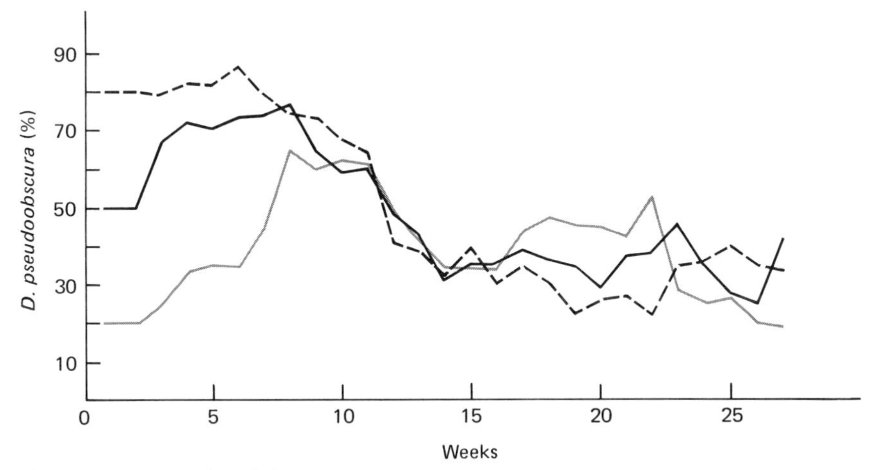

Figure 5–20 Results of the competition of strains of *Drosophila pseudoobscura* and *D. willistoni* in three experimental populations begun at different initial frequencies. (Source: F. J. Ayala, 1971, *Science, 171:* 820–824.)

bility properties when more than a few competitive pathways are considered, since the number of possible sets of equilibrium equations becomes very large.

POPULATION REGULATION OF MICROTINE RODENTS

In this final section on population regulation, we will briefly survey the variety of regulating mechanisms proposed for complex population fluctuations and oscillations in the well-studied microtine rodents (voles, mice, and lemmings). The population cycles of small rodents have always been a classic problem in population ecology, and because population cycles represent an ideal situation in which to study population regulation, intense interest has centered on the possible mechanisms behind the rise and fall of these rodent populations.

Generated by Elton's (1924) original documentation of regular population fluctuations in some animals, the first approach to the problem of population cycles was to emphasize the regularity of the cycles, attempt to determine the precise period of these cycles for each species, and search for some phenomenon capable of generating cycles. The second approach was to turn attention away from the periodicity, which as Cole (1951, 1958) pointed out could be interpreted as essentially random fluctuations with some serial correlation between successive years, and concentrate on the causes of each case of population growth and decline. This approach suggests that "the problem of cyclic length will be solved once the mechanism of these cycles is understood" (Krebs, 1964).

Three general facts have emerged from the work on fluctuations in microtine rodents to date (Krebs, 1964; Krebs et al., 1973). First, a cycle is typically a three- or four-year fluctuation in numbers which is characterized by high body weights of adults during the summer of a peak year. Second, fluctuation in numbers occur in many species of microtine rodents (lemmings, voles, and mice) in many different genera; no well-studied population has been found to be stable in numbers from year to year. Third, these periodic fluctuations are found in a variety of north temperate to arctic ecological communities: lemmings on the tundras of North America and Eurasia, red-backed voles in the boreal forests of North America and Scandinavia, meadow voles in New York, and field voles in coastal California, New Mexico, Indiana, Britain, Germany, and France. While no cyclic fluctuations have been described for tropical rodents or for temperate South American species, population studies on these species are almost totally lacking. Because of the widespread nature of population cycles in rodents, and the fact that demographic events are similar in a variety of species living in quite different plant communities under different climatic regimes, Chitty feels that these cycles are a single class of events and that a common explanatory mechanism underlies all rodent cycles. Most recent workers have adopted this viewpoint in proposing their theories of mechanism—a simpler assumption to handle and test than the concept that all these rodent cycles have a different explanation.

The principal controlling factors which have been proposed to stop populations from increasing are *extrinsic agents* such as weather, predators, food supply, or disease and parasites, and *intrinsic agents* involving the effects of one individual upon another, particularly those of stress, behavior, and genetics. The last three hypotheses have received particular emphasis in recent work.

Weather

The problem of synchrony of cycles among populations scattered over large areas of country has long intrigued workers. Ignoring cosmic theories such as radiation fluxes, solar flares or sunspot cycles, weather seems to be the only reasonable variable that could account for this synchrony (Krebs, 1964). If weather was not involved and only an intraspecific factor controlled cycles, we would expect to observe non-synchronous fluctuations (Chitty, 1952). Weather thus serves to synchronize increase or decline in the cyclic rodent species, but undoubtedly it cannot be a sufficient cause of the cyclic increase or decline because we would not normally get three- or four-year cycles if this were true. (Weather cycles of this periodicity have not been observed in these regions.)

In lemmings, it appears that winter weather (but not summer

weather normally) may be a partial cause of the increases and declines. For promoting synchrony, Chitty (1952, 1960) postulates that the effect of weather on a population will be determined in part by the past history of the population. Populations in a low state would increase during a mild winter, while the decline of peak populations would be retarded under such conditions, thus tending to bring scattered populations into phase.

Predation

Relatively few workers (Lack, 1954; Pitelka et al., 1955; Pearson, 1966) today believe that the lemming cycle is caused by predators, although Pitelka (1959) has observed that jaegers and owls tend to dampen the fluctuations of the lemmings in northern Alaska. The strongest evidence for the role that carnivores might play in the population cycles of microtine rodents has been offered by Pearson (1966) in studies of meadow mice (*Microtus californicus*) living in Tilden Park near Berkeley, California. Population peaks occurred in this mouse population in 1947, 1951, 1958, 1961, and 1963, and the role of carnivores (feral cats, raccoons, gray foxes, and skunks) in a complete microtine cycle on this 35-acre study area was followed for three years (Figure 5–21). The curve for the number of *Microtus* eaten in each moth lagged behind the curve for the number of *Microtus* present (Figure 5–21).

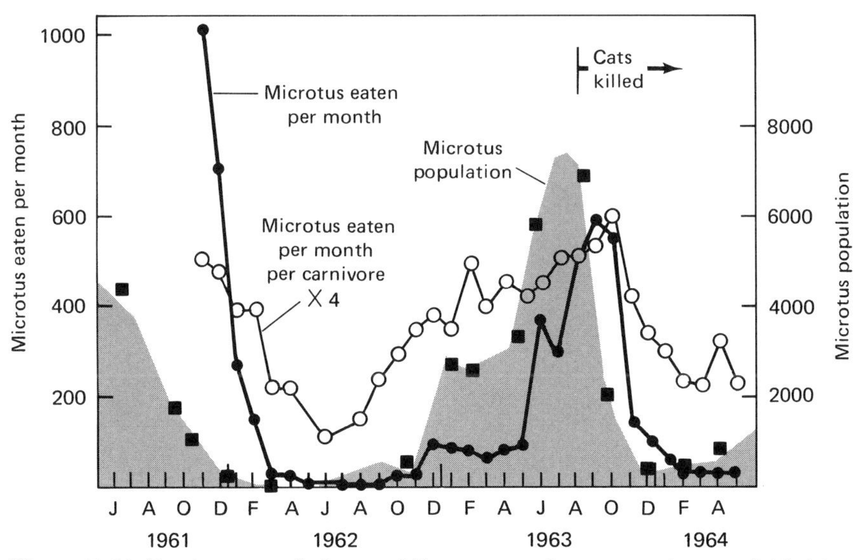

Figure 5–21 Carnivore predation on *Microtus* on a 35-acre study area (14 ha) in California during three years. Divide left-hand vertical scale by four to obtain number of *Microtus* eaten per month per carnivore. (Source: O. P. Pearson, 1966, *J. Animal Ecology, 35:* 217–233.)

During the collapse of the 1961 peak population, for example, the monthly toll taken by carnivores becomes minimal two or three months after the low point of the microtine cycle. At this time, the carnivores turned to secondary prey such as gophers, harvest mice, wood rats, and rabbits. The *Microtus* population started rebuilding rapidly in November 1962, but the toll taken by carnivores did not increase markedly until six months later. During the 1963 collapse, the decline of the carnivore catch followed the mouse decline by at least one to two months, but the situation was influenced at this time by the park officials shooting the feral cats. Pearson estimates that carnivores ate about 88 percent of the 1961 *Microtus* peak population and 25 percent of the 1963 peak. Because there were fewer carnivores present during the second peak, predation pressure on the 1963 population did not become as severe, and the *Microtus* numbers were not reduced as much.

The nature of carnivore predation during the cycle observed by Pearson is summarized in graphic form in Figure 5–22. Pearson (1966) notes that if carnivore predation were immediately and completely responsive in a constant-proportion manner to *Microtus* numbers, the percentage capture during a *Microtus* cycle would follow a course like configuration A in Figure 5–22; all points would fall in a vertical distribution above some appropriate monthly percentage. However, if carnivore predation removed a constant number of mice each month throughout the cycle, the points on the graph would follow a distribution like B in Figure 5–22. In a third case, if carnivore predation were to exert, through compensatory mortality, a stabilizing effect on the mouse population during a cycle, the points would fall along some line similar to configuration C.

The distribution that Pearson found (Figure 5–22) shows that the carnivores tended to capture a relatively constant percentage of the population (5 percent) after a crash and during the subsequent increase (months 7–20), and a much greater percentage during a crash. The postulated curve (dashed line in Figure 5–22) is intended to serve as an estimate of conditions during a typical crash, and accordingly falls between the data for the first crash (months 1–6, in 1961), which are clearly too high because some of them exceed 100 percent and the data for the second crash (months 21–25), which are too low because many carnivores had been filled. Of special importance is the observation that the data can in no way be construed to have a slope like that of insert C, which would implicate a stabilizing effect through increased percentage of predation when mice were abundant. The lack of compensatory control by carnivore predation undoubtedly contributes to the great amplitude of the *Microtus* cycle.

In Pearson's view, the importance of carnivore predation to the microtine cycle of abundance lies in the low phase of the cycle, when car-

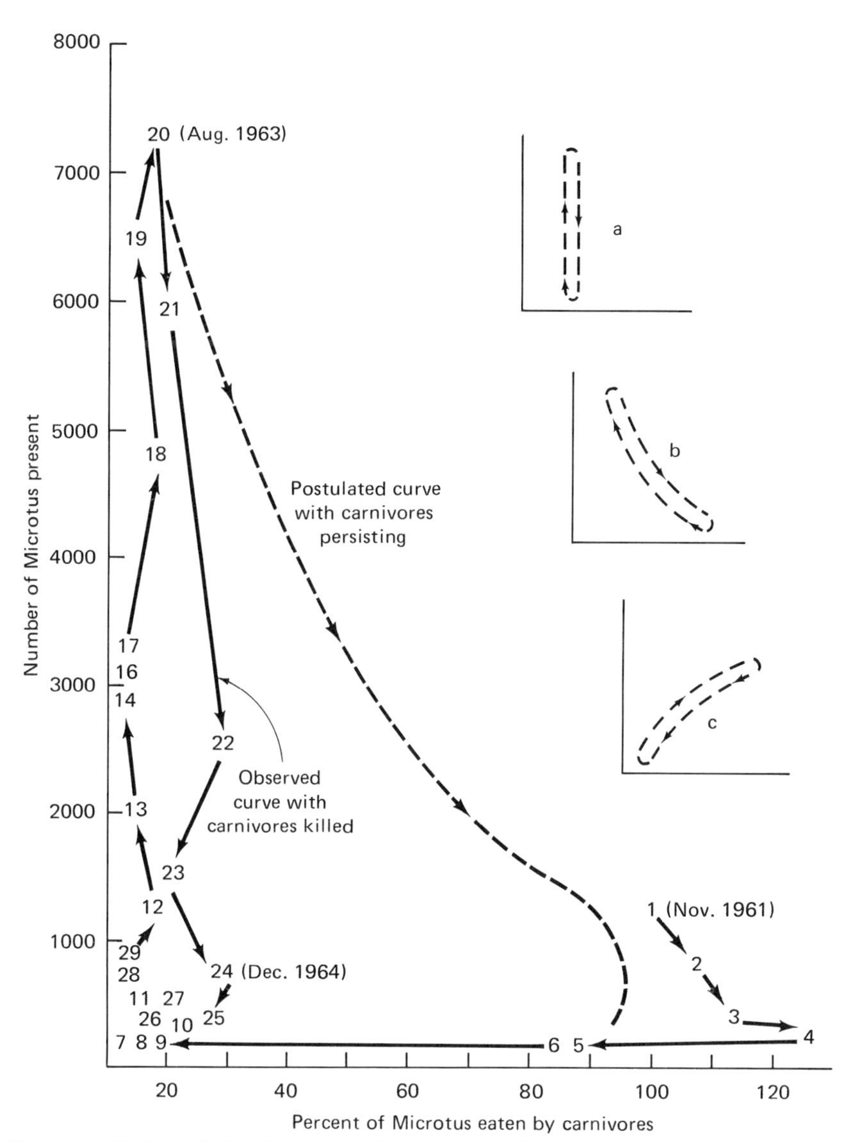

Figure 5–22 Correlation between the number of *Microtus* existing on the 35-acre study area (14 ha) in each month and the percentage of them eaten by carnivores. The datum for each month is represented by a number to enable one to follow the impact of carnivores during different stages of the microtine cycle, such as the first crash (1–5) and the second crash (20–25). The three insets represent the theoretical distribution and sequence of data under three kinds of mortality without any time lag: (a) constant percentage; (b) constant number; and (c) density-dependent or compensatory. (Source: O. P. Pearson, 1966, *J. Animal Ecology, 35:* 217–233.)

nivores retard premature recovery of the mouse population. The critical role played by the predators, therefore, is one of suppression and delaying explosive population growth. The timing of the cycle at three to four years is a function of the amount of alternative prey available and the reproductive capacity of prey and predator. Although the characteristic timing of the cycles may be the result of predation, Pearson (1966) emphasizes that the assistance of other factors is needed in reducing peak populations by increasing mortality or decreasing reproduction. Predation by carnivores cannot control or destroy the rodent population at high densities because predators with their rate of reproduction (one or two litters per year and their slow sexual maturity) is too slow to keep up with the mice until other factors limit the amplitude of the cycle and start a decline from peak levels. As Pearson (1966) summarizes, "It is because the carnivores are unable, without the help of these other factors, to control that part of the cycle that people have expected them to control [i.e., the peaks] that carnivore control of microtine cycles has received so little attention."

Food supply

The food-supply hypothesis, first suggested by Lack (1954) and supported by studies on Alaskan lemmings by Pitelka (1958, 1964) and Thompson (1955), holds that the cause of the decline of lemming populations is a qualitative and quantitative change in the forage (Figure 5–23). The bulk of the evidence (Thompson, 1955; Pitelka, 1958) consists of an observed association between lemming declines and extensive forage utilization.

Chitty (1964) and others consider this to be inadequate evidence. Schultz (1964), however, found that fluctuations in primary production, forage quality, and decomposition rates were correlated with the lemming cycle in northern Alaska. Chemical composition (calcium, phosphorus, and protein) varies significantly during a lemming cycle, the quality improving as the peak phase of the cycle approaches and sharply declining during and following the population decline phase. The variation may be explained partly by botanical composition changes resulting from selective grazing and partly by changes in the available soil nutrient pool. Grazing and manuring activities of the lemmings cause important direction changes in the soil system. Grazing decreases the insulating quality of the vegetative cover; two or three years of no grazing allows the insulation to build up again. Hence, there is a three- to four-year cycle in depth of the active soil layer. Excreta produced during the peak years add a sudden influx of available nutrients to the substrate and perhaps modify the decomposer flora quantitatively as well as qualitatively.

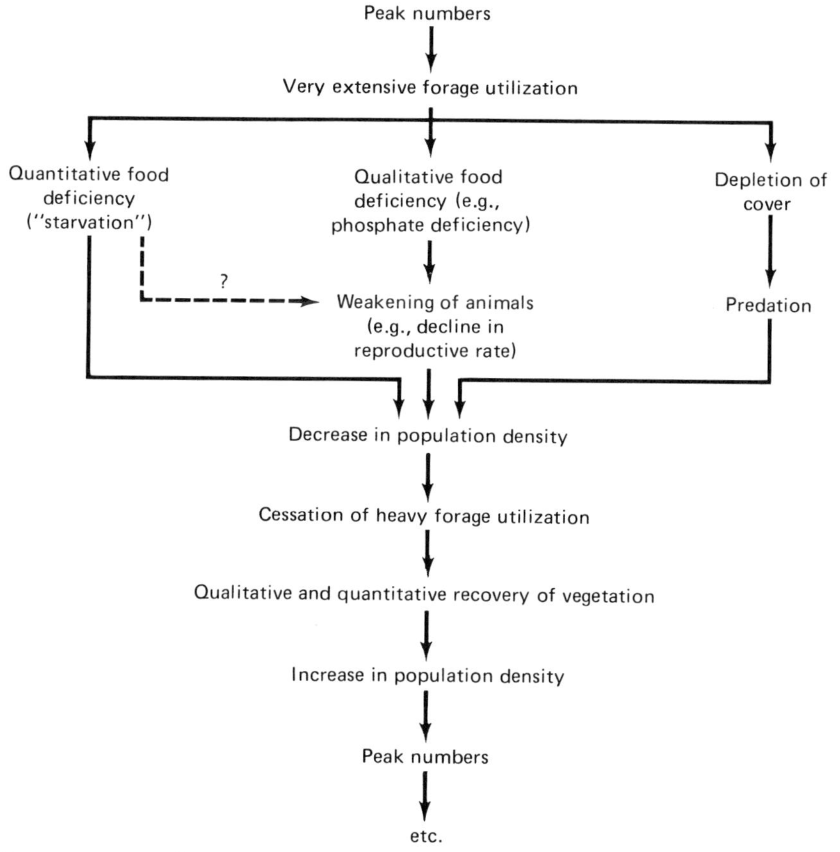

Figure 5–23 Pitelka's food supply hypothesis, one of the proposed regulatory mechanisms in microtine rodent populations. (Source: C. J. Krebs, 1964, *Arctic Inst. N. Amer. Tech. Paper, 15:* 50–67.)

Disease and parasites

Available evidence indicates that disease is a local factor of variable intensity and occurrence and not an essential part of the cyclic process (Krebs, 1964). Elton (1942) and Chitty (1954) show that disease alone was an insufficient cause for a decline in numbers in well-studied microtine rodent populations.

Christian's stress hypothesis

Christian (1950) proposed that cycles in microtine populations were caused by stress under conditions of crowding and that declines could be associated with changes in adrenal–pituitary functions and "shock disease" (Figure 5–24). Christian's general hypothesis is that all

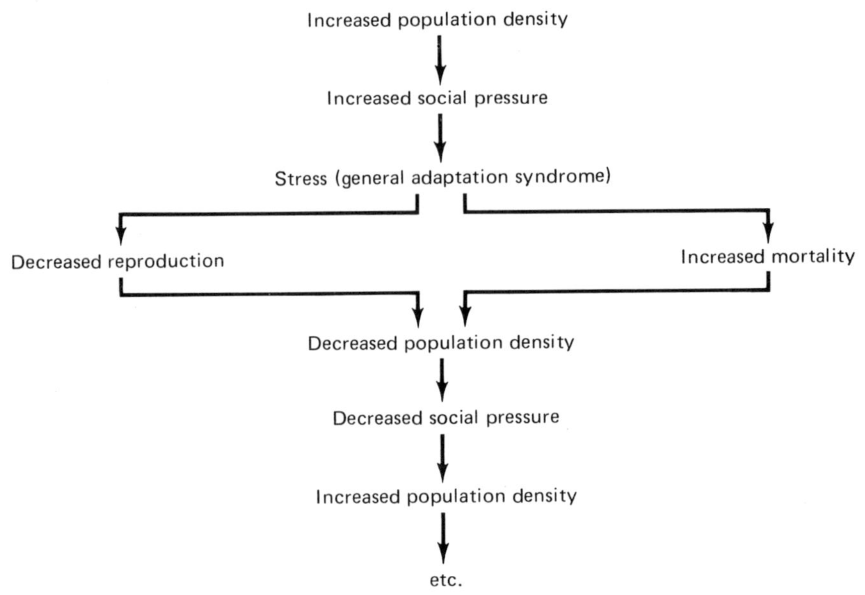

Figure 5–24 Christian's stress hypothesis of population regulation. The system is purely phenotypic and operates through the General Adaptation Syndrome. (Source: C. J. Krebs, 1964, *Arctic Inst. N. Amer. Tech. Paper, 15:* 50–67.)

mammals limit their own densities by a combination of behavioral and physiological changes. The specific aspect of his ideas is that the mechanism of this limitation involves the General Adaptation Syndrome and is purely phenotypic; that is, the reactions take place rapidly by physiological changes and there is no genetic component.

The theory includes the concepts that stimulation of pituitary–adrenocortical activity and inhibition of reproductive function would occur with increased population density. In addition, increased adrenocortical secretion from aggressive competition would increase mortality indirectly through lowering resistance to disease, through parasitism or adverse environmental conditions, or more directly through "shock disease." The evidence demanded by this specific hypothesis is that there must be increased adrenal activity and decreased reproductive activity at high densities, and that this increased adrenal activity must cause an increased death rate (Krebs, 1964). Christian and Davis (1964) amass considerable evidence from both laboratory and wild microtine populations that such behavioral and physiological changes occur. In assessing the effects of behavioral factors on adrenal function, they found that the important point is the number of interactions between individuals rather than density of population *per se*. Therefore, age, sex, previous experience, local distribution, and other factors may be critical in producing adrenal effects. Very brief encounters were found to produce development of the adrenal responses. As few as two 1-min exposures per day resulted in a 14 percent increase in adrenal weight, and eight exposures daily resulted in a 29 percent increase. Corticosterone hormone levels in the blood plasma increased by 67 percent. Most critics of Christian's theory (Chitty, 1960; Krebs, 1964) feel that the laboratory situations do not correspond to anything that goes on in nature, and field data on adrenal weights are ambivalent in some reported populations, showing no correlation with population density. The controversy remains unresolved.

Chitty's polymorphic behavior hypothesis

Chitty (1952, 1958, 1960) proposed that mutual antagonism associated with high breeding densities in lemmings causes selection for a

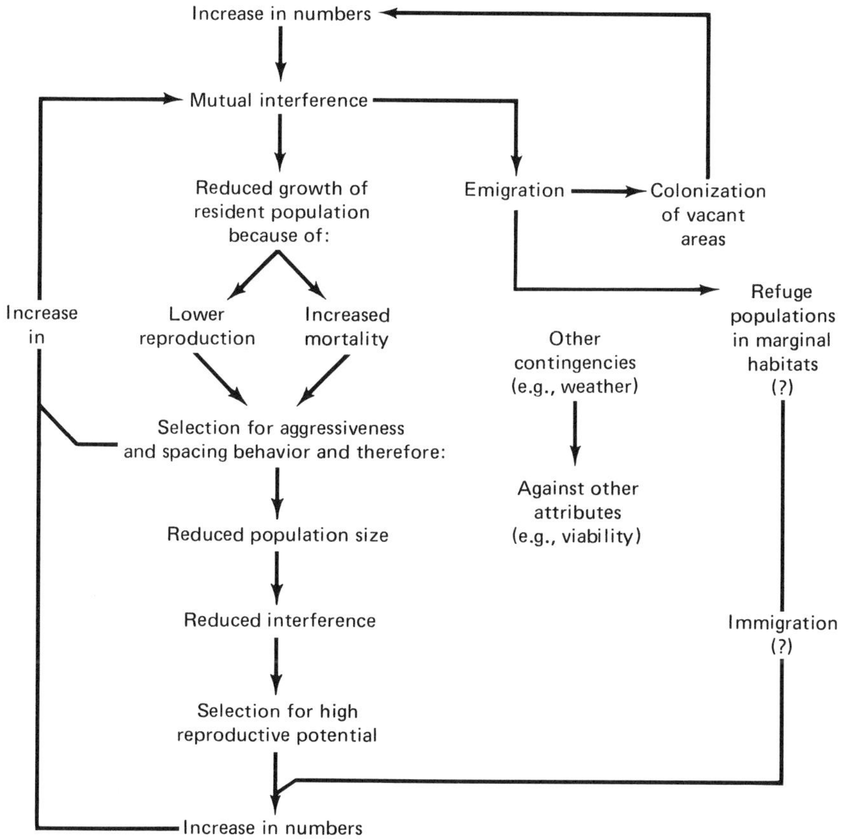

Figure 5–25 Modified version of Chittys' genetic and behavioral hypothesis to explain population fluctuations in small rodents. Density-related changes in natural selection, especially through emigration, are central to this hypothesis. (Adapted from C. J. Krebs et al., *Science, 179:* 35–41.)

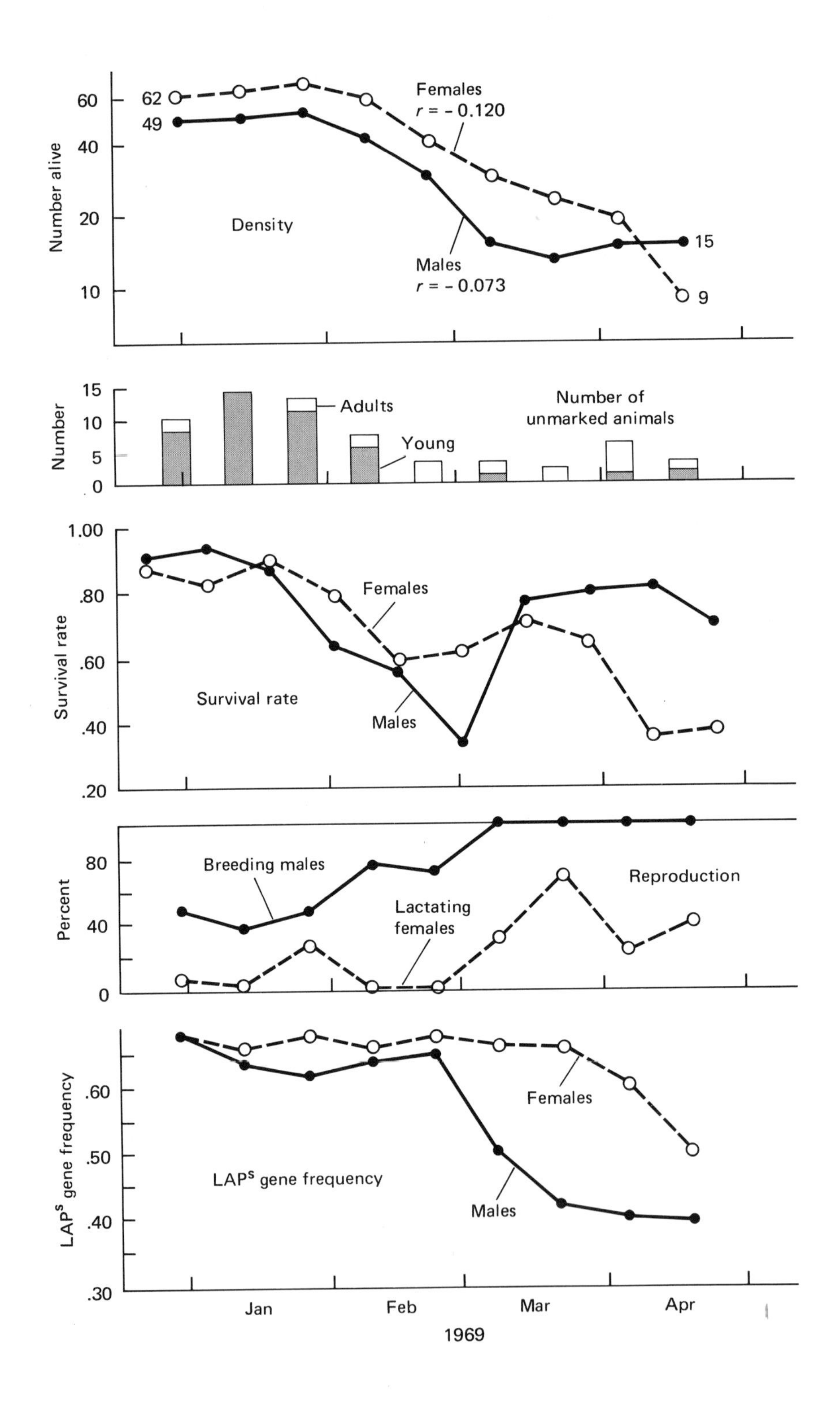

Number alive
62
49
Females
r = - 0.120
Density
Males
r = - 0.073
15
9
Number
Adults
Number of
unmarked animals
Young
Survival rate
Females
Survival rate
Males
Percent
Breeding males
Reproduction
Lactating
females
LAP^s gene frequency
Females
LAP^s gene frequency
Males
Jan
Feb
Mar
Apr
1969

change in the properties of the individuals in the contemporary population, and of the subsequent generations, which become less resistant to the normal source of mortality (Figure 5–25). This selection to withstand social strife is at the expense of other properties, which renders the animals less able to withstand other stresses in the environment. The relevant changes produced by mutual antagonism might involve (1) changes in maternal physiology which affect the offspring (that is, similar to the nongenetic effects of the stress hypothesis of Christian, wherein social interactions affect the endocrine balance of the female and thus affect the young), or (2) changes in the genetic composition by selection. The latter idea would also provide a means whereby weather changes could bring about synchrony of population cycles. Chitty (1958, 1960) rejected the first mechanism after extensive laboratory experiments indicated that, whereas adults could be greatly affected by mutual antagonism, the offspring did not show the changes in quality found in natural populations.

No direct evidence on the importance of genetic events in demographic trends of microtine populations was reported until 1973. Krebs and his co-workers (Krebs et al., 1973) studied the relationships among population dynamics, aggressive behavior, and genetic composition of field mice populations in southern Indiana during 1965 to 1970. *Microtus pennsylvanicus* fluctuates strongly in numbers with peak densities recurring at intervals of two or three years. Polymorphic serum proteins, including the genes *Tf* (transferrin) and *LAP* (leucine aminopeptidase), were used as genetic markers to study the possible role of natural selection in population fluctuations of this *Microtus*. The electrophoretically distinguishable forms of the protein products of these genes are inherited as if controlled by alleles of single autosomal loci.

Large changes in gene frequency at these two loci occurred in association with population changes. Figure 5–26 shows the detail breakdown of a population decline in *M. pennsylvanicus* during the late winter and early spring of 1969. The survival rate of males during this decline dropped to a minimum in early March; female survival dropped six weeks later. These periods of poor survival coincided with the onset of sexual maturity in many adult males and the approximate dates of weaning first litters in adult females. The frequency of the LAP^S allele (distinguished by slow electrophoretic mobility)

Figure 5–26 Detailed breakdown of a population decline in *Microtus pennsylvanicus* during the spring of 1969. The critical observation is the difference in timing of male losses (highest in early March) and female losses (highest in mid-April). This timing is reflected in the gene frequency changes shown on the lowest graph. In the upper graph, r is the instantaneous rate of population increase, which is negative during the decline. (Source: C. J. Krebs et al., 1973, *Science, 179:* 35–41.)

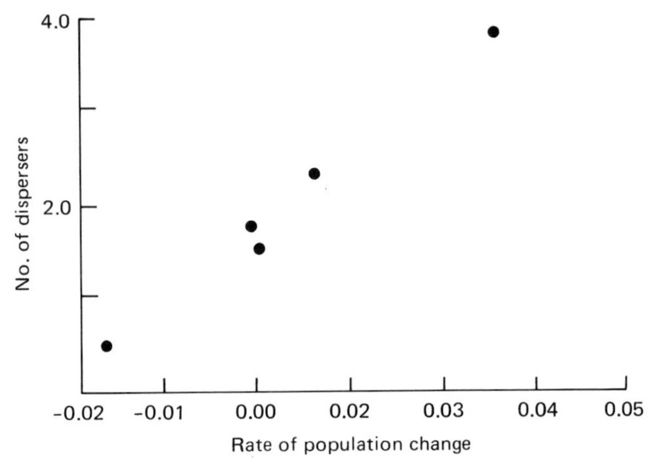

Figure 5–27 Rate of population change in *Microtus* control population from southern Indiana in relation to dispersal rate from that population during 1968 to 1970. Rate of population change is the instantaneous rate of change per week, averaged over eight-month "summer" and four-month "winter" periods during the three years. Dispersal rate is the mean number of mice dispersing from the control population (to a trapping grid) per two weeks, averaged over the same time periods. Populations increasing rapidly show the highest dispersal rates. (Source: C. J. Krebs et al., 1973, *Science, 179:* 35–41.)

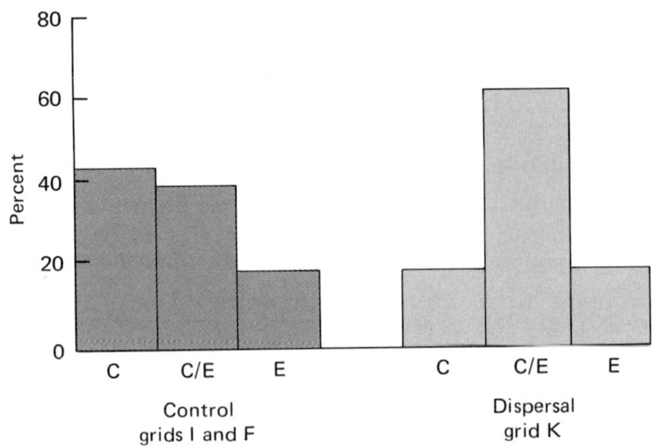

Figure 5–28 The increase phase of *Microtus pennsylvanicus* in fall, 1969. Transferrin (*Tf*) genotype frequencies of dispersing females on trapped grid (N = 39) compared with those of resident females on control grids immediately adjacent in the same grassland (N = 224). C, C/E, and E represent the three transferring genotypes. Dispersing mice in the increase phase are not a random sample from the control population, but show a distinct preponderance of the heterozygous genotype. (Source: C. J. Krebs et al., 1973, *Science, 179:* 35–41.)

dropped by about 25 percent in the males beginning at the time of high losses, and four to six weeks later its frequency declined an equal amount in the females. This observation strongly supports the hypothesis that demographic events in *Microtus* are genetically selective and that losses are not found equally among all genotypes. Without knowledge of linkage effects and how selection is acting, the mechanisms and therefore the cause and effect of the associations shown (Figure 5–26) cannot be assigned.

Dispersal from the population increases significantly in the increase phase of *Microtus* population cycles (Figure 5–27). Thus, Krebs et al. (1973) looked for qualitative differences in the *Tf* and *LAP* variation between dispersing *Microtus* and resident animals. Figure 5–28 shows the genotypic frequencies at the locus of the *Tf* gene for control populations and dispersing animals during an increase phase. Heterozygous females (Tf^{C}/Tf^{E}; where *C* and *E* are alleles of the *Tf* gene) are much more common among dispersing *Microtus* than in resident populations, and 89 percent of the loss of heterozygous females from the control population during the population increase phase was due to dispersal (Krebs et al., 1973). Along with an earlier report on the same genetic characteristics in these *Microtus* populations (Myers and Krebs, 1971), this was the first demonstration that certain genotypes in a natural population show a tendency to disperse. The genetic differences between dispersing and resident populations support the frequently advanced hypothesis (for example, Howard, 1960; Mayr, 1963) that a genetic polymorphism could cause a polymorphism in the dispersal behavior of animals. The association demonstrated between increasing populations and abundant dispersal agrees with Lidicker's (1962) and Wynne-Edwards (1962) theory of population regulation by emigration in a homeostatic or self-balancing system and Chitty's hypothesis that a behavioral polymorphism regulates population density.

1 2 3 4 5 6 7 8 9 10 11

Dispersion, Dispersal, and Population Structure

A population of plants or animals is made up of many individuals reacting to each other and to features of the environment. The spacing patterns that result will lead to particular patterns of encounters for potential mates in animal populations and for potential crossing in plant aggregations. The numbers of individuals, life cycle length, and age and sex distribution within the population also affect its organization. As such, a population may be said to have a definite *structure,* which refers to *all the factors that govern the pattern in which gametes from various individuals unite with each other* (Ehrlich and Holm, 1963). We shall examine in subsequent chapters the many factors besides distribution that affect population structure. Our present goal is to examine concepts of spacing patterns and movement in animal and plant populations from the perspective of their proximate causes, their ecological consequences for the population, and their ultimate evolutionary or adaptive significance (Brown and Orians, 1970).

THE NICHE CONCEPT AND ENVIRONMENTAL GRAIN

Organisms tend to distribute themselves according to particular ecological requirements. Hence, we use the general word *habitat* to describe the characteristic physical location that is typically inhabited by a particular species. In the Kaibab Plateau region of northern Arizona, a Kaibab squirrel is always found in a coniferous forest habitat. It does not descend to the geographically adjacent but arid desert situation near the floor of the Grand Canyon. A lionfish will be found along coral reefs and not in the open oceans of the South Pacific. The characteristic habitat implies a generally restricted occurrence for each species.

The early ecologists, even in Darwin's day, realized that the organism usually seemed to respond to more than just physically suitable environments in its choice of a place to live. Joseph Grinnell (1924, 1928) was the first to employ the word *niche* to an animal's ecological position in the world. He defined (Grinnell, 1928) the ecological niche as "the concept of the ultimate distributional unit, within which each species is held by its structural and instinctive limitations, these being subject only to exceedingly slow modification down through time." Grinnell emphasized the *potential* nature of the niche; it represented the idealized distribution of individuals over a geographical area as if there were no interactions with other species. Little consideration was given to the importance of food supply or other biotic components as limiting factors. Vandermeer (1972) has termed Grinnell's conception of niche as being preinteractive in nature, and points out in his excellent review of niche theory that this potential niche concept would later be equated with the modern notion of fundamental niche.

Almost simultaneously with Grinnell, Charles Elton (1927) was developing a post-interactive concept of the niche—it was the actual place of the organism in nature as opposed to its potential place in nature that was significant (Vandermeer, 1972), and this was defined in large part by the food habits of a species. Thus, the organism's relationship to other organisms became an integral component of the niche definition: "the 'niche' of an animal means its place in the abiotic environment, its relations to food and enemies," and "the niche of an animal can be defined to a large extent by its size and food habits" (Elton, 1927). This post-interactive or realized niche was later to be equated with the modern notion of realized niche (Vandermeer, 1972).

The next 30 years saw a merging of these niche concepts by Grinnell and Elton so that textbooks in ecology generally defined the ecological niche as an animal's "occupation" or way of life, as well as physical placement, in a community. The only major biological "law" in ecology was promulgated during this period by G. F. Gause (1934), who demonstrated experimentally that two species cannot occupy the same niche simultaneously. If they are very closely related and have similar food and other resource requirements, competition will be great in any areas of habitat overlap. Gause's principle (or Gause's theorem) generated a large body of field work and laboratory experiments with sympatric species that has continued to the present day. The usual findings are some degree of competition owing to niche overlap in one or more factors, along with subtle but significant ecological differences in other requirements.

Then in 1957, Hutchinson initiated a revolution in niche theory by developing the formal concept of the ecological niche as a hypervolume of space. If one considers two independent environmental variables, say temperature (x) and annual rainfall (y), which can be measured along ordinary rectangular coordinates, then we can assume there are limiting values (minimal x_1 and maximal x_2, minimal y_1 and maximal y_2) that permit a species to survive and reproduce. The niche area defined by these variables, assuming that they are independent in their action on the species, will be a rectangle (Figure 6–1a). At any combination of rainfall and temperature falling within this range of environmental states, the species is in its optimal niche and can exist indefinitely.

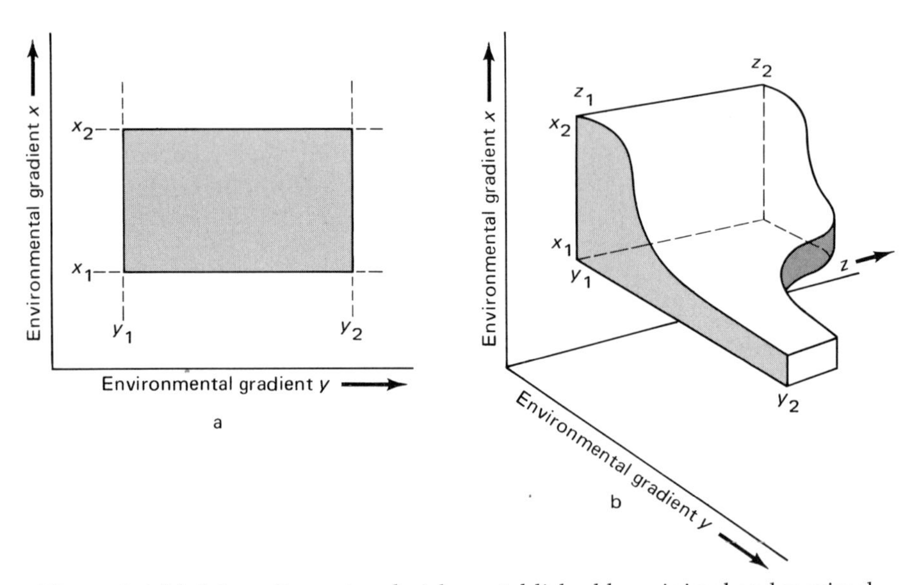

Figure 6–1 (a) A two-dimensional niche, established by minimal and maximal values for each of two independent environmental variables; for example, x could equal temperature and y could equal annual rainfall. The species can only exist within the shaded areas defined by these parameters. (b) An example of a three-dimensional volume or Hutchinsonian niche, involving only three variable environmental components (x, y, and z). The species is maximally tolerant of minimal conditions of x, y, and z, but the niche narrows rapidly with increasing values of any of the three components. The odd-shaped volume that results is the species' niche. If we added more variables on still more axes, we could have an example of Hutchinson's n-dimensional hypervolume niche.

We may now introduce a third variable, say humidity (z), that influences this species, and this time let us assume that the three variables are not independent in effect. Hence, in Figure 6–1b, we obtain a volume corresponding to the environmental state where the species may persist. If we added further variables until all of the ecological factors relative to this particular species have been considered, we define an n-dimensional hypervolume, which can be called the *fundamental niche* of the species. The result is best described in Hutchinson's (1957) original words:

> It will be apparent that if this procedure could be carried out, all X_n variables, both physical and biological, being considered, the fundamental niche of any species will completely define its ecological properties. The fundamental niche defined in this way is merely an abstract formalization of what is usually meant by an ecological niche.

It ought to be evident that this formal abstraction can never actually be measured for a species in the real world because we would have to know practically everything about the plant or animal concerned.

The *realized* niche itself is difficult to define, even with its reduction in size from the fundamental niche due to competition and displacement from other species in the first species' area, as in Figure 6–2. The realized niche is also subject to considerable change with time, as physical conditions and biotic competitors and food supplies vary on both a short-term basis and a long-term evolutionary basis.

Implied in the preceding discussion are several additional recent concepts of the niche, including the notions of *niche breadth* and *niche overlap*. Niche breadth, also termed "niche size" and "niche width," refers to the extent that the realized niche spreads across a set of niche

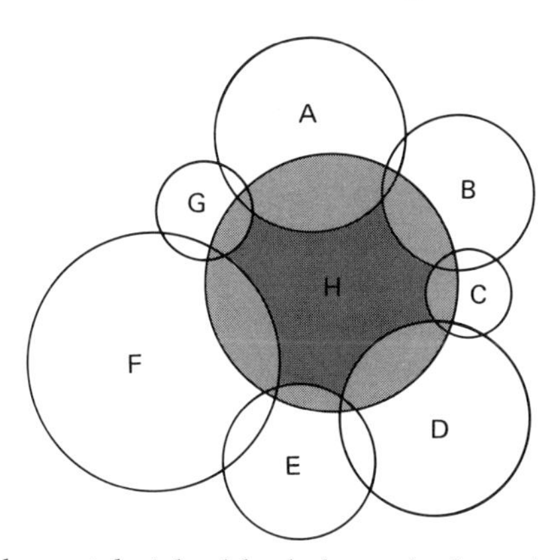

Figure 6–2 The fundamental niche (shaded areas) of species H, and its realized niche (darker shade), a subset of the fundamental niche which results from competition and competitive displacement with the various neighboring competitors, species A, B, C, D, E, F, and G. (After E. R. Pianka, 1974, *Evolutionary Ecology*, Harper & Row, New York, p. 195.)

dimensions; in other words, the effective size of the Hutchinsonian hypervolume. A generalized species which has broad tolerances and can feed in a variety of habitats is said to be a broad-niched organism; it is a jack-of-all-trades. Such species tend to build up large populations in relatively equal numbers through all occupied habitats. A specialized species has narrow tolerances, is restricted in distribution to one or several habitats, and is normally not very abundant. More specialized species persist because they are specifically adapted and are hence more efficient in survival and reproduction in their chosen niche than are the generalist species that may invade the boundaries of that niche. Niche overlap represents the degree to which two species co-occupy a specified habitat or food resource or other parameter. It usually is taken to mean the extent of interspecific competition over a particular resource. Thus, in Figure 6–3 the two species of insectivorous birds specialize on prey of different sizes and lower competitive overlap for this particular niche-gradient parameter.

These considerations bring us to the concept of *environmental grain,* first introduced by MacArthur and Levins (1964) and expanded further by Levins and MacArthur (1966) and MacArthur and Wilson (1967). Environmental grain refers to the way a species perceives and utilizes its environment. If the members of a population tend to remain in one specific habitat for long periods of time, the population is said to be coarse grained with respect to its resources. If individuals tend to sort out randomly and readily move from habitat to habitat, the population is basically fine-grained with respect to its resources. There are many concepts of environmental grain presently in the literature, but these definitions as proposed recently by Vandermeer (1972) perhaps reflect those which are most generally useful. The organism that regards its

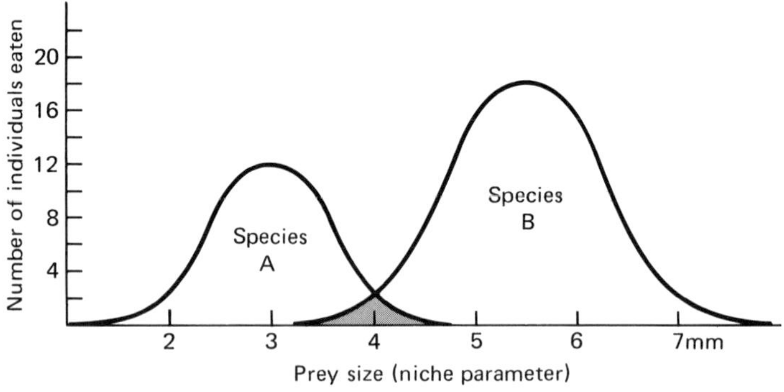

Figure 6–3 Two hypothetical species of flycatchers preferentially take prey of different sizes. Species A specializes on insects averaging 3.0 mm in size, whereas species B has a greater niche breadth and captures insects ranging primarily from 4.5 to 6.5 mm in size. Competition reduces the survival of individuals that capture insects around 4.0 mm in size; hence, genes for characters determining such prey choice are eliminated, minimizing niche overlap.

resources as "coarse-grained" is a specialist that searches out and selects the grains of environment that contain a particular kind of resource (say 3.0-mm-sized insects). In contrast, "fine-grained" species may encounter and select its food more or less indiscriminately over a broad range of parameters and take them in the proportion in which they occur. MacArthur and Wilson (1967) emphasize that a fine-grained generalist may out-compete several coarse-grained specialists if the latter must waste a considerable amount of time searching an area for food or other resource particles. If resources come in a fine-grained mixture, there will be an upper limit to the number of resource-limited or specialist species the habitat will hold. In Levin's (1968) view of graininess, the coarse-grained environment presents itself to the individual as alternatives and if the environmental patches are large enough (or the grain is sufficiently coarse), the individual can spend his whole life in a single patch, specializing in one particular kind of resource. The fine-grained generalist will wander among many patches during his foraging forays, and he will exhibit maximum fitness by utilizing a variety of foods or other resources. Colwell and Futuyma (1971), Pielou (1972), and Vandermeer (1972) develop several elegant analyses of the problems in precisely assessing the fine points of environmental grain and measuring niche breadth and overlap.

TYPES OF DISPERSION PATTERNS IN POPULATIONS

In our overall theme of spacing patterns in populations, it is clear that niche theory and environmental grain may play a useful, indeed key role in interpreting the proximate and ultimate causes[1] of spacing. Let us now examine in more detail the types of distributional patterns found in plant and animal populations, and the factors that cause this observed spacing.

Dispersion refers to the distribution or physical location of individuals within a population at a particular moment in time. In V. C. Wynne-Edwards' treatise on the subject, *Animal Dispersion in Relation to Social Behavior* (1962), dispersion is defined as comprising "the placement of individuals and groups of individuals within the habitats they occupy, and the processes by which this is brought about." It is important to note that this distribution should not be confused with *dispersal*, which refers to the movement of animals (or plant parts, especially seeds) from one place to another. Dispersal movement commonly involves spreading from an area of high concentration to one of lower density. We shall spend most of the remaining part of this

[1] An *ultimate* cause is grounded in the evolution of the species; genotypes with particular kinds of spacing behavior were more successful in reproduction than other genotypes. A *proximate* cause is an "immediate" cause of certain spacing behavior; it may be the rainfall pattern that year, the abundance and spacing of an insect's host plant, and so forth.

a

chapter in discussing how different types of dispersal affect population structure and dispersion patterns.

If we consider dispersion as the internal distribution pattern of individuals within a population, three general patterns of spatial relationship may be expected. With *random dispersion,* individuals are scattered over an area without regularity or any degree of affinity for each other. The probability of finding an individual at a point in the spatial arena of the population is equal for all points. Random distribution is relatively rare in nature, and it occurs only where the environment is quite uniform, with resources spread evenly throughout the area of the population's distribution. There must also be no interactions to repel one another, or social attractions causing a tendency to aggregate. Because a nonrandom dispersion of resources is usually encountered in natural environments, organisms are almost never random in distribution throughout their lives. In Figure 6–4a the marine iguanas (*Amblyrhynchus cristatus*) on the uniformly rocky shoreline of Hood Island in the Galapagos Archipelago show a more or less random distribution as they bask in the morning sun. Later in the year, the males in this population would develop territorial behavior and space themselves out more regularly.

In *regular dispersion* (also called even or uniform distribution) the individuals tend to be as far apart from each other as it is possible for them to be with that particular population density. At the start of a breeding season, the male marine iguanas on Hood Island (and other

b

Figure 6–4 Dispersion in Marine Iguanas (*Amblyrhynchus cristatus*) on islands in the Galapagos Archipelago. (a) Random dispersion on shoreline of Hood Island; (b) clumped dispersion on Punta Espinosa, Narborough Island. (Photos by Boyce A. Drummond and T. C. Emmel.)

islands of the Galapagos) space themselves out rather evenly along the rim of the coastal cliffs and establish breeding territories. At the height of the breeding season in January and February, the Hood Island males develop very colorful red, orange, turquoise, and black mottling on their bodies and exhibit agressive behavior in displays and actual attack against any adult male intruders. When male encounters result in actual contact, this takes the form of head butting which can last for many minutes and even hours (Carpenter, 1966). Nesting sea-bird colonies typically show a regular spacing of breeding pairs throughout the nesting grounds. The Blue-footed Boobies (*Sula nebouxi*) that nest on Hood Island throughout the year exhibit rather uniform spacing (at least within cleared areas, ignoring brushy or excessively rocky habitats) between established pairs that have begun the long period of courtship, incubation of the eggs, and rearing of the chicks. The evenly spaced pairs in Figure 6–5a are in the courtship phase. Plants may exhibit regular dispersion in areas where there is high competition for water and other scarce resources, as in deserts of the world and the chaparral areas around the Mediterranean Sea and the western Australia or southern California hills. Creosote bush (*Larrea divaricata*) and a great many other plants, from microorganisms

a

releasing penicillin to black walnut trees releasing plant toxins, limit the nearby distribution of their own or other species by excreting antibiotic poisons or *allelopathic agents* from their roots, or even aerially through their leaves (Figure 6–6). We shall discuss chemical interactions and competition between species in more detail in Chapter 11.

The most common type of distributional pattern in nature is clumped or *aggregated* dispersion. This may result from nonuniformity of the habitat or attraction to other individuals. Aggregated dispersion may vary in intensity and may even be present only at certain times in the annual cycle or in the organism's entire lifetime. The Galapagos marine iguanas on Narborough Island aggregate in huge numbers on the tiny tip of Punta Espinosa (Figure 6–4b) where a few square yards of sand and rock seem to have the most appropriate exposure for basking and raising body temperatures to a level necessary for digestion of their algal food. Seeds from a perennial or annual plant will normally drop rather near the parent and thus clumps of genetically related individuals may characterize the distribution of that species. Many annual plants are quite colonial in distribution and are restricted to relatively limited sites where they generally recur each year. Species of the western North American genus *Clarkia* (in the Onagraceae, or evening primrose family) have colonies ranging in number from several to innumerable individuals which may carpet several square miles of grassland slopes. The usual colony, however, numbers several hundred to several thousand individuals and covers

b

Figure 6–5 (a) Uniform dispersion of pairs of the blue-footed booby (*Sula nebouxi*) on Hood Island in the Galapagos; (b) male (left) and female blue-footed boobies are essentially indistinguishable except for the female's enlarged, dark-pigmented area in the iris around the pupil. (Photos by T. C. Emmel.)

two to ten square meters, showing relatively little change in number from year to year at a particular locality (Lewis, 1953).

Proximate factors that result in clumped dispersion are in part those that determine local habitat differences, such as temperature, light, water, mineral content of soil, and wind velocity and direction. The best combination of environmental factors may vary seasonally. During dry periods of the year in south Florida, populations of birds and other wildlife move from the open marshlands and scattered hammocks to feed and congregate around "gator holes," depressions in the limestone substrate of watercourses excavated by alligators. These gator holes retain water long after the moving sheets of water covering the everglade marshlands have dried up. Groups in animal species also result from the demands of reproductive activity, where at least one male and female cohort must aggregate. Often, breeding and rearing of young is best accomplished in a whole colony situation. Feeding aggregations in the form of schools of fish, flocks of birds, and herds of mammals (Chapter 1) make efficient utilization of time and energy in finding and securing food; consequently, social attraction promoting clumping is adaptively advantageous to them, as we shall see in cases cited in more detail later in this chapter.

Figure 6–6 Widely spaced creosote bushes (*Larrea divaricata*) on the western Mojave Desert, near the Calico Mountains of California. (Photo by T. C. Emmel.)

ULTIMATE FACTORS CAUSING DISPERSAL INTO PARTICULAR DISPERSION PATTERNS

The ultimate reasons for particular types of dispersal behavior and dispersion patterns must be sought in their adaptive or evolutionary value. Two opposing tendencies are found in populations of organisms: (1) to spread through the available habitat; and (2) to remain close enough in physical distribution so that breeding is not hindered and feeding or other vital activities will be optimized in relation to available resources, requiring flocking in some species and generally solitary existence in other species.

In the evolution of types of dispersal behavior to be examined, these opposing tendencies have led to a remarkable diversity of outcomes. We may say, however, that the ultimate results of dispersion are threefold.

Density is optimized by dispersal in relation to the distribution of resources within the population area. Wynne-Edwards (1962) believes that optimum density is the central form of all dispersal behavior and that such an optimum is in fact usually achieved. In an ideal sense this ecologist believes "the habitat should be made to carry everywhere an optimum density, related to its productivity or capacity," without overcrowding any parts of the habitat or leaving some areas underpopulated and thus wasting resources. The critical resources that are necessary to achieve optimal population density are said to be (1) food supply, (2) the habitat itself (all the physical environmental factors), and (3) minimal biological competition from individuals of the same or different species.

Genetic variability is introduced by dispersal through interchange between populations. Individuals that successfully cross unsuitable intervening habitat between two populations have the potential possibility of bringing new variant genes into the recipient population if the dispersing adult is able to reproduce there. This may require short-range migration between geographically separated areas.

A frequent result of dispersal is that the range of the species is increased, which probably enhances the chances of genetic survival over evolutionary time. Dispersal behavior usually *maintains* a particular population structure and should solely involve movements within the breeding area of a population. But the process of dispersal or dissemination intrinsically involves a scattering of individuals in a population over wider areas than those formerly occupied by their forebears. Thus, the range of species is potentially increased with each generation, although the actual extent, if any, of this increase is not usually significant unless the propagules reach unoccupied and otherwise suitable habitats. Species that have evolved migratory behavior are often able to "spin off" small resident populations at either end of their route, but the great majority of individuals continue to include movement from one place to another as part of their daily, monthly, or annual cycle. Note that the migratory species may have two entirely different kinds of population structure at each location, exhibiting considerably different dispersal behavior. Migratory waterfowl, for instance, breed and nest singly at the "summer" temperate zone end of the migratory range, where they become highly territorial. At the other ("winter" or tropical) edge of their range they may feed most of the time in communal flocks.

DISPERSAL IN PLANTS

From the point of view of the population biologist, plant dispersal mechanisms have been well studied throughout most of the seed-plant groups and certain fungi. Developing and mature plants are essentially immobile, of course, and must disperse primarily by means of their seeds and spores. A few groups exhibit other types of local dispersal, such as via runners of vines or joints of *Opuntia* cactus pads, breaking off from the parental plant and rooting in a nearby favorable location. Vegetative reproduction, however, seldom offers an opportunity for long-distance dispersion by seed plants, and it represents essentially a method of multiplication and colonization within a local area (Harper et al., 1970). In contrast, dispersal of seeds or spores represents a means for the offspring to escape competition with parents and siblings in the local parental environment. Just as significant, dispersal presents a means for the sessile seed to escape predation by animals (Janzen, 1971). Clearly, the overall significance of dispersal in the strategy of plant populations involves an evolutionary balancing of many physiological processes and reproductive patterns, from a single dispersal move to seed size, chemistry, and parental structure and behavior, affecting both degree of dispersion and protection from seed predation.

Seed dispersal by wind is common among plants, especially in temperate zone forests (e.g., pines, maples, beech) and in grasses and weedy or colonizing species. The distribution of a species of this type is often directly related to its dispersal ability, and there is high selection for smaller, more numerous, readily dispersed seeds. The four species of British poppies (*Papaver* spp.) all inhabit open weedy fields and disperse seeds by wind from a seed capsule on the end of a long stalk. Salisbury (1942) found that the more common, geographically widespread *Papaver* species have far greater reproductive capacities in seed number than do the two rare species (Table 6–1). But the seed weights of the four species are identical. It is the relative height of the seed stalk above the ground that largely influences the effective dispersal distance of the seeds. Thus, the two common species have seed stalks nearly twice the height of the two uncommon species, and the greater dispersal capacity generated by this structural feature increases their ubiquitous distribution.

Seed dispersal mechanisms are well known to be correlated with habitat (Harper et al., 1970). Wind dispersal usually occurs in canopy

Table 6–1 The relationship between dispersal ability, as seen in seed output and height of seed capsule pores aboveground, and geographical range in four species of poppies (*Papaver* spp.) in Britain.

SPECIES OF *Papaver*	MEAN NO. CAPSULES PER PLANT	MEAN NO. SEEDS PER CAPSULE	MEAN PERCENTAGE GERMINATION	MEAN REPRODUCTIVE CAPACITY (SPROUTING SEEDS FROM ONE PARENTAL PLANT)	MEAN HEIGHT OF SEED CAPSULE PORES ABOVEGROUND (RANGE IN CENTIMETERS)
			UNCOMMON SPECIES		
P. argemone	6.81	314	63	1,347	35 cm. (14–65)
P. hybridum	7.28	230	91	1,529	26 cm. (16–43)
			COMMON (WIDESPREAD) SPECIES		
P. dubium	6.83	2,008	42	5,757	56 cm. (21–98)
P. rhoeas	12.5	1,360	64	10,928	58 cm. (32–88)

SOURCE: Salisbury (1942).

forest trees or open habitat species. Coastal plants tend to have large seeds, often containing air cavities to enhance water dispersion. The ground floor vegetation of temperate forests often bear hairy or hooked seeds which catch readily on fur and are dispersed opportunistically by small mammals. Herbs of the shrub layer in temperate and tropical forests very commonly bear attractively colored, fleshy or juicy fruits which are collected by birds and animals, the seeds later being passed out in excreta. Climax forest species tend to bear heavy, large seeds with greater food reserves, and dispersal of these seeds are dependent upon the specialized foraging of particular animals.

When a plant species moves into a new habitat, selection for dispersal ability may change. This is particularly true in insular species that develop from mainland progenitors. The loss of dispersibility in Compositae seeds of the Pacific Islands has been elegantly documented by Sherwin Carlquist (1966, 1974). One of the best examples is the genus *Bidens,* a member of the sunflower tribe, which is well represented by endemic species on remote islands in the Pacific. The widespread tropical weed, *Bidens pilosa,* is probably representative of the original mainland stocks of *Bidens.* The small, dry, one-seeded fruits (called achenes) of *B. pilosa* bear widely spread bristles (called awns) at their ends, and these awns are relatively long compared to the achene (see upper left achene in Figure 6–7). With stiff barbs and upward-pointing hairs on the achene, this seed fruit will readily adhere to fur or feathers. *Bidens* probably reached these remote islands by chance adherence to sea-bird feathers. Only a few of the South Pacific species of *Bidens* retain these well-developed dispersal features today, and these occur largely in the coastal habitats. The species with least dispersibility are those which have penetrated furthest into wet upland forests. The two forest species from the island of Raiatea (center, left side of Figure 6–7) show the greatest loss of dispersibility among South Pacific *Bidens* species, for they lack any hairs or effective appendages.

The loss of dispersibility in Pacific Compositae can be partly attributed to a negative selective premium on dispersal mechanisms which remove large numbers of seeds from the islands to fall into the sea. Because *Bidens* and other genera are not adapted to wind dispersal and the number of sea birds on some of the islands is minimal, three other principles may influence the observed loss of dispersibility (Harper et al., 1970). Ecological opportunity may be quite different in cases where one species is living in pioneer habitats, necessitating strong dispersal at each generation, and another species may be in a stable rain forest habitat where the seeds can germinate almost anywhere. Species that penetrate the shady forest habitats may need to put more energy into large seed size or storage of starch for rapid growth of the embryo under conditions of competition; hence, the former premium

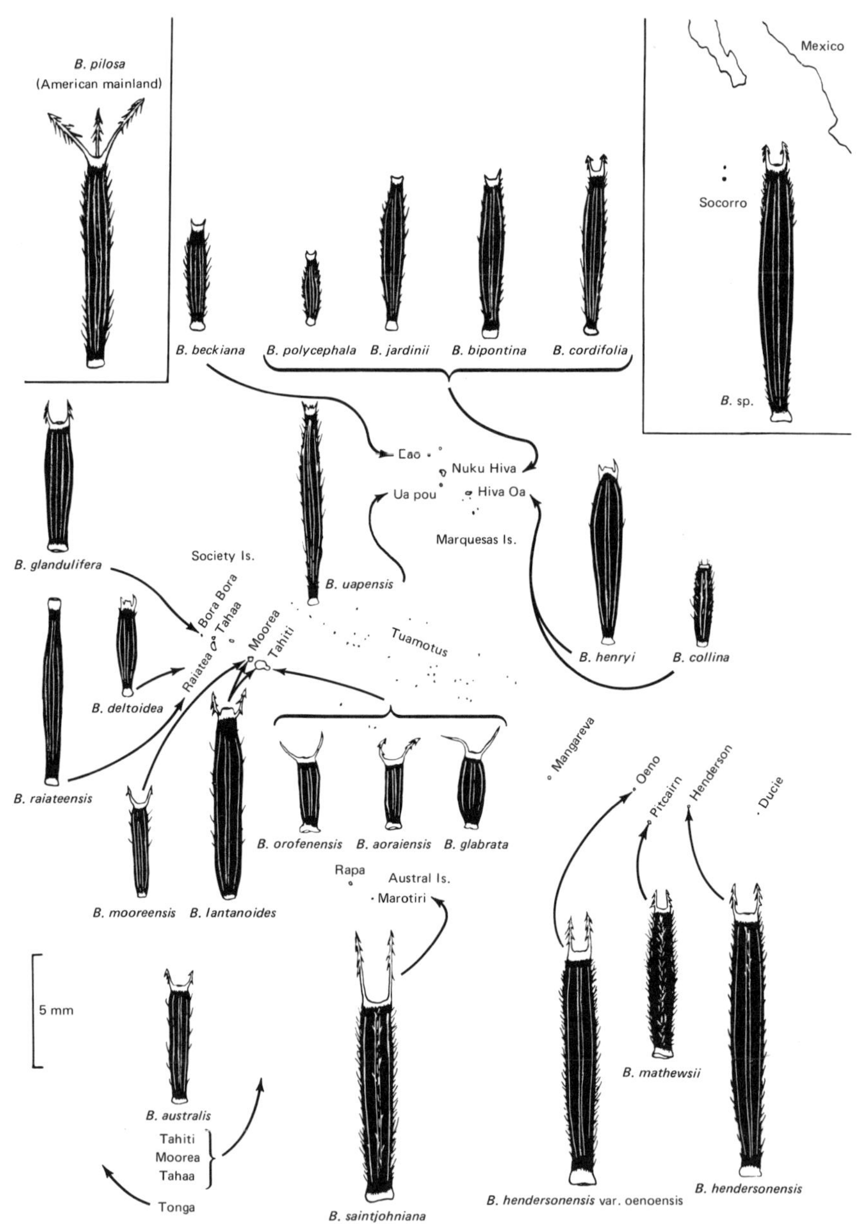

Figure 6–7 The dispersal-linked characteristics of fruits (achenes) of endemic species of *Bidens* (Asteraceae, in the Compositae) in the islands of southeastern Polynesia. The inset at the top left shows a typical achene of *B. pilosa*, a common tropical mainland species. The inset at top right shows an unnamed species native to Socorro Island off Mexico; like many of the island species in Polynesia, it has lost its probable ancestral condition of long and widespread barbed bristles. (Source: S. Carlquist, 1974, *Island Biology*, Columbia Univ. Press, New York, p. 444.)

on elaborate dispersal appendages is overshadowed. Perhaps the largest seeds in the Composite are found in a forest *Fitchia* species, *F. speciosa*, with achenes 50 mm long and 15 mm broad. Finally, the loss of the relationship between the dispersal mechanism and the mainland animal dispersal agent existing in the mainland population areas would likely influence the loss of dispersibility. Over time, selection would likely not maintain the channeling of energy into unneeded dispersal mechanisms in insular species.

The dispersal patterns exhibited by plants may also be heavily influenced by seed predation, which can be either predispersal or postdispersal in nature (Janzen, 1971). Many dispersal agents swallow seeds while eating fruits and excrete them later, but we refer here to animals that specifically predate on seeds, restricted so far as known to insects, mammals, and birds. Some species of all plant families experience seed predation, and there exists a tremendous variety of adaptive responses to counter such predation and ensure successful dispersal of part of the seed crop. Daniel H. Janzen reviews these in an exceptionally fine analysis (Janzen, 1971) and points out that following any predispersal seed predation (such as predation on foliage, flowers, ovaries, green fruits, or even mature seeds), the dispersal agents generate a seed shadow around the parent with the remaining viable seeds. Postdispersal predation on these seeds and resultant seedlings should be a function of parental proximity, and we would normally expect that the seed shadow should be most intense near the parent for gravity- or wind-dispersed seeds and for some mammal-dispersed seeds. Predation intensity ought to be concentrated near the parent where fruit or seeds are primarily dropping, and decrease with distance and decreasing density of seeds. Seed survivorship is thus enhanced with increasing distance from the parental plant, and at some point a "crater rim" of successful seed survivorship in maximum numbers will result (Figure 6–8). Local heterogeneity of the habitat and competitive interactions with members of the seedling's own and other species will result in a patchy distribution of new individuals in the approximate "rim" area of maximum realized survivorship.

Sometimes, two species of flowering plants may have coordinated dispersal. The self-incompatible annual herb *Orthocarpus densiflorus* (owl's clover, in the family Scrophulariaceae) is an obligate parasite of the other flowering plants. Root grafts are made by contact between seedling roots of young *Orthocarpus* and a suitable host, such as the composite *Hypochoeris* (cat's ear). Atsatt (1965) found that in natural populations of *Orthocarpus* and *Hypochoeris* near Santa Barbara, California, the seeds of both species mature at approximately the same time. The *Hypochoeris* achenes fall to the ground with the upward-directed pappus of barbed bristles forming a dense trap, in common with neighboring achenes, for the small falling seeds of *Orthocarpus*.

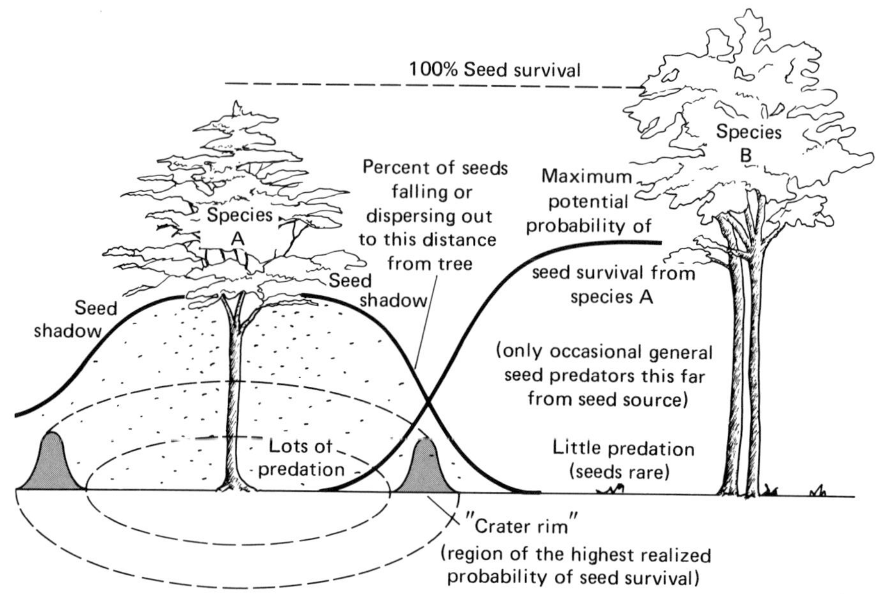

Figure 6–8 Seed dispersal and postdispersal predation on these seeds in a tropical forest. Tree species A has maximum realized survival where the decreasing "seed shadow" curve (indicating seed quantity dispersal per unit distance) intersects the increasing survivorship curve well away from the concentrated predation around the parental source of seeds. (Data after D. H. Janzen.)

The *Orthocarpus* seeds are much smaller and are surrounded by a rigid net of outer integument. This net readily catches on an achene bristle (Figure 6–9). Atsatt points out that the seeds of host and parasite fit together so that the seeds may be clustered and their dispersal coordinated. A self-incompatible plant like *Orthocarpus* requires clusters of individuals to accomplish outbreeding. Hence, as Atsatt states, "The attachment of more than one *Orthocarpus* seed to a single *Hypochoeris* achene not only provides the potential for long-range migration and establishment, but at the same time produces a degree of clustering within populations of these self-incompatible annuals."

Finally we should note that some species of fungi have evolved quite sophisticated solutions to the problem of dispersal. Most fungi that grow on dung have wind-carried spores that simply land and germinate. *Pilobolus* has a much more elaborate method of spore discharge and dispersal that is very effective in its particular environment. The mature spores remain in the thick-walled sporangium, which adheres to grass blades due to a slimy outer wall and sticky bottom (Figure 6–10). The grass is eventually eaten by an herbivore and the spores germinate to form a mycelium after the dung is depos-

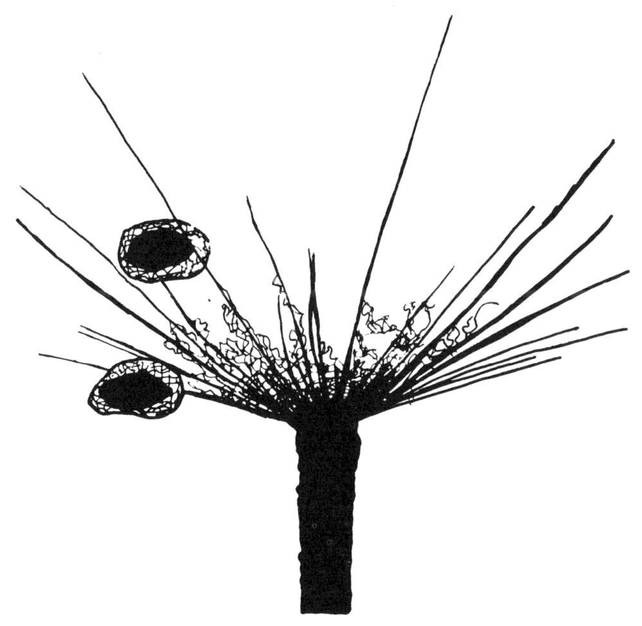

Figure 6–9 Two seeds of the root parasite *Orthocarpus densiflorus* attached to an achene of its composite host, *Hypochoeris glabra*. The net of the lower seed is lanced by one of the lateral bristles of the host and would presumably be dispersed with the host achene. (Drawn from photograph by P. R. Atsatt, 1965, *Science, 149:* 1389–1390.)

ited and the sporangial wall ruptures. Because this dung is lying at ground level, some distance from leaves where the newly formed sporangia on the mycelium could be eaten by another animal and dispersed again, the plant must move its sporangia by some mechanism. The evolutionary solution to this problem is a remarkable fungal cannon that shoots an entire sporangium toward sunlit, presumably open areas (Figure 6–10). The sporangiophore tube of *Pilobolus* stands up to a centimeter above the surface of the dung and ends in a transparent bulb directly below the black sporangium. This transparent lenslike bulb concentrates the light on a basal photoreceptive region which bends the stalk toward the maximim source of light. In the morning, the turgor pressure in the sporangiophore builds up and the bulb finally bursts just below the sporangium attachment point, shooting the whole sporangium over several yards. The sticky sporangium base catches hold of an obstacle like a leaf and flips the sporangium over, with the dark protective sporangium wall on top. Now the sporangium will be eaten along with the grass blade or leaf and be dispersed with its load of spores to a new area by the herbivore.

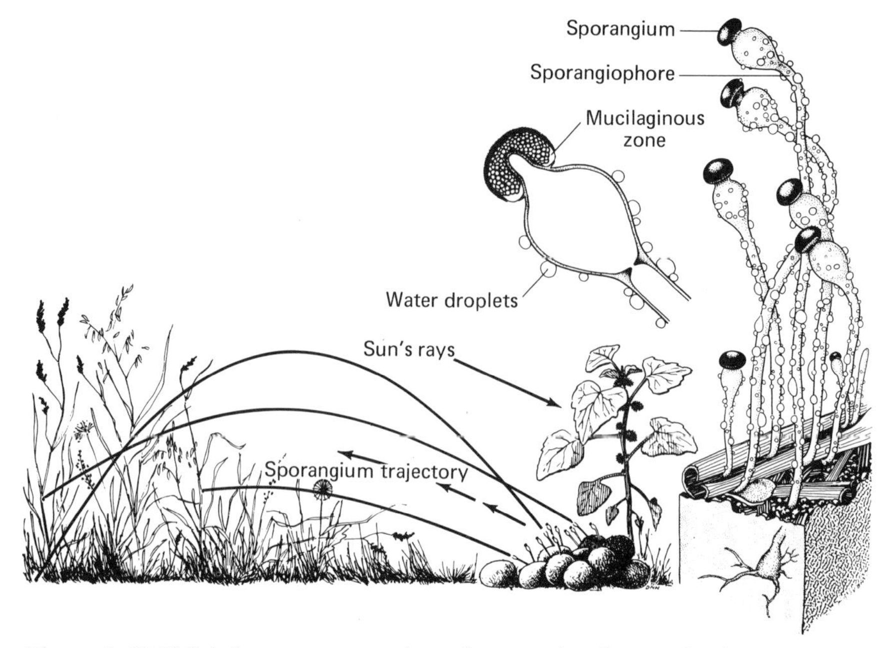

Figure 6–10 *Philobolus,* a common dung fungus, that has evolved a pressure-cannon dispersal mechanism for shooting its sporangia toward the light. (Source: W. A. Jensen and F. B. Salisbury, 1972, *Botany: An Ecological Approach,* Wadsworth, Belmont, Calif., p. 342.)

DISPERSAL IN ANIMAL POPULATIONS

Although most or all stages in the life cycle of an animal are mobile, the usual situation is that one stage has become specialized as the dispersal phase. The variety of dispersal patterns observed in animal populations is considerably more complex that that exhibited by plants, and ranges from species with extremely sedentary behavior, often exhibiting intrinsic barriers to dispersal, to those with homing abilities, migratory capabilities, and elaborate territorial systems maintained by unique dispersal behavior. Coupled with these kinds of dispersal in the mobile animals are internal population changes relating to dispersal behavior, which may exhibit even daily or hourly dispersal patterns correlated with changing environmental conditions. Let us look at these problems in more detail now.

Apparent absence of dispersal

In a particularly coarse-grained environment or in very strong territorial systems, some species show virtually no dispersal at all. A situation involving reproductively isolated family groups within a local population has been described by P. K. Anderson (1964) for populations of the house mouse, *Mus musculus,* near Calgary, Canada. On one farm near Calgary, he looked at the distribution of lethal *t* alleles in mice inhabiting 13 adjacent buildings, separated by only a few meters. The numbers of mice were small, with 1 to 84 mice per inhabited building. By experimentally analyzing the gene distribution of these *t* alleles between the buildings over two years' time, Anderson discovered that dispersal between these small family groups was very rare or nonexistent, despite the closeness of the populations. Other workers have shown that such family groups in *Mus* and Norway rats maintain reproductive isolation from other groups through defense of communal territories.

Standard dispersal behavior of freely moving individuals

Because of the element of choice in their mobility, the majority of animals tend to move about more or less continuously in their daily search for food and other environmental-grain components. Flying insects are particularly noteworthy in this regard, and dispersion of this type has been rather well studied in several groups of economic or evolutionary interest (Johnson, 1969).

The fruit fly family Tephritidae includes injurious fruit flies such as the oriental fruit fly and Mediterranean fruit fly, which attack commercial orchards. Surprisingly, these flies often disperse without regard to availability of hosts. Apparently, dispersal is influenced mostly by the wind after the adults hatch. Populations of the olive fruit fly (*Dacus oleae*) in Greece, were marked with radioactive P^{32} and a more or less continuous movement of these flies was found from olive groves on mountain slopes to adjacent plains. Within 24 hr, flies were captured 2.5 km from the release point, and 4.3 km away after 10 days. Within olive groves, individuals will disperse up to 2 km in several days. Some flies remain within a few hundred meters of the place where they hatch. Young flies tend to emigrate more than do older ones. Coupled with these observations is the fact that female tephritids have a relatively long preoviposition period of up to 14 days. Infestations of another tephritid fly, *Rhagoletes completa,* have been observed to advance at a rate of 0.5 to 4.8 km per year. The overall picture is that dispersal in the population structure of these flies seems to be mainly employed toward increasing the dispersion of the genotype to new areas rather than equalizing density within a small area.

Often, dispersal behavior is keyed to the timing of daily flight activity. The fruit fly family Drosophilidae has been rather well studied from this viewpoint. Many species of *Drosophila* exhibit a bimodal flight curve near a breeding source, one peak occurring at dawn and another at dusk. The onset of each crepuscular flight peak is determined endogenously by the external cues of light and temperature, which vary during the overall flight season. The adaptive significance of these flight peaks is subject to diverse interpretations, but they probably minimize desiccation problems (because direct sun is absent and the levels of relative humidity are higher at dawn or dusk), and maximize feeding opportunity (flower nectar and yeast activity in oozing sap are most abundant at those hours). These crepuscular peaks also serve as dispersal control mechanisms because these small insects avoid flight when the wind currents are strongest.

Dispersal in various species of United States and European *Drosophila* has been studied by Dobzhansky and Epling (1944) and others, using mark-release–recapture methods. Diverse species move at rates ranging from 4 to 100 meters per day. The rate of dispersal depends upon both air temperature and age. The flies move considerably more during the first day after release than later. In Drosophilidae as with many other insects, dispersal seems to occur as an incidental byproduct of generalized flight activity associated with feeding and oviposition.

Internal population changes in dispersal behavior

Dispersal behavior commonly varies during daily and seasonal cycles. This variation may be cued directly by either exogenous or endogenous factors, and normally represents an adaptive response to insure maximum exposure to favorable conditions for the particular activity involved. Mourning doves (*Zenaidura macroura*) migrate through the arid desert and irrigated farmland country of the hot Imperial Valley of southeastern California in the fall of each year, and reside a few weeks before continuing on to wintering grounds in Mexico. While in the Valley, they follow a daily cycle of moving into feeding areas (e.g., stubble fields) and watering holes in the early morning and evening hours, while in the very hot mid-day hours and at night, they fly to roosting sites in sheltered groves of trees.

The melon fruit fly (*Dacus cucurbitae*, family Tephritidae) exhibits a fairly elaborate cycle of dispersal behavior that changes the basic dispersion pattern of gravid females in a population no less than four times a day. Nishida and Bess (1950) found that *D. cucurbitae* breeding in a tomato field flew upon emergence as adults into the surrounding trees where they stayed until they had matured sexually. Then each day, gravid females would move into the tomato field early in the morning to oviposit. Near mid day they flew back into the trees. During mid-afternoon, these females moved back into the tomato field to lay more eggs on their food plants. In the evening, they returned to the shelter of the trees.

Other changes in the environment besides temperature and light intensity may cue changes in dispersal behavior. McFarland and Moss (1967) have found that the relative positions and dispersal of fish within traveling schools change with variations in environmental oxygen. These structural changes within schools of fish result from the behavioral responses of the individual fish to the environmental consequences of group metabolism. Thus, in schools of striped mullet (*Mugil cephalus*) the individuals sense when dissolved oxygen concentrations are reduced and carbon dioxide increases as pH is low-

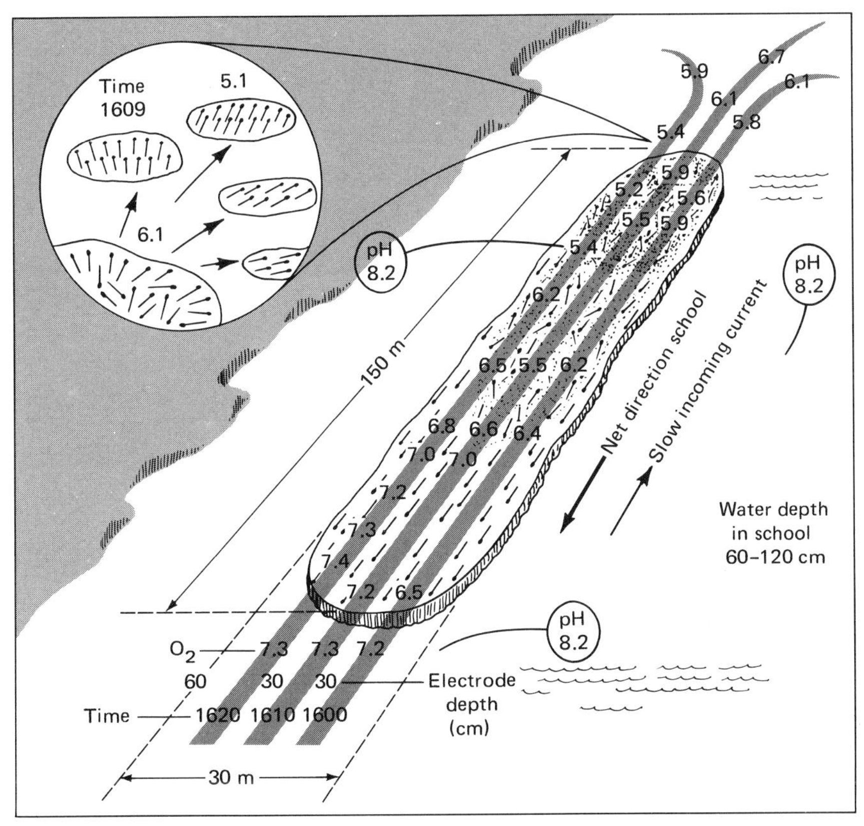

Figure 6–11 The relationship of school structure and behavior in striped mullet to metabolic modification of environmental oxygen and pH. The dot (head) and line (body) signify the orientation of individual fish in various parts of the school. Fish were dense throughout the school, but greatest density occurred in the rear, as indicated by stippling. Fish in the back of the school were actively roiling the water surface. The inset indicates breakup of the rear portion of the school into individual groups. Oxygen reported in milligrams per liter of seawater. (From W. N. McFarland and S. A. Moss, 1967, *Science, 156:* 260–262.)

ered because of the passage of the front portion of the school population. The mullet change their swimming direction or orientation, their spacing, and even their swimming velocity. The overall result of this change in social behavior from individual responses to physiological parameters is that the exposure of individual fish to less favorable conditions (low O_2, high CO_2, and low pH, or both) is shortened. Thus, Figure 6–11 shows that the oxygen reduction during the passage of a striped mullet school ranged (on three traverses) from 22.6 to 28.8 percent, and in the rear portion of the school, several small schools broke off temporarily and swam into less affected water. Constant interchange acts to equalize the time of exposure of each member of a school to the metabolic impact of the group (McFarland and Moss, 1967).

Seasonal changes in dispersal behavior within populations are commonplace. Foremost is the aggregation caused by pairing during the courtship and mating seasons and the dispersal following maturation of the young. Dispersing members of a population, such as in the field voles *Microtus pennsylvanicus* and *M. ochrogaster,* may differ genetically from resident members and the polymorphism can be a method of regulating population density (Myers and Krebs, 1971). In some groups of mammals males commonly stay apart from females except during the breeding season. In the rhesus monkey (*Macaca mulatta*), males remain with females in a particular group all year, except during the mating season when part of the adult male contingent in populations in northern India change groups (Lindburg, 1969). This action thoroughly disrupts and destabilizes the social structure of these groups, though the males conform in foraging, resting and travel activities to the habits of the new group they have just joined. At the end of the mating season, which lasts about six weeks, the transfer males return to their original groups in almost all cases. The situation suggests that selection has favored this "exchange migration" of males because of its effect in reducing inbreeding in the normally small, behaviorally isolated troops of rhesus monkeys. Genetically, this type of dispersal behavior promotes gene exchange among rhesus monkeys from diverse genetic backgrounds.

Intrinsic barriers to dispersal

The lack of dispersal shown in the population structure of the house mice studied by Paul Anderson (p.193) seemed to be due to the presence of closed family units maintaining exclusive territories. Now we shall look at another type of situation involving lack of dispersal in birds and butterflies and other species that could normally be considered highly vagile organisms, able to cross ordinary physical barriers. Many of these organisms are actually quite sedentary, and in a well-studied checkerspot butterfly species, Ehrlich (1961) has termed this type of dispersal behavior as being controlled by *intrinsic barriers to dispersal.*

This particular butterfly, *Euphydryas editha,* exists in well-defined colonies throughout cismontane California and higher mountains in the western United States. Ehrlich (1961, 1965) and numerous co-workers have worked intensively since 1960 with a series of three adjacent populations in a colony located on Jasper Ridge, a grassland and chaparral-covered range of hills with serpentine outcroppings on the Stanford University campus in central coastal California. Starting with the spring flight season in 1960, individual adults were marked in coded numbers with felt-tipped pens. During that 25-day flight period, 185 individuals were given distinguishing marks and released. Of these individuals, 97 were recaptured at least once, accounting overall for a total of 224 separate recapture events. (Sampling was conducted only once a day, so all recaptures of any given individual were on different days.) From this procedure, the total population size was estimated at 250–400 adult butterflies in 1960. This mark-release–recapture procedure was repeated in 1961, 1962, and 1963. Results (Table 6–2) clearly showed that the overwhelming majority of marked butterflies stayed in their area of previous capture, hence showing no dispersal. The changes in relative abundance of adults in the three adjacent populations C, G, and H on Jasper Ridge are shown in Figure 6–12. In 1964, a capture–recapture program was not done, but a total of 1444 adults were captured in areas C and H (see Figure 6–12), with only one found in area G. If movement of individuals between areas was frequent, a considerable number should have been captured in area G, even if the resident population was essentially extinct.

The lack of movement outside of their original area is curious because no insurmountable physical barriers separate the various sections of the colony; the butterflies are fully capable of flying over the low shrubby chaparral surrounding their grassy habitat if they are chased. In addition, the colony sections are *not* totally isolated because the chaparral and woodland growth around each area is discontinuous. The locations of main concentrations of individuals within each population area remain static through the season and from year to year

Table 6–2 Recaptured *Euphydryas editha* butterflies on Jasper Ridge in four different flight seasons (1960–1963).

YEAR	RECAPTURES IN AREA OF PREVIOUS CAPTURE	TOTAL RECAPTURES	PERCENT IN AREA OF PREVIOUS CAPTURE
1960	216	224	96.4
1961	411	425	96.7
1962	159	164	97.0
1963	235	235	100

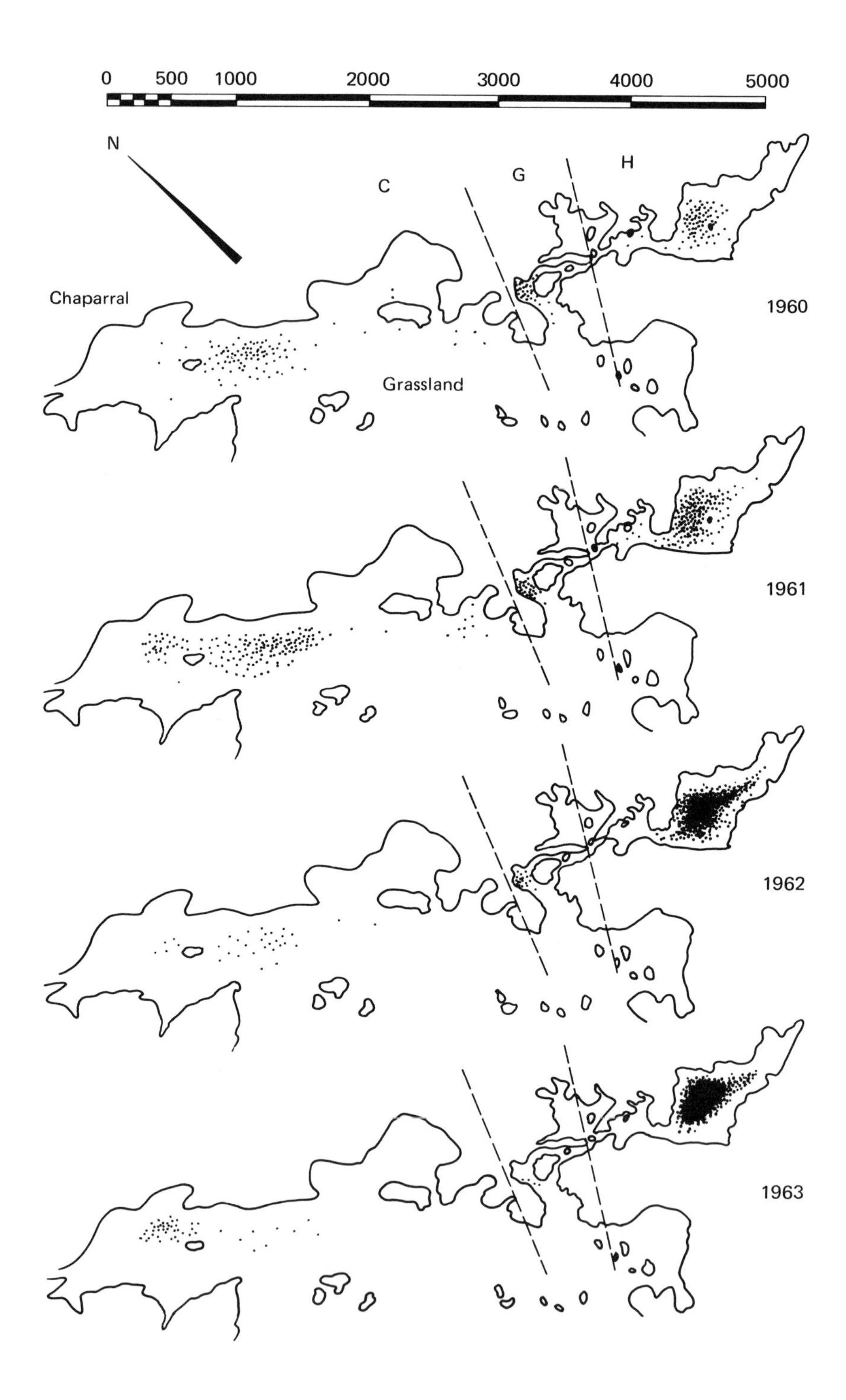
0
500
1000
2000
3000
4000
5000
N
C
G
H
Chaparral
Grassland
1960
1961
1962
1963

(although population density varies annually). Yet within areas C and H there is abundant open space with apparently appropriate food plant growth, immediately adjacent to these principal concentrations. There was no explanation, then, in 1961 for the reasons for this lack of dispersal, but the butterflies clearly seem to "choose" to remain within a particular area—in other words, the barriers to dispersal are *intrinsic* and are genetically determined within the organism.

With further work on the biology of this butterfly (Labine, 1964; Ehrlich, 1965; Singer, 1972), dispersal behavior was shown to be intimately related to the selective factors regulating reproductive success in the serpentine grassland on Jasper Ridge. Females are mated within a short time after emergence and are "plugged" by the first successful male, to prevent further matings. Males that might change areas are less likely to find a virgin female than those that stay home, and since the foodplant dries up toward the end of the flight period, the larvae from eggs laid by late-mated or transfer females will not have sufficient time to reach the size for successful summer diapause. Thus, transfer adults are at a disadvantage and movement among areas is rare. Long-term dispersal has been sacrificed for short-term gains in maintaining high reproductive success in present populations. Singer (1972) has also pointed out that the adult insects do indeed occupy all the habitat area available to them when one considers both adult and larval resources. For survival to the diapause state, most larvae require a site where two species of food plants are intermingled. Ovipositing adults can respond to one species of food plant only. Females that disperse will leave fewer offspring because the codistribution of the two larval food plants is greater in these presently inhabited areas on Jasper Ridge than elsewhere.

Homing and home range

The home range of an animal is *that area regularly traversed by an individual in search of food and mates, and caring for young* (Kendeigh, 1961). It is not a territory unless it is defended against intruders. Often, a central part of a home range serves as a defended territory. *Homing* is the behavior of returning to the home area when the animal is displaced. It seems increasingly apparent that most adult animals have home ranges, at least during the breeding season, and that this

---

Figure 6–12 Distribution of initial captures of adult *Euphydryas editha* butterflies during 1960–1963 in the grassland area of Jasper Ridge, California. The lines represent the edge of chaparral–oak woodlands. The letters in the top map refer to divisions (population areas) of serpentine grassland recognized by the butterflies in their dispersal behavior. See text for further explanation. (Source: P. R. Ehrlich, 1965, *Evolution, 19:* 327–336.)

influence on dispersal behavior considerably affects gene exchange and population structure in such species.

Goddard (1967) studied home range characteristics in two populations of the black rhinoceros (*Diceros bicornis*) in northern Tanzania during 1964 to 1966. One population (94 animals) inhabited the caldera of Ngorongoro, a 102-square-mile area of mostly open grassland, with 26 inches of annual rainfall. The estimated density was one rhino per 1.2 square miles. The second population (70 animals) inhabited a 170-square-mile area in the vicinity of Olduvai Gorge, with 16 inches of annual rainfall, sparse water supplies, and mostly thornbush vegetation, and had an estimated density of one rhino per 2.5 square miles. Goddard found that the black rhinoceros is a solitary and very sedentary species, showing no territorial behavior, and having a home-range size governed by several factors. In the central Ngorongoro crater, with open grassland and seasonal marshes, the adult rhinos had a mean home range of 6.0 square miles. In the dry thornbush country of Olduvai, the mean home range expanded to 11.7 square miles. Besides the influence of the availability of food and surface water, the size of the home range varied considerably according to the age of the animal (Table 6–3) and the season. The immature animals wander notably more than the adults. The mother drives off her old calf when a new one is born, and the immature animal generally joins up with a neighboring adult on a temporary basis and travels into new areas with him. Interestingly, the proportion of the home range which is utilized by the animals is considerably greater in the wet season, when more herbs are green and palatable. In the dry season, the rhinos remain close to a water source and do not search afar for food.

An animal that is so extremely sedentary would be quite susceptible to the effects of inbreeding if this small home range of the adult black rhinoceros were true throughout the animal's life. However, the two factors of (1) *the larger home range of the immature animal,* and (2) *the intolerance of the mother for her old offspring after a new calf is born,* afford some measure of population dispersal and thus serve an evolutionaly function in enabling a certain amount of outbreeding in the population structure of the species. The advantages to the individual

Table 6–3 The relation of sex and age class to mean home range size in two Tanzanian populations of black rhinoceros.

AGE AND SEX	NGORONGORO POPULATION (SQ MI)	OLDUVAI POPULATION (SQ MI)
Adult male	6.1	8.5
Adult female	5.8	13.7
Immature male	13.9	14.5
Immature female	10.7	8.4

SOURCE: Data from Goddard (1967).

animals in terms of maximizing their reproductive success are of course clearly evident in this behavioral pattern.

Homing as a dispersal-control mechanism has received particular attention in social insects, such as ants. Maintenance of distinct population structure in adjacent colonies would rapidly break down and be impossible, in fact, if it were not for the homing ability of the workers that forage for food. Stimuli used to control dispersal and return include chemical, mechanical (vibrational), and visual cues. Most ants use chemical pheromones to label trails to food sources and to the nest. Predatory, solitary-hunting species of ants apparently use visual stimuli in the normal course of events. *Cataglyphis bicolor,* a very vagrant species of ant that lives in North Africa and the Near East, depends exclusively upon visual orientation and terrestrial landmarks. When individuals wander long distances from the nest, beyond the known home range, they use time-compensated orientation to the sun's position in order to locate their nest (Wehner and Menzel, 1969). Pattern recognition and learning abilities are thus very highly developed in this ant.

Giant toads (*Bufo marinus*) also use visual orientation in their local homing activities. On Barro Colorado Island in the Canal Zone, Brattstrom (1962) found that homing in *Bufo marinus* in the non-breeding season (when no vocalizations were made) is probably largely due to the use of visual cues and the retention of previously learned topographic cues. These kinds of homing modification of dispersal behavior enable a species to retain a particular population structure over a considerable area of home range.

Long-range homing and migration

Population structure in species with long-range homing and periodic or nonperiodic migratory movements can become an immensely complicated analytical problem for the population biologist. Often, the tracking of individuals over the considerable distances involved may present almost insurmountable technical problems to solve before any "hard" biological data may be obtained. Animals must be tagged with radioactive isotopes, colored plastic or metal bands, painted numbers or codes on parts not exposed to excessive wear, miniature transmitters and other electronic devices, mutant genes with easily recognized phenotype effects, clipped toes, or other persistent and easily recognizable coding systems. Southwood (1966) offers an excellent recent summary of methods of marking animals. Once a significant portion of the population has been marked, an investigator can observe or recapture sufficient individuals to establish the extent of dispersal behavior and the operation, if any, of homing mechanisms and migratory movements.

It is difficult to distinguish "homing" and "migration" in many species. Homing may be an integral part of the migratory dispersal mechanism in a great many cases where an animal moves from one specific area to another specific area and back again. Examples may be found in the movements to and from breeding ponds by many ambystomid salamanders, which are terrestrial most of the year and go to specific ponds for breeding and oviposition in March and April. In eastern Massachusetts, individual adults of *Ambystoma maculatum* returning to ponds to breed will use the same track as they used when they first left the ponds as newly metamorphosed juveniles in a previous year, and they then use the same track in leaving the breeding pond (Shoop, 1965). These migratory-homing movements are usually carried out at night in fog, rain, or under cloudy skies when visible celestial cues may not be available. The sensory basis for this remarkable orientation ability is unknown.

From 1953 to 1967, Victor C. Twitty and his students have carried out exceptionally comprehensive studies of the extraordinaly migratory movements of the salamander *Taricha rivularis* on a 14,000-acre ranch in Sonoma County, central California. Twitty (1966) has given an excellent overview of these studies and the influence of these movements on the population structure of this California newt. The general habitat occupied by these newts is rolling mountains with steep-sloped valleys, in which a few small streams such as Pepperwood Creek flow; from early summer to early winter, conditions are extremely dry and the streams may dry up in the lower elevations. The mature adult newts that breed in the creek descend in the early spring from the adjoining mountain slopes and begin to enter it in late February or early March after subsidence of the winter floods. The adults mate and spawn by early April, barring too-frequent interruptions by rains and consequent flooding of the breeding pools. By May almost all the adults have left the stream. During the dry and hot summer months they hide in underground cavities where they can find moisture. With the first heavy autumn rains they begin to emerge in the more humid evening hours and forage for insects on the forest floor. During late winter, the newts begin working their way back to the streams.

Twitty and his co-workers analyzed this regular population flux of dispersal movements from 1953 to 1965 by marking (with toe-clipping of combinations of the four front-foot toes and the five back-foot toes) over 20,000 individuals at specifically recorded sites of 50-yard stretches along Pepperwood Creek. Only *one* of these salamanders was ever found in a stream in a neighboring canyon. These marked adults were found to return to Pepperwood Creek from the surrounding mountains year after year for as long as 10 to 12 years. Twitty decided to test the apparent homing ability of these newts by displacement experiments. In 1960, he moved 564 male *Taricha rivularis* downstream 1 mile from their capture points. Some 65 percent of these were recaptured in the site of origin in the following five years. Also in 1960 he displaced 692 newts from an experimental stretch along Pepperwood Creek to two other creeks which lay over 2 miles away with an intervening 1000-ft-high mountain ridge. In the following five years, 81 percent returned to the original site! Some of these newts were observed to wait a year or two before returning to their site of origin.

In 1963, Twitty's group displaced 730 males approximately 5 miles to another creek. As before, an overland return route was involved, but this time a *third* creek lay between the original creek and the site of release. Drift fences and can traps were put out on intervening ridges to check the course of movement of the marked salamanders. By 1965, about 40 percent of the total number displaced had moved home, or were known (because of their presence in traps) to be en route to that destination. A very few lingered in the intervening creek, at least for 1 year. Besides the long-range accuracy of this homing movement, this series of experiments showed also that the homing journey is not prompted solely by the urge to breed. "Some of the males captured en route home, for example in the traps along the experimental stretch, showed no development whatever of the secondary sexual characteristics always associated with impending breeding activity" (Twitty, 1966). Dissection of some of these animals showed the testes were undeveloped, and that the animals would not be breeding for at least another year. Hormonal factors associated with breeding are clearly not directly implicated in the motivation or "release" of homing activity. As Twitty has aptly put it, to newts, *home* seemingly means "the right place to live," not merely *"the right place to breed."*

The homing behavior exhibited by *Taricha rivularis* is *oriented migration* and does not represent a random search or series of wanderings until the right place is accidentally encountered. This oriented migration is not by *vision,* as displaced blinded newts are perfectly capable of returning to their home area. It is not by *mechanical stimuli* through the sense of touch, getting cues from the texture and composition of the soil and the topography of the land, because newts did not respond to such cues on a huge outdoor tilting table platform. It does not occur by hearing of *auditory stimuli.* As salamanders are voiceless and make no calls, they could not assemble in one place by reason of audible communication. There also would be no reliable reference sounds provided by the creek because of great seasonal and yearly changes due to fluctuating volume of flow and shifting course of the actual stream bed. The only known senses left are those of *olfaction* or sensory response to related chemical stimuli. The best evidence for the operation of this sense in homing comes from marked experimental animals where sections of the olfactory nerves were destroyed surgically and the individual then displaced. Of 692 normal animals displaced, 171 were found back at the home site the first year and 260 the second year. But of 607 surgically treated animals displaced (with destroyed olfactory nerve trunks), 0 homed the first year, 5 returned to the home site the second year, and 10 the third year. When Twitty's group dissected these 15 animals that had successfully homed to their original site, all 15 were found to have *regenerated* their olfactory nerves. More evidence is needed to prove that homing depends principally upon odor detection (presumably characteristic odors from particular mixtures of soil, vegetation, and humus). In these experiments there is the disturbing possibility that the lack of sense of smell may prevent most animals from surviving the dry summer months, perhaps through interference with their ability to detect moisture and find suitable estivation chambers in the ground.

After examining this remarkable example of long-distance dispersal and homing in a weak-limbed animal normally not thought to be capable of moving very far, we may well ask what the significance of homing is with regard to the population biology of this animal. Why should association with a particular segment of a particular stream be of such compelling importance to newts? There are at least two major selective benefits accruing to a species with such a dispersal system. First, the distribution of individuals is stabilized and equalized within the area occupied by a population, say upper Pepperwood Creek, so that competition for food and other environmental resources, such as suitable terrestrial hiding places during the dry summer months, will be minimized. Second, and probably of greater benefit, the accurately homing individuals are *protected against reproductive wasteage.* Many of these smaller coastal streams in central California are semipermanent, and dwindle and disappear during the dry summer season. The upper part of Pepperwood Creek runs all year, but the lower portion sinks underground in summer. This lower meadow stretch is fully suitable for spawning earlier in the year with the water runoff from winter and spring rains, but *T. rivularis* rarely enters it. Any adults choosing the lower stretch for spawning would leave few, if any, heirs, for most of the tadpoles would die during the summer drought. The species is *protected* against such errors of judgment by the homing instinct, which ensures that spawning will be confined to waters of proved suitability—proven by the fact that the spawners themselves were born and survived there (Twitty, 1966).

Such fidelity to a successful breeding site is shown by many birds, mammals, and even marine animals like the giant sea turtles. The green turtle (*Chelonia mydas*) makes transoceanic migrations of remarkable accuracy to nesting beaches on either the mainland or remote oceanic islands. This species has been studied for many years by Archie Carr and his students at the University of Florida. An example of great dispersal and subsequent orientation to a mainland nesting ground is provided by the *Chelonia mydas* population nesting at Tortuguero Beach in northeastern Costa Rica. After hatching, the turtle swims off into the Caribbean where it grows to maturity and then returns to the ancestral nesting beach where it hatched, to mate in the offshore surf. If the turtle is a female, she will clamber ashore at night on her next "home" visit and dig a pit in the sand, laying about a hundred eggs (Figure 6–13). This terrestrial nesting habit has allowed Carr and associates to tag these females and maintain a record of every observed subsequent nesting visit. Tagging of both males and females caught offshore as well as onshore has resulted in hundreds of recapture observations by native fishermen throughout the Caribbean, as the metal shell tags include mention of a reward to the finder if the tag is returned to Carr in Florida! Even in the first decade of work on Tortuguero (Carr, 1964), with over 3000 turtles tagged at Tortuguero, the extensive tag returns from international points demonstrated that the Tortuguero adults disperse to many points over a huge range, returning to their home beach on the average only every three years.

A population of green turtles, then, can be defined as those animals originating at a particular nesting beach, which may represent 8 to 10 miles of beach or more, with tens of thousands of eggs laid every year by returning female adults (a third of the total adult female population). The highest percentage of mortality during the green turtle's

Figure 6–13 A nesting female green turtle (*Chelonia mydas*). (Photo by Archie Carr, Univ. of Florida, Gainesville.)

lifetime occurs during the nesting period when predators dig up the eggs or kill the hatching young as they scramble from the sand dunes to the sea. Population structure for green turtles is ill-defined outside of the group meeting on this common offshore mating and beach nesting ground. When the young hatch, they do exhibit unconscious cooperation when freeing themselves from their sandy prison nearly two feet below the surface of the ground. They all hatch nearly simultaneously and then wiggle upward through the sand together, causing a collapse of the roof into the lower cavity of the nest. From that point on, however, the population seems to be a collection of independently dispersing individuals as each makes his way to the sea and swims out into the Caribbean to mix with individuals from other nesting-beach populations in the Sargasso Sea or wherever they may go to grow to maturity. Every three years, that particular age class, or mixture of three-year-internal age classes (including those that hatched three years, six years, etc., earlier), returns to the home population area for mating and nesting.

This is population dispersal at its extreme (Figure 6–14). Throughout the entire temporal existence of the population, except for several days or weeks off the nesting beach every three years, the individuals seem to be independently and widely dispersed. The remarkably precise homing shown in their migratory return to the nesting beach from which they hatched (and that therefore their parents used) seems to be because of the same general adaptive principles hypothesized for the *Taricha* salamander populations. The turtles are returning to the remote and isolated place that has fulfilled their requirements generation after generation, thereby removing the uncertainties of (1) searching for a new nesting beach every breeding season, (2) encountering males that had come to that new area likewise by chance, and (3) discovering a new location equally suitable in sun exposure, sand consistency, moisture content, etc., for the optimal incubation of their eggs, as well as (4) having a site reasonably free of resident predator concentrations (until the appearance of man on the scene). In dispersing widely as immatures and adults, individual *Chelonia mydas* can graze on turtle grass throughout the Caribbean. The selective disadvantage to individuals would be considerable if the turtles remained in the pastures near the nesting beach, where overgrazing would soon result and insufficient food would be present.

Even certain marine invertebrates show regular migratory dispersal movements. In the shallow waters off Bimini, Bahamas, and the Florida east coast, spiny lobsters (*Panulirus argus*) are normally active at night, feeding on annelid worms and small mollusks, and seclusive during the day, hiding in crevices. But each fall, all the lobsters in a region—comprising thousands of individuals—form parallel single-file queues and migrate *diurnally* in tremendous mass movements

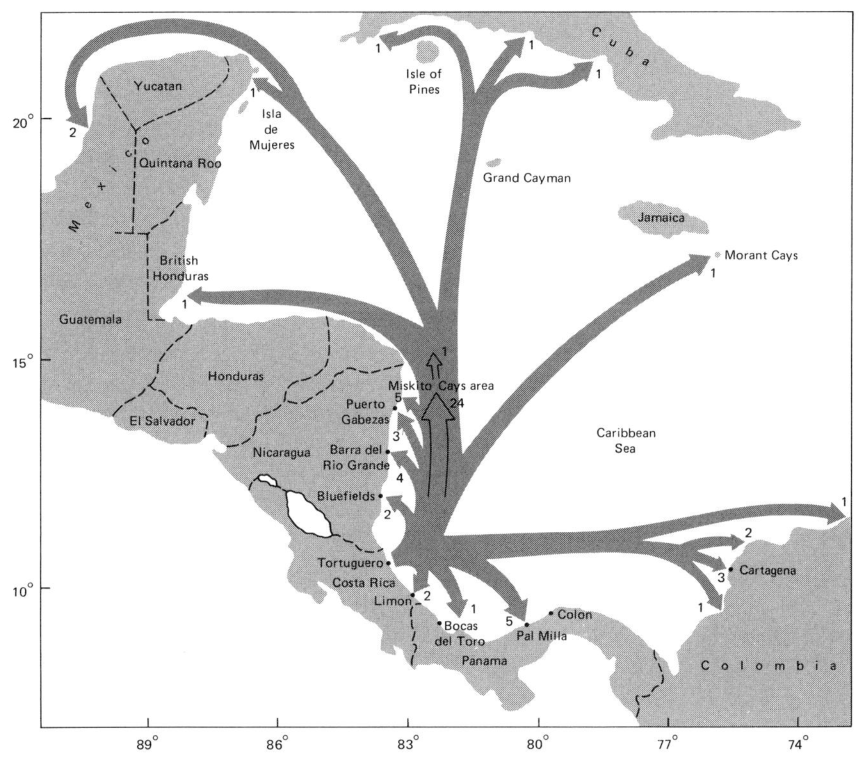

Figure 6–14 A map showing general dispersal distances covered by mature female green turtles (*Chelonia mydas*) marked at Tortuguero, Costa Rica, and based on recoveries of tags from 2000 originally marked individuals. The arrows only indicate spread from the tagging locality, and are not intended to suggest routes. (Source: A. Carr, 1962, *Amer. Scient., 50:* 359–374.)

which are oriented in a particular direction, characteristic for a given population (Herrnkind, 1969, 1970). Individuals link up by touching their anterior appendages (antennae or the front pair of legs, or both) to the posterior end of the next individual (Figure 6–15). Vision is used in the initial linkup, but tactile cues are most important in maintaining the migratory chains. The individuals in the queues maintain constant contact and the same speed and direction, even when detouring around an obstruction. Herrnkind picked up some marching lobsters off Boca Raton, Florida, that continued marching around the perimeter of a plastic seawater pool for almost five weeks and an estimated 500 miles before the "migration" halted. Some idea of the magnitude of these movements in the field may be gained from the observations of a column that was one-quarter mile in length, with more than 1000 spiny lobsters, and of a 1969 migration of many columns near Bimini which probably involved at least 200,000 lobsters. Experimental displacement of marked lobsters showed that the lobsters have

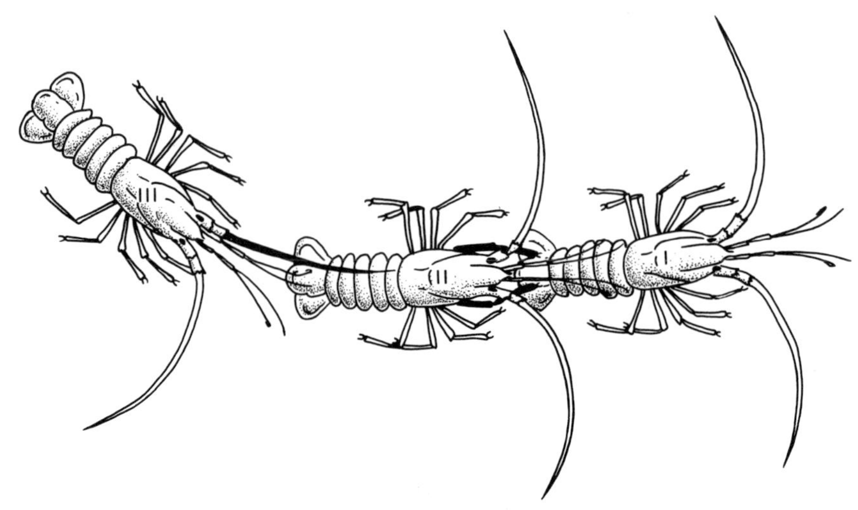

Figure 6–15 The use of appendages in queuing dispersal behavior of the spiny lobster, *Panulirus argus*. Appendages involved are shown in solid black; they are sensitive to tactile stimulations. Lobster III joins the queue after touching lobster II with an extended antenna and turning until both antennular inner rami touch the sides of II. Lobster II maintains queue alignment either by frequent intermittent flicks of the antennules against lobster I, by grasping or touching I with extended anterior pereiopods, or both. (Source: W. F. Herrnkind, 1969, *Science, 164:* 1425–1427.)

a remarkable homing and orientation ability and use a guidance system independent of vision or current and bottom substrate conditions. The adaptive significance of these mass single-file autumnal migrations cannot be explained as yet. A "return" migration is still unknown. This dispersal behavior is not connected with reproduction since females do not carry spermatophores until spring. Possible functions are local dispersal, reduction of population pressure, attainment of maximal shelter for molting, and attainment of seasonally better feeding grounds. The function of the queuing behavior appears to be linked to defensive purposes when relatively open areas are crossed, as the spinous cephalothorax of all but the last trailing individual covers the vulnerable abdomen of each preceding lobster. Under normal nonmigratory conditions, the soft abdomen is protected in a rocky crevice during the daytime. Perhaps the queuing is an example of reciprocal altruism, or even kin selection if these lobsters are closely related (Chapter 3).

Coastal populations of the northern deep-sea lobsters (*Homarus americanus*) exhibit localized movements, but populations found on the outer continental shelf undertake extensive seasonal migrations that seem as profound a navigational task as any that the better-known migratory birds attempt. In tagging some 5710 lobsters from depths of 70 to 400 meters, Cooper and Uzmann (1971) found that the lobsters moved shoalward in spring and summer, and returned to the edge of the shelf in fall and winter, apparently motivated by seasonal change in optimum bottom temperature. The lobsters were able to move up to 10 km/day, and while 21 percent of the recaptured individuals had moved distances of less than 16 km, some 58 percent migrated between 16 and 80 km, and the remaining 21 percent in excess of 80 km. Ten tagged lobsters had moved farther than 160 km and one had migrated 338 km shoalward (Figure 6–16). Females seemed to migrate farther and in greater numbers than males. Apparently, the role of this extensive, regular and seasonal dispersal behavior in the population structure of the deep-sea lobster is to allow adults to molt, mate, and extrude eggs, as well as permit eggs to hatch, at the sufficiently high water temperatures which are present seasonally in the shoal areas but absent in the deeper continental shelf habitats. Thus the

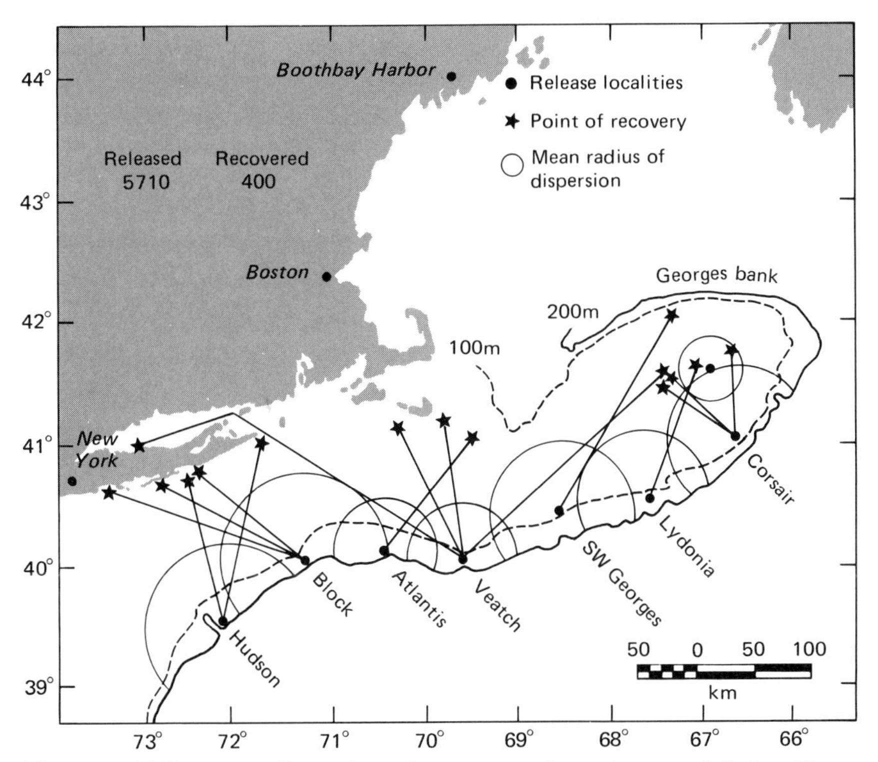

Figure 6–16 The mean dispersion of tagged northern deep-sea lobsters (*Homarus americanus*), and examples of the longest shoalward migrations. (Source: R. A. Cooper and J. R. Uzmann, 1971, *Science, 171:* 288–290.)

individual maximizes its reproductive contribution to the next generation of lobsters, and benefits in a selective sense.

Territoriality and dispersal behavior

Simply defined, a *territory* is *any defended area.* Territoriality represents "the defense of an area by one or more individuals against the invasion of that area by other members of the same species" (Sexton, 1960). Animal species that exhibit territorial behavior have characteristic dispersion patterns because of these defended areas, and these seem to be primarily a response to temporal and spatial distribution of exploitable resources (Brown and Orians, 1970). For this reason, territoriality often plays an important role in population regulating systems (Wynne-Edwards, 1962), and we considered this aspect of it in depth in Chapter 5. For the present, let us emphasize that territorial behavior is related to time and energy budget considerations and that the dispersal behavior of territorial species is generally intended to equalize spacing and is frequently linked to population regulation. A single example will suffice here.

The troops of the black and white colobus monkey (*Colobus quereza*) maintain well-defined territories in the Budongo Forest, Uganda. Marler (1969) found that the average group of eight animals maintained a territory of 0.062 square miles (0.317 km^2) in area, which virtually coincided with the troop's home range. Male roaring and visual displays served to maintain intertroop distance and this territorial organization resulted in a distinctive uniform dispersion of colobus troops through the forest. Dispersal is controlled within the troop boundaries by this strong territorial defense. Marler (1969) points out that forest-living monkeys in the Old World are generally territorial with small home ranges, while the non-territorial primates such as baboons and the patas monkey (*Erythrocebus patas*) tend to invade more open habitats and exhibit a great increase in the size of home range. In the latter cases, large home ranges become impractical to defend effectively as territories and the home ranges of adjacent troops often overlap extensively. At very large home-range sizes, perhaps around 1 square mile, territoriality, and hence territorial controls on dispersal behavior, may disappear because intertroop contacts become rare and reinforcement of the territorial boundaries becomes insufficiently frequent.

Finally, note that in some groups such as lizards and odonates (dragonflies and damsel flies), territorial behavior influences the dispersal of individuals in relation to mate selection far more than to resource exploitation in home range areas. Johnson (1964) shows that the evolution of territoriality in dragonflies and damsel flies has been related to mate acquisition and the male's breeding behavior rather than density regulation, for the males do most of their feeding in other areas. Species-specific bright colors in territorial males and their potential mates have likely been favored by selection because of their role in reducing unproductive flights toward other species and conspecific males, and favoring flights to proper mates. The small defended territories result in reduced interference by other odonates to mating and oviposition, and offset the disadvantages of frequently having a high male and female density around a suitable but limited body of water.

DISPERSAL BEHAVIOR AND POPULATION STRUCTURE

Because population structure is related to the factors that influence reproductive contacts and dispersion patterns between individuals in a population, the structure of a population can be highly dependent upon dispersal behavior. Each individual's genotype contains genes that affect dispersal in some way (strength of locomotion, behavioral tendencies to remain sedentary or to wander, and so forth). Although they vary in genotype, the members of a species share a dispersal strategy for spreading through the available habitat, but still keep them close enough together that reproduction is not hindered. These two opposing tendencies, to spread through the habitat and to stay close for breeding, have resulted in the tremendous range of simple to complex forms of dispersal behavior that we have observed in the preceding pages.

The general ecological and evolutionary effects of ordinary dispersal behavior are to distribute the population optimally with respect to resource availability, and to affect genetic variability by increasing contact and gene flow between separate populations. The latter will always be increased by greater dispersal (in terms of introducing new alleles), but occasionally the former will not, and this as well as other factors may lead to the establishment of intrinsic barriers to dispersal. The checkerspot butterflies studied by Ehrlich are a good example. Here, the short season during which suitable food concentrations were present made immediate mating and oviposition obligatory, and thus there was no time or selective advantage for dispersal. This set of factors resulted in noticeable genetic isolation between populations, but it insured optimal temporal distribution and survival of larvae.

Homing and migratory behavior, closely related forms of dispersal behavior, are two ways of insuring the return of individuals which have spread out too far for efficient breeding or which must reproduce in a particular area which has proved suitable in the past at the proper season. Migration in particular allows regular utilization of separate and diverse habitats for breeding and food resources that are unavailable, at least to support the same population densities, on a permanent basis. Homing and territorial dispersal behavior tend to reduce genetic contact within or between populations, but these traits normally insure better distribution relative to appropriate resources. The role of dispersal in maintaining appropriate population densities was specifically examined when we discussed regulatory systems in populations (Chapter 5).

1 2 3 4 5 6 **7** 8 9 10 11

Population Structure: Age and Sex

The life-span or longevity of an individual and the proportion of the sexes are complementary features that affect the total reproductive potential and hence the structure of a population, as we have defined it (Chapter 6). In addition, the age structure and proportion of sexes at different times may themselves be properly considered integral descriptive components of population structure. We can gain some appreciation of the importance of the interrelationship between age and sex in affecting frequency of breeding by looking at the Pepperwood Creek population of the California newt, *Taricha rivularis,* already famed for its homing prowess (Chapter 6).

The actual longevity of this small terrestrial salamander may amount to several decades or more; just how long is not yet known because the Stanford investigators pursuing this project through 1965 (see Twitty, 1966) were still recapturing surprisingly high percentages of the groups marked during their first seasons of work at Pepperwood Creek. In 1953, the first year of the project, two series of adult males were marked by amputation of an entire hind limb, one large group numbering 1835 and a relatively small one of 262. Eleven years later, during the 1964 breeding season, Twitty and his co-workers recaptured 28 percent and 30 percent, respectively, of the members of these two series. These captures represented only part of the survivors from 1953, since not all males enter the stream for breeding each year and some that do undoubtedly escape recapture. A distinguishing mark was given to these 1964 recaptures, so that they could be distinguished from any new 1953 newts encountered. Additional captures in 1965 raised the total numbers of recaptures in 1964 and 1965 to 38 percent and 36 percent, respectively, for the two series. The additional members of the groups captured in 1966 meant that about 40 percent of the adults marked in 1953 were still alive 11 years later in 1964, the first year of the three-year census. In a third experimental series, comprising 636 animals first marked in 1955 by amputation of a hind foot and one front foot, the recaptures during 1964 and 1965 showed a survival of 41 percent.

In all of these series the percentages of recaptures decrease very little from year to year (Twitty, 1966), implying remarkably long lifespans. These extraordinarily small annual decrements in recaptures are even more remarkable when one considers the extensive leg amputation initially employed to "mark" the animals, and that for several years after first marking the three series these appendages were reamputated each year the animals were recaptured during Twitty's patrols. Thus, once *T. rivularis* reaches adulthood it appears to enjoy a rather long period of immunity to the ravages of time and environmental hazards, during which it can engage in reproduction for perpetuation of its genotype.

Male *rivularis* have a relatively regular frequency of breeding. A given male will commonly, but definitely not always, breed every year. Thus, only about 60 percent of marked males are found in the stream again the following year.

Female *rivularis* breed much less frequently. Twitty and co-workers marked all females since 1960 with a distinctive toe clipping for each year. As shown in Table 7–1, the pattern of subsequent recaptures has been closely similar in each year-group. In the first year following

Table 7–1 The nonannual frequency of breeding in *Taricha rivularis* females at Pepperwood Creek, as shown in the pattern of subsequent recaptures of marked females.

FEMALES MARKED		PERCENTAGE OF FEMALES RECAPTURED				
YEAR	NUMBER	1ST YEAR	2ND YEAR	3RD YEAR	4TH YEAR	5TH YEAR
1960	1273	0.7	14	19	9.2	10.2
1961	836	2.3	13	16.5	10	—
1962	1158	1.0	16	19	—	—
1963	1295	1.0	18	—	—	—
1964	1025	2.3	—	—	—	—

SOURCE: Twitty (1966).

marking, the number of females returning to the water to breed is uniformly low in all year-groups, from 0.7 to 2.3 percent. In the second and subsequent years, the number increases sharply, but the total female captures for the oldest (1960) series are still, after five years far short of the total originally marked. The number of new (previously unmarked) females was about the same each year. Thus it appears that females wait for intervals of from one to several years for successive breedings. Not more than 2 to 3 percent of females breed in immediately successive years. This arhymthmicity in breeding explains the skewed sex ratio in favor of the males at the breeding sites. As many as 20 or more males may be seen attempting to mate with a single female.

The difference in breeding frequency between the two sexes, and especially the great variability in the interval between successive breedings of a female, are perhaps unique among amphibians and other vertebrates (Twitty, 1966). The low average frequency of breeding for females and the sex ratio at the breeding sites mean that the total production of eggs is considerably less than we might have expected from the long life-span of the species and the huge numbers of newts (mostly males) in the stream area. Yet the Pepperwood Creek populations are apparently at equilibrium with their environment, neither increasing or decreasing significantly in size for at least 13 years (1953 to 1965 seasons). Thus, the average lifetime production of a female must be only two offspring, one of each sex, which survive to breeding maturity. Natural selection must have led to a balance between life-span and frequency of breeding in *rivularis* that assures a suitable adjustment between population size and the capacity of the environment to accomodate it. A long-lived species need breed less frequently in order to guarantee survival of its offspring in a population at equilibrium, a life cycle type resulting from selective processes known as *K* selection, which will be discussed in more detail in Chapter 8. Selection favoring the wide variability in frequency of female spawning, as well as adult longevity for increasing breeding opportunities, likely results from the precariously variable nature of the stream habitat in these California hills. The eggs of *rivularis* females, which are mostly attached to small stones or boulders in the stream bed, are vulnerable to the silting and scouring forces of strong spring floods. In years when heavy late rains have occurred, Twitty (1966) found few surviving larvae during the summer. Hence, it seems possible that wide variability in the frequency of female spawning might help assure that cataclysmic floods would never destroy the progeny of more than an allowable fraction of the total pool of females in the population. Males, on the other hand, place a comparatively low energy investment in their sperm and can afford, in a selective sense, to attempt to breed every year and place their genes in that year's reproductive sweepstakes entry. Their intense competition to breed with the many fewer available females at the streams suggests the value of yearly breeding journeys on their part (see Chapter 9 for a more general discussion of sexual selection). Thus, we see strong selective pressures influencing the total population structure through the effects of the component structural features of age and sex characteristics in the population. We shall now look at each of these components of population structure in more detail.

LIFE-SPANS AND AGE STRUCTURE

Once an individual is born, it begins to age immediately and at some point it dies, or to put it demographically, the individual suffers mortality. This truism compresses the many provocative theories of aging and the numerous factors reported to influence life-span (Table 7–2) into a simple statement which explains nothing, but does reflect the fact that the life-span of an organism is both intrinsically and extrinsically determined. The intrinsic component includes the ways in which the genetic constitution might determine the life span. Each species has a characteristic and specific maximum longevity; this potential biological upper limit of age is genetically determined as much as any other species character. This limit may be affected detrimentally by cellular mutation rates, physiological disorders caused by parentally inherited traits, and speed of maturation, among other intrinsic factors, in individuals of a single species. Extrinsic factors include predation, accident, environmental change, and so forth. These factors combined cause mortality in the individual (discussed already in Chapter 5) and thus influence the average life-span for members of a population.

The results of these intrinsic and extrinsic determinants of life-span are sometimes categorized as two types of longevity. *Physiological longevity* is reflected by the total life-span of an individual under the most optimal environmental conditions, and is a genetically fixed trait. *Ecological longevity* refers to the longevity of individuals under environmental conditions which are less than optimal. The extent to which these types of longevity approach equality is a measure of the closeness of environmental conditions to the optimum state.

We can describe life-span and age structure in a population of plants or animals with more precision by considering life tables, survivorship curves, and age pyramids. *Life tables* provide an enumeration of age groups in the population, which allows inferences to be drawn as to the rate of survivorship after birth and, conversely, the specific chances of mortality at any particular age. *Survivorship curves* are a graphical expression of the same data, plotting the number of individuals left in a particular age cohort versus time. *Age pyramids*

Table 7–2 Some factors reported to influence life-span in organisms.

GENERAL FACTOR	SPECIFIC CHARACTERISTIC
1. Age at parenthood	Age of father Age of mother (including age at time of ovulation, oviposition)
2. Mating pattern and fertility	Immediate versus delayed mating regimen Assortative mating Order of birth Breeding pattern: inbreeding vs. outbreeding Fertility Hybrid superiority
3. Physiological and biological characteristics	Body size Brain and body weight "Build" of females and their mortality Sex Loss of protoplasm Onset of disease, lesions Growth and metabolic rate Rapid maturation Blood groups Immunological reactions
4. External conditions	General environment Climatic conditions Radiation and ionizing radiation Diet, nutrition, protein intake Male–female differences in dietary enhancement Temperature
5. Population characteristics	Density of population Size of brood nest in relation to number of nursing bees Sex ratio of reproductives to nonreproductives Social classes in man Socioeconomic factors and occupation in man

SOURCE: Summarized from Cohen (1964).

show diagrammatically the proportion of each age class in the population. We can now look in more detail at these demographic methods analyzing population age structure.

Life tables

Mortality varies with age groups and sex and hence greatly influences population structure according to the age structure and sex ratio present. The life table expresses in account-book fashion the pattern of deaths and survivorship at different ages within a population. Because of the influence of sex on survivorship, males and females are almost

Table 7–3 Life table for human males in the 1966 United States population.

AGE (YEAR CLASS) (x)	NO. ALIVE AT TIME x (BEGINNING OF INTERVAL) (l_x)	NO. DYING IN THE INTERVAL (d_x)	MORTALITY RATE $q_x = d_x/l_x$ (q_x)	SURVIVAL RATE $s_x = 1 - q_x$ (s_x)	LIFE EXPECTANCY (MEAN LIFE EXPECTANCY OF INDIVIDUALS IN THIS AGE CLASS) (e_x, IN YEARS)
0	1000	26	0.02576	0.97424	66.75
1	974	4	0.00405	0.99595	67.51
5	970	2	0.00253	0.99747	63.78
10	968	3	0.00260	0.99740	58.93
15	965	7	0.00730	0.99270	54.08
20	958	10	0.00992	0.99008	49.46
25	949	9	0.00938	0.99062	44.93
30	940	10	0.01088	0.98912	40.33
35	930	14	0.01520	0.98480	35.74
40	916	21	0.02345	0.97655	31.26
45	894	33	0.03716	0.96284	26.94
50	861	51	0.05956	0.94044	22.88
55	810	75	0.09216	0.90784	19.16
60	735	97	0.13260	0.86740	15.84
65	637	124	0.19505	0.80495	12.86
70	513	137	0.26772	0.73228	10.35
75	376	132	0.35064	0.64936	8.22
80	244	115	0.47188	0.52812	6.33
85+	129	129	1.00000	0.00000	4.75

SOURCE: Data from Keyfitz and Flieger (1971).

always tabulated separately. The standard form of a life table is shown in Table 7–3. It consists of a series of columns headed by certain standard notations. These include, in order: x, the year class or other units of age for a representative group (usually adjusted to 1000 individuals) starting at birth or hatching; l_x, the number of individuals in a cohort group that survive to age x (if expressed as a fraction of 1000, this column can also be read as the probability at birth of an individual surviving to age x); d_x, the number of individuals in a cohort that die during the age interval from x to $x + 1$. The next column, q_x, is an age-specific mortality rate, which can be calculated by dividing the number of individuals (d_x) dying in the interval x to $x + 1$ with the number of animals alive (l_x) at the beginning of age x. The survival rate (s_x) of that age group can then be calculated as $1 - q_x$, and this figure can be used to calculate e_x, the life expectancy at the end of each age interval. Life expectancy is simply the probable number of years left to live for members of a population in a particular age group, given an empirically determined survival rate.

Life tables are of several kinds: dynamic, time-specific, or composite. The *dynamic life table* follows the survivorship of a particular age cohort until all members have died. This practice requires having a large group of plants or animals all germinated or born at the same time. Because it is usually difficult to obtain such a cohort, especially in the terrestrial vertebrates, most population biologists utilize two other approaches to constructing a life table.

The *time-specific life table* is constructed from a single sample of animals in a population taken over a given period of time, usually not more than a year. One obtains the population age ratio (assuming that the sampled animals of each age class have been taken in proportion to their numbers in a population) and determines the mortality rate from the existing age ratio, as was done in Table 7–3 from data on the age structure of the United States male population in the year 1966. This type of life table has three principal assumptions: (1) the population size is stationary; (2) the recruitment rate and age-specific mortality rates are constant, leading to a stable age ratio; (3) there is no ingress or egress of individuals which is age-group specific. Each age class in the life table represents a composite of many cohorts. Thus the time-specific life table presents only a crude generalization of the probable age structure of a population at the time the sample was collected.

The *composite life table* advantageously uses the principal features of the dynamic and time-specific life table approaches by combining data from several years to produce a practical life table. Thus, wildlife biologists will band a number of newly hatched waterfowl over several years, and after following the fate of each year class (too small in themselves to construct a dynamic life table), they treat all of the marked birds as a single cohort sample in a life-table analysis. The assumptions involved in the composite life table are that: (1) the marked sample is a representative sample from the population in terms of age groups; (2) mortality and survivorship in the age classes of the recaptured marked cohort are similar in their distribution to the unreported mortality and survivorship in the banded or marked cohort. In waterfowl, for instance, band losses tend to be more common in older birds due to wear, and the younger age groups are more susceptible to recovery (by hunters) than the experienced older ducks which have migrated for several seasons. Murie (1944) constructed a life table for the age at death of mountain sheep in Mount McKinley National Park, Alaska, by collecting skulls of 655 animals in the wild. The records were obtained by estimating the age of mortality from tooth wear and the size of the sheep horns, which persist for many years after the death of the animals. The total number of deaths was used as the initial composite number of live animals (Table 7–4), and q_x and s_x were calculated as previously outlined. From this composite life table (with a dynamic analysis), it appears that if a sheep can survive into its second year of life, it has a good chance of attaining what for this animal is old age. The mean length of life for these mountain sheep is a little

Table 7–4 Life table for Dall Sheep in Mount McKinley National Park, Alaska, based on age grouping of skulls of animals dying of natural causes.

AGE INTERVAL (YEARS) (x)	NUMBER ALIVE AT START (l_x)	NUMBER OF DEATHS (d_x)	ANNUAL MORTALITY RATE (q_x)	ANNUAL SURVIVAL RATE (s_x)
0–1	655	41	6.3%[a]	93.7%
1–2	614	117	19.1	80.9
2–3	497	10	2.0	98.0
3–4	487	9	1.8	98.2
4–5	478	9	1.9	98.1
5–6	469	23	4.9	95.1
6–7	446	34	7.6	92.4
7–8	412	37	9.0	91.0
8–9	375	48	12.8	87.2
9–10	327	79	24.2	75.8
10–11	248	92	37.1	62.9
11–12	156	88	56.4	43.6
12–13	68	64	94.1	5.9
13–14	4	1	—	—
14–15	3	3	—	—
Total:	(5239)	655	12.5%	87.5%

SOURCE: Original data taken from Murie (1944).
[a] Percentages of q_x carried one place beyond statistical significance.

more than seven years; the average annual mortality rate is 12.5 percent, and therefore the annual survival rate is 87.5 percent (Table 7–4). Sources of error in this application of the composite life table come from the fact that one is less apt to find the skulls of younger animals, where there is less ossification, and hence more deterioration from exposure. Not all vertebrates, of course, are as long-lived as mountain sheep. Tinkle and his colleagues have studied population turnover and survivorship of the lizard *Uta stansburiana stejnegeri* in western Texas by marking studies. From 1961 to 1965, every juvenile lizard in two 2.07-acre study areas was marked as soon as possible after hatching in order to obtain accurate information on natality. Multiple recaptures of these individuals until their disappearance provided reliable survivorship data (Tinkle, 1967). The data on population turnover and on mortality are summarized in Table 7–5. An average of only 9 percent of the resident adults and only 20 percent of the young survive for more than one breeding season. The juvenile survival estimate is almost certainly high because some lizards are not captured for the first time until they are approaching adult size, when their death rate is much lower. Consequently, a greater proportion of these late-marked individuals survive; many may be immigrants rather than offspring produced in the two study areas.

More reliable data may be derived by determining the rate at which *Uta* lizards marked at hatching disappear from the population

Table 7–5 Percentage of juvenile and adult *Uta stansburiana* surviving from one season to the next. Both groups must have lived to the onset of the reproductive season in March to qualify as survivors, and juveniles must have been of reproductive size by that time.

YEAR:	1960		1961		1962		1963		1964	
AREA:	I	II	I	II	I	II	I	II	I	II
Total adults	46	—	64	72	54	64	39	57	81	110
% surviving to next season	9	—	8	10	7	9	8	11	6	9
No. of young produced	184	—	455	489	238	383	318	498	269	310
% surviving to maturity	19	—	22	12	17	18	28	24	19	18

SOURCE: Tinkle (1967).

Table 7–6 Life-table data for *Uta stansburiana stejnegeri.*[a]

AGE INTERVAL (WEEKS)	FEMALES l_x	d_x	q_x	MALES l_x	d_x	q_x
0–1	1000	350	0.35	1000	380	0.38
1–2	650	52	0.08	620	56	0.09
2–3	598	90	0.15	564	73	0.13
3–4	508	76	0.15	491	83	0.17
4–5	432	43	0.10	408	57	0.14
5–6	389	66	0.17	351	70	0.20
6–7	327	20	0.06	281	45	0.16
7–8	307	25	0.08	236	21	0.09
8–9	282	11	0.04	215	19	0.09
9–10	271	19	0.07	196	18	0.09
10–11	252	20	0.08	178	16	0.09
11–12	232	5	0.02	162	2	0.01
12–13	227	9	0.04	160	0	0.00
13–14	218	7	0.03	160	6	0.04
14–15	211	13	0.06	154	5	0.03
15–16	198	2	0.01	149	0	0.00
16–17	196	8	0.04	149	1	0.01
17–18	188	8	0.04	148	1	0.01
18–19	180	13	0.07	147	3	0.02
19–20	167	2	0.01	144	4	0.03
36–37	131	12	0.09	103	4	0.04
43–44	86	3	0.04	72	6	0.09
50–51	56	7	0.12	43	9	0.20

SOURCE: Tinkle (1967).
[a] The q_x was computed from actual time between hatching and disappearance from the population for all lizards marked at a length of 24 mm or less.

and then calculating the age-specific rather than simply crude mortality (Tinkle, 1967). These data can be summarized in the form of a composite life table (Table 7–6), and the mortality rate (q_x) determined from the assumed time of death (time of last capture) for 482 males and

585 females marked at hatching over three generations (1961–1964). Females lay the first of three clutches of eggs at 36–37 weeks of age. From these data, approximately 13 percent (131/1000) females survived to maturity at age 36–37 weeks; according to the crude mortality method of estimation in Table 7–5, 20 percent of both sexes of young did so. It is also clear that males have a higher mortality rate than females, perhaps because of their more frequent long-distance emigration. Other survivorship data indicate that the greatest known longevity of utas is 100 weeks for a single female and 97 weeks for a single male. Less than 1 percent of the males live 57 weeks; less than 1 percent of females live 60 weeks. The mean life expectancy for hatchlings is about 18.5 weeks, the median only about four weeks (Tinkle, 1967). As can be observed in the q_x column of Table 7–6, mortality is about constant for age, with the exception of the first few weeks when mortality is high. A nearly complete annual turnover in the population is the rule.

Survivorship curves

Age-specific mortality can also be expressed by plotting the number of individuals in a particular age cohort against time, in the form of a *survivorship curve.* If all animals or plants of a particular species could express their physiological longevity uninhibited by environmental or deleterious intrinsic factors, they would have an extremely convex survivorship curve with a sharp terminal angle at maximum life-span, like that in Figure 7–1. Starved fruit flies of the same genotype raised

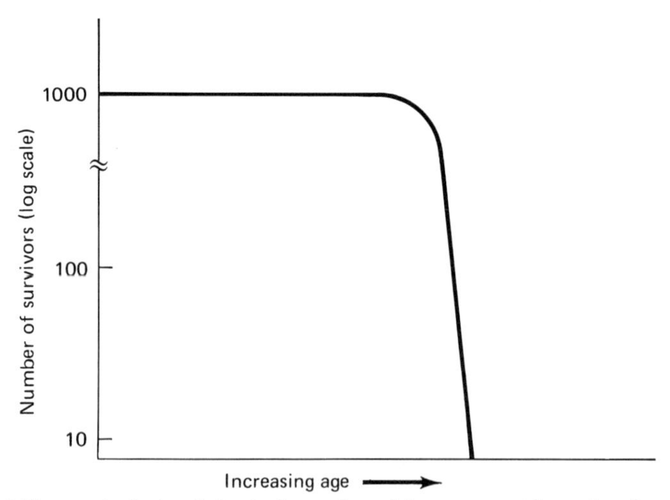

Figure 7–1 Theoretical physiological survivorship curve with optimal environmental conditions. All individuals in the cohort survive until maximum physiological longevity is reached. This curve ignores juvenule mortality, which is almost always substantial in both animal and plant populations.

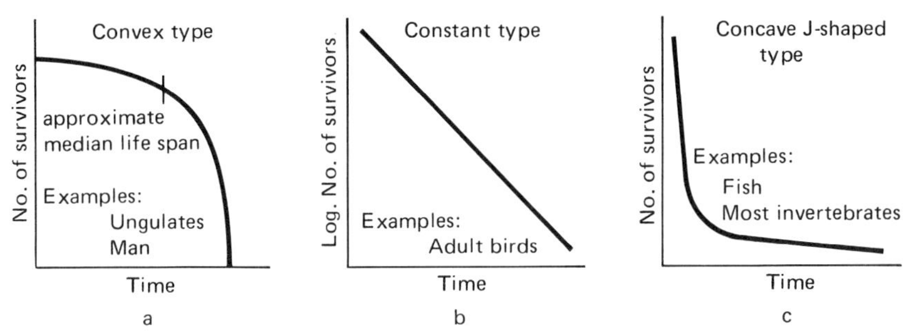

Figure 7–2 The three basic types of survivorship curves. The arithmetic or logarithmic number of survivors on the vertical axis is plotted against age on the horizontal axis. (a) Convex type, in which the organisms live out most of the full physiological life span of the species. Life expectancy declines with age. (b) Constant type (or concave, if plotted arithmetically), in which the rate of survival tends to be constant at all age levels and life expectancy does not decline with age. (c) Concave J-shaped type, wherein there is high mortality early in life and then life expectancy increases with age of the survivors.

from similarly aged batches of eggs die after almost the same interval of time. Ideal conditions are rarely met and thus we find different types of ecological survivorship curves in nature.

Natural survivorship curves fall into three main types (Figure 7–2), with all degrees of intermediacy. In the *convex* type, there is high survivorship through most of the possible life-span, then a sudden increase in mortality as maximum physiological longevity is approached. The older the animal becomes, the less the life expectancy, as the latter declines with age. Man, mountain sheep, and other mammals typically show such a curve.

In the *constant* type of survivorship curve, the rates of survival for each age class are similar and thus the cohort numbers show an exponential rate of decline. Plotted arithmetically, the survival curve is concave. The lizard *Uta stansburiana* shows such a curve when life-table data (Tinkle, 1967) are graphed (Figure 7–3). The adult stages of birds and many rodents also demonstrate this type of survivorship curve; once the fledging stage is passed or the babies are weaned, mortality may occur at an equal rate at any phase of development—a straight-line survivorship relationship if the logarithm of the number of survivors is plotted against increasing age.

In the *concave J-shaped* survivorship curve, mortality rates are extremely high during early life, producing a sharp rate of population decline. Then the survivorship curve tends to level out, and life expectancy increases with the older members of the population. Invertebrates such as oysters and vertebrates such as fish and frogs tend to show this type of survivorship.

The relationship of these three types of survivorship curves to establishing a particular age structure in a population may be best seen

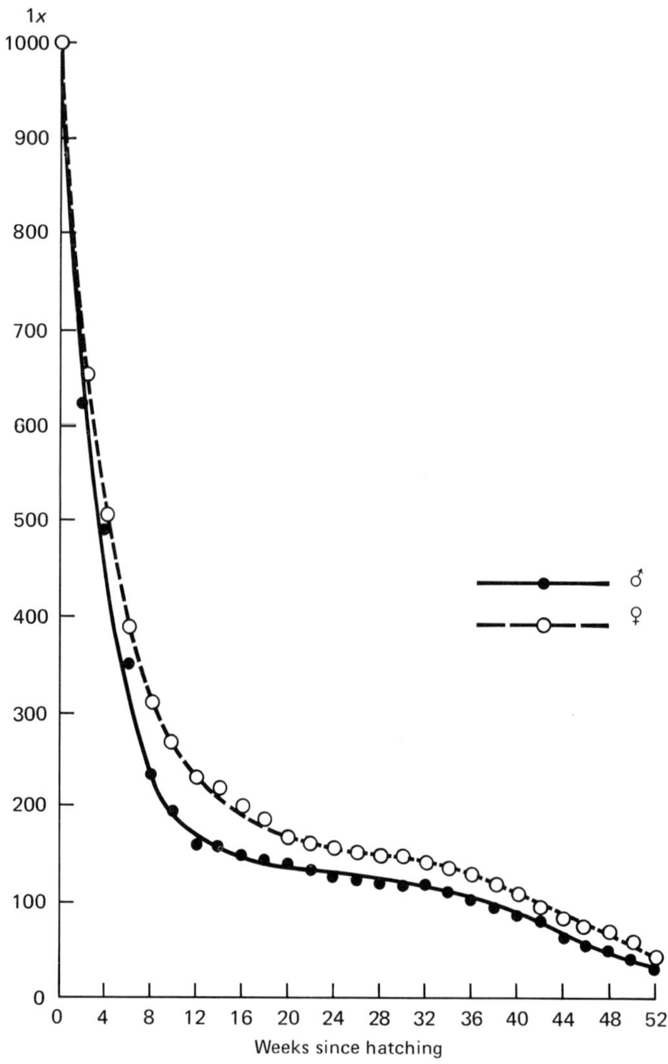

Figure 7–3 Survivorship curves for males (solid line and circles) and females (dotted line and open circles) of the lizard *Uta stansburiana* in western Texas, based on life table data in Table 7–6. (Source: D. W. Tinkle, 1967, *In* W. W. Milstead, ed., *Lizard Ecology: A Symposium*, Univ of Missouri Press, Columbia, p. 22.)

if we carry out life-table analyses of the three generalized cases. We will consider three hypothetical populations of 100,000 animals. From each, we shall take a 10 percent sample of 10,000 individuals and we shall look at its mortality and survivorship using the time-specific and the dynamic life-table analyses. With the dynamic approach, the mortality data are generated by returns (e.g., bird bands) coming from the sample of the number dying. As will be seen in a comparison of

Table 7–7 with Tables 7–8 and 7–9, the only occasion when one can use the dynamic approach to the sample of living individuals is with a population showing a constant rate of mortality (Table 7–7). In the time-specific analysis, the morality data can be estimated from the sample of living individuals at each age interval. However, as just stated, the application of the dynamic life table to the sample of living individuals gives incorrect estimates of mortality in all cases except constant survivorship (Table 7–7), such as with adult bird populations where it has been successfully applied by Lack.

The age structure of a hypothetical population with a *constant yearly* q_x of 60 percent is given in Table 7–7. This population would have an exponentially declining (straight-line) survivorship curve,

Table 7–7 Age structure of a hypothetical population with a constant annual q_x of 60 percent, giving an exponentially declining survivorship curve.

AGE INTERVAL (x)	NUMBER LIVING	NUMBER DYING	10% SAMPLE: NUMBER LIVING	10% SAMPLE: NUMBER DYING
0–1	100,000	60,000	10,000	6,000
1–2	40,000	24,000	4,000	2,400
2–3	16,000	9,600	1,600	960
3–4	6,400	3,840	640	384
4–5	2,560	1,536	256	154
5–6	1,024	614	102	61
6–7	410	246	41	25
7–8	164	98	16	10
8–9	66	40	7	4
9–10	26	26	3	3

LIFE-TABLE ANALYSES

AGE INTERVAL (x)	LIFE-TABLE ANALYSIS FROM SAMPLE OF LIVING: TIME-SPECIFIC l_x	d_x	q_x	LIFE-TABLE ANALYSIS FROM SAMPLE OF LIVING: DYNAMIC l_x	d_x	q_x	LIFE-TABLE ANALYSIS FROM SAMPLE OF DYING: DYNAMIC l_x	d_x	q_x
0–1	10,000	6,000	60%	16,665	10,000	60%	10,000	6,000	60%
1–2	4,000	2,400	"	6,665	4,000	"	4,000	2,400	"
2–3	1,600	960	"	2,665	1,600	"	1,600	960	"
3–4	640	384	"	1,065	640	"	640	384	"
4–5	256	154	"	425	256	"	256	154	"
5–6	102	61	"	169	102	"	102	61	"
6–7	41	25	"	67	41	"	41	25	"
7–8	16	9	"	26	16	"	16	9	"
8–9	7	4	"	10	7	"	7	4	"
9–10	3	3	"	3	3	"	3	3	"
	16,665	10,000	60%	27,757	16,665	60%	16,665	10,000	60%

SOURCE: Data from James J. Dinsmore.

Table 7–8 Age structure of a hypothetical population with a convex-shaped survivorship curve.

AGE INTERVAL (x)	NUMBER LIVING	NUMBER DYING	10% SAMPLE NUMBER LIVING	10% SAMPLE NUMBER DYING
0–1	100,000	5,000	10,000	500
1–2	95,000	5,000	9,500	500
2–3	90,000	5,000	9,000	500
3–4	85,000	15,000	8,500	1,500
4–5	70,000	15,000	7,000	1,500
5–6	55,000	15,000	5,500	1,500
6–7	40,000	20,000	4,000	2,000
7–8	20,000	20,000	2,000	2,000
8–9	0	0	0	0

LIFE-TABLE ANALYSES

AGE INTERVAL (x)	Sample of living, time-specific l_x	d_x	q_x	Sample of living, dynamic l_x	d_x	q_x	Sample of dying, dynamic l_x	d_x	q_x
0–1	10,000	500	5%	55,500	10,000	18%	10,000	500	5%
1–2	9,500	500	5.3%	45,500	9,500	20.9%	9,500	500	5.3%
2–3	9,000	500	5.6%	36,000	9,000	25.0%	9,000	500	5.6%
3–4	8,500	500	5.9%	27,000	8,500	31.5%	8,500	500	5.9%
4–5	7,000	1,500	21.4%	18,500	7,000	37.8%	7,000	1,500	21.4%
5–6	5,500	1,500	27.3%	11,500	5,500	47.8%	5,500	1,500	27.3%
6–7	4,000	2,000	50.0%	6,000	4,000	66.7%	4,000	2,000	50.0%
7–8	2,000	2,000	100%	2,000	2,000	100%	2,000	2,000	100%
	55,500	10,000	18.0%	202,000	55,500	27.5%	55,500	10,000	18.0%

SOURCE: Data from James J. Dinsmore.

which would be concave if the life-table data in the lower part of the table were plotted like Figure 7–3. A fixed proportion of the animals dies each year.

The age structure and life-table analyses of a hypothetical population with a *convex-shaped survivorship curve* are outlined in Table 7–8. Mortality is very low (5–5.9 percent) until the end of the fourth year (except in the erroneously applied dynamic analysis of the sample of living individuals). Then it increases dramatically in the following four years and the cohort goes extinct two years before the population shown in Table 7–7, which had the seemingly tremendous but *constant* annual mortality rate of 60 percent. Yet this present population with a convex-shaped survivorship curve had only an average annual mortality rate of 18 percent.

Finally, the age structure and life-table analyses of a hypothetical population with a *concave J-shaped survivorship curve* is treated in Table 7–9. Again, the application of the dynamic life-table analysis to the sample of living individuals produces erroneous results in the absence of constant survivorship. Although a tremendous number of

Table 7–9 Age structure of a hypothetical population with a concave J-shaped survivorship curve.

AGE INTERVAL (x)	NUMBER LIVING	NUMBER DYING	10% SAMPLE NUMBER LIVING	10% SAMPLE NUMBER DYING
0–1	100,000	90,000	10,000	9,000
1–2	10,000	8,000	1,000	800
2–3	2,000	1,000	200	100
3–4	1,000	500	100	50
4–5	500	200	50	20
5–6	300	100	30	10
6–7	200	50	20	5
7–8	150	50	15	5
8–9	100	25	10	3
9–10	75	75	8	8

LIFE-TABLE ANALYSES

AGE INTERVAL (x)	LIFE-TABLE ANALYSIS FROM SAMPLE OF LIVING: TIME-SPECIFIC			LIFE-TABLE ANALYSIS FROM SAMPLE OF LIVING: DYNAMIC			LIFE-TABLE ANALYSIS FROM SAMPLE OF DYING: DYNAMIC		
	l_x	d_x	q_x	l_x	d_x	q_x	l_x	d_x	q_x
0–1	10,000	9,000	90%	11,433	10,000	87.5%	10,000	9,000	90%
1–2	1,000	800	80%	1,433	1,000	69.8%	1,000	800	80%
2–3	200	100	50%	433	200	46%	200	100	50%
3–4	100	50	50%	233	100	43%	100	50	50%
4–5	50	20	40%	133	50	38%	50	20	40%
5–6	30	10	33%	83	30	36%	30	10	33%
6–7	20	5	25%	53	20	38%	20	5	25%
7–8	15	5	33%	33	15	45%	15	5	33%
8–9	10	2	20%	18	8	44%	10	2	20%
9–10	8	8	100%	10	10	100%	8	8	100%
	11,433	10,000	87.5%	13,862	11,433	82.5%	11,433	10,000	87.5%

SOURCE: Data from James J. Dinsmore.

individuals die during the first three years (90 percent of the population in the first year and 99 percent by the end of the third year), the remaining adults actually persist very well until the maximum life-span is reached at 10 years. Greater numbers (75) survive out of the original 100,000 in the population than with either of the other two age structures (Tables 7–7 and 7–8). The mortality rate drops steadily (from 90 to 20 percent) until the last year of life, when it of course becomes 100 percent as the original cohort becomes extinct.

We should also take special note here of the relationship between *age distribution* and *population growth*, discussed briefly in Chapter 5. Age distribution in a population reflects the direction of population dynamics, for the growth of populations must be a function of (1) an increased rate of recruitment or (2) a decrease of age-specific mortality in younger cohorts. Age ratios are functions of birth rates. If mortality increases, the population decreases but the age ratios remain the same. The three general types of age distributions that appear in stationary, increasing, and decreasing populations are depicted in Table 7–10. Stable age distributions are achieved by any population in a steady environment regardless of whether the population is increasing, decreasing, or holding steady in size. Under particular environmental conditions, each population has its own stable age distribution.

The *life table* or *stationary* age distribution appears in the population with balanced birth and death rates, having a zero rate of increase (Table 7–10, left). Fifty percent of the hypothetical population depicted in Table 7–10 is in the first-year age class, and each successive age class has half as many members up to the maximum life-span. Populations with life-table age distributions will have a more even distribution of individuals in the several early age classes than will those with Malthusian age distribution.

The *Malthusian* or *stable* age distribution appears in the population where the birth rate exceeds the death rate and the population is increasing in size (Table 7–10, center). A very high percentage of individuals are in the early age group compared to older groups. Although the birth and death rates are fixed, they are not balanced. Yet a stable age distribution is maintained with that particular rate of increase. If the rate of increase changes, the age ratio will stabilize at a new frequency. A majority of the population is quite young.

In a *decreasing* population, the age distribution involves a greater number of individuals in the middle age classes than in the earliest and oldest class (Table 7–10, right). We can see this more readily in graphical representations of age structure in a population.

Table 7–10 Age distribution in a representative sample of 1000 individuals drawn from three hypothetical populations: stationary, increasing, and decreasing in size.

	STATIONARY OR "LIFE-TABLE" POPULATION[a]		INCREASING OR "MALTHUSIAN" POPULATION[b]		DECREASING POPULATION[c]	
AGE (YEARS)	NUMBER IN YEAR-CLASS	% OF TOTAL POPULATION	NUMBER IN YEAR-CLASS	% OF TOTAL POPULATION	NUMBER IN YEAR-CLASS	% OF TOTAL POPULATION
1	500	50	666	67	333	33
2	250	25	222	22	222	22
3	125	13	74	7	148	15
4	63	6	25	2	99	10
5	31	6	8	2	66	13
6	16	6	3	2	44	13
7	8	6	1	2	29	13
8	4	6	1	2	20	13
≥9	3	6	0	2	39	13

SOURCE: Data from James J. Dinsmore.
[a] Assuming that $R_b = 1.0$/year, $R_d = 1.0$/year, $R_i = 0.0$/year.
[b] Assuming that $R_b = 2.0$/year, $R_d = 1.0$/year, $R_i = 1.0$/year.
[c] Assuming that $R_b = 0.5$/year, $R_d = 1.0$/year, $R_i = -0.5$/year.
In all three examples, mortality between age classes was assumed to be identical.

Age pyramids

Age distributions can be represented diagrammatically by *age pyramids* (Figure 7–4). The pyramid is a compilation of bar graphs of the ratio of one age group to another in the population, expressed as a percentage of the total number of individuals. Often, the proportion of males to females in each age group of a population are shown to either side of a center line in the pyramid (males on the left, females on the right).

A pyramid with a broad base and a narrow top indicates a large juvenile age group and a very small old age group, a condition characteristic of young, rapidly growing populations. As increasing numbers enter the reproductive period, rapid population growth would be predicted.

When the age structure of a population has a medium percentage of young individuals in the population and the ratios of one age group to another are about the same except for the oldest classes, we have a stable population. No significant population growth will occur.

If a large number of individuals are in the reproductive age groups but few are present in the juvenile and oldest age classes, we have a declining population. With few individuals being born and therefore few being available to contribute to future population growth, the population will continue to increase its proportion of old individuals and hence will decrease in numbers.

The age structure of plant populations has been rarely studied to the same degree as animal populations, a notable exception being the excellent work by Harper (e.g., Harper and Sagar, 1953) on buttercups and other higher plants. The concept is widely but indirectly employed in forestry, where tree stands are analyzed by diameter classes

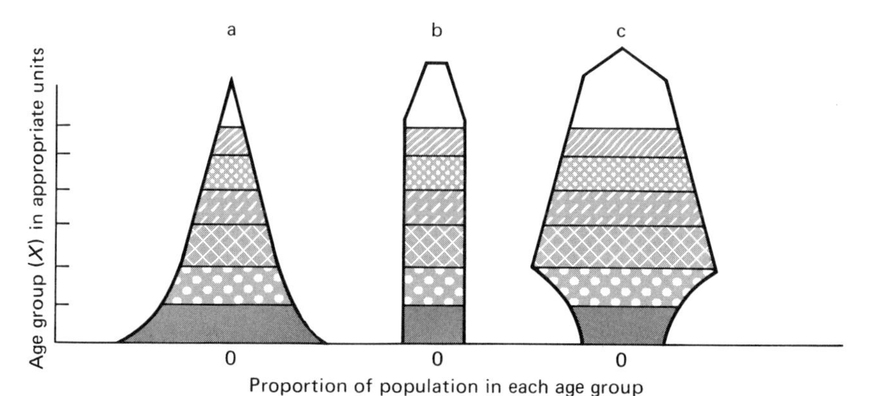

Figure 7–4 Age structures of (a) a rapidly increasing or "Malthusian" population; (b) a stationary or "life table" population; and (c) a declining population. (Source: J. T. Giesel, 1974, *The Biology and Adaptability of Natural Populations,* Mosby, St. Louis, p. 61.)

and divided into even-aged stands and uneven-aged stands on that basis. Unfortunately, there is little actual correlation between trunk diameter and age. Levin (1973) used age structure to good effect in analysis of interspecific hybridization of *Liatris* composite species in Illinois. The age structures of three species of *Liatris* and their hybrids at Zion Prairie were assessed by making radial sections of the corms of large samples of these perennial plants. Growth rings could be readily scored. The age structures of the three species have several features in common (Figure 7–5). The plants begin flowering at about seven years of age and few live beyond 20 years. Most of the plants are in their early teens. The mean ages of the species are *L. spicata,* 13.08 years; *L.*

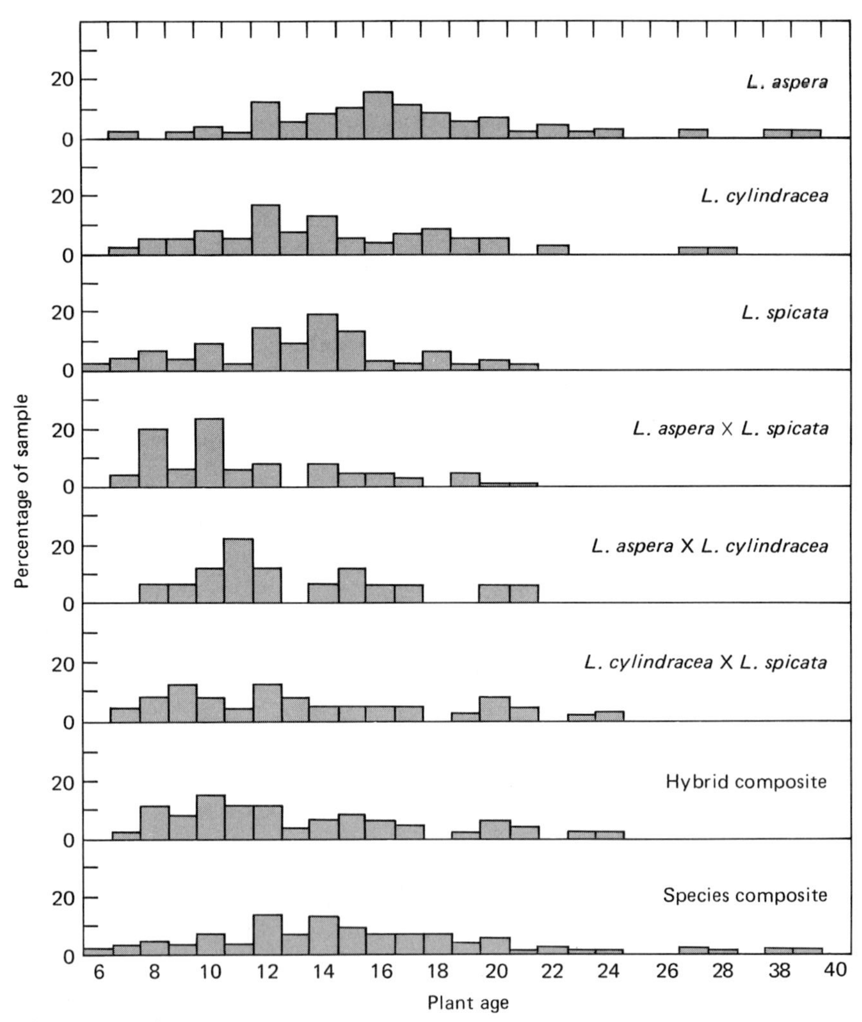

Figure 7–5 The age structure (in years) of three *Liatris* plant species and their hybrids at Zion, Ill. (Source: D. A. Levin, 1973, *Evolution, 27:* 532–535.)

Table 7–11 Summary of the age structure of a hybrid swarm in *Liatris* plants at Zion, Illinois.

PARAMETER	*L. aspera* × *L. spicata*	*L. spicata* × *L. cylindracea*	*L. aspera* × *L. cylindracea*	*L. spicata*	*L. cylindracea*	*L. aspera*
Sample size	46	26	18	86	100	98
Mean age (years)	11.02	12.76	13.50	13.08	14.10	16.15
Age range (years)	7–19	7–24	8–21	6–21	7–28	7–40
Variance (years)	11.71	39.06	19.58	10.01	17.34	27.34

SOURCE: Levin (1973).

cylindracea, 14.10 years; *L. aspera,* 16.15 years (Table 7–11). The difference between *spicata* and *aspera* is statistically significant, whereas other interspecific differences are not. The age structure of reproductive hybrids is also depicted in Figure 7–5. The similar age profiles yield preteen modal classes, and the mean ages of all the hybrids are less than any of the parental species (Table 7–11).

The data in the composite histograms for the age structures of the species and hybrids (Figure 7–5, bottom) show that the modal age class for reproductive hybrids is 10 years, whereas that for the species is 12. Some 56 percent of the hybrids are preteen, as opposed to 39 percent of the species. On the average, then, hybrids reach reproductive maturity sooner than their parents and this earlier flowering seems to be a manifestation of general precocity and hybrid vigor. However, further inspection of the mean and maximum ages shows that the longevity of the pure species individuals is greater than that of the hybrids. Without further information on seed production as a function of age and the stability of the age structure and population size, it is impossible to draw more specific conclusions on the fitness of adult hybrids relative to their adult progenitors. But this study does indicate the potential usefulness of the age-structure and life-table approach in offering insights into the dynamics of population structure and hybridization phenomena in evolutionary biology.

THE EFFECT OF SEX RATIO ON A POPULATION

Just as age structure reflects population growth and mortality, sex ratios in a population change with age and differential mortality. The *primary sex ratio* exists at the time of conception, whereas the *secondary sex ratio* is that proportion of males and females which exists at the time of birth. The *tertiary sex ratio* exists among the young juvenile age groups and is now often called "the ratio of juveniles." The *quaternary sex ratio* exists among the adult population. Sex ratios are altered by differential mortality between males and females at different times in the life history, and the same level of sex ratio (e.g., quaternary) may differ between populations of the same species.

In West and East Africa, the butterfly *Acraea encedon* occurs in well-defined populations that are often predominately female (Owen, 1966). Thus in field samplings during 1963–1965 in the Kampala-Entebbe area of Uganda, Owen found six populations that were from 0.6 percent male to 38.6 percent male—some colonies having less than one male per 100 females. Elsewhere, the sex ratio was near 1:1. The field samples were accurate indicators of male frequency in these populations, for rearings from wild-caught females (including copulating pairs) and wild-collected eggs and larvae yielded 1.1 percent males among the reared adults. A collection made in this same Ugandan area in 1909 to 1912 contains about a 1:1 ratio of males and females. As this butterfly has two to five generations per year, the change in sex ratio could have arisen during the past 100–250 generations (Owen, 1966). Later sampling (Chanter and Owen, 1972) showed that fairly rapid changes from high-female to normal sex ratio could be found in population samples taken a year apart, but the low sex ratios of some of these populations are now known to have persisted for at least 45 breeding generations (Owen et al., 1973a). The species is not parthenogenic, and to produce eggs a female must mate with one of the rare males, a process which lasts several hours (Owen et al., 1973b). By breeding the butterflies Owen showed that there are two kinds of female, one producing females only, the other producing males and females in the expected 1:1 ratio. The best explanation at present (Chanter and Owen, 1972; Owen et al., 1973a) is that the sequence of emergence in normal broods could be partly responsible for the maintenance of stable sex-ratio equilibria in predominantly female populations. It has been suggested by Owen (1966) that the predominantly female populations are of recent origin and have perhaps resulted from hybridization between previously isolated populations brought about by human disturbance of the environment, and especially by the rapid spread of agriculture. One evolutionary effect of the rarity of the males is that those populations remain genetically isolated, and each population is characterized by a particular frequency of color forms in this polymorphic species.

Sometimes a biased sex ratio is clearly compensated for by increased mating expectancies. In most Odonata (dragonflies and damselflies), females are more numerous than males in both larval and adult stages. Johnson (1964) showed that the female preponderance among breeders of the Arizona damselfly *Enallagma praevarum* (where the male percentage ranges from 20.8 to 42.9) was offset by a greater mating expectancy among males. The daily survival rates were 0.80 and 0.82, corresponding to average life expectancies of 5.0 and 5.5 days for males and females, respectively. The similarity of the values suggests that individuals of one sex could not mate more frequently than could individuals of the opposite sex by virtue of greater longevity. Relative mating expectancies (the number of matings for individuals of each sex during their average life expectancies) were estimated by determining the number of different (marked) males and females contributing to 251 successful matings. The estimates were 3.0 matings per male and 1.9 per female. The population sex ratio was therefore suggested to be a function of mating expectancies, such that the low male proportion might not be disadvantageous under these circumstances.

Sex ratio changes with age are particularly well documented in populations of mammals and birds. As a general rule among mammals, the sex ratio in older cohorts becomes more imbalanced favoring females, while in birds the sex ratios in older cohorts favor males. These general trends are shown in the sex ratios of mammal and bird species listed in Tables 7–12 and 7–13. Thus, whereas both groups start with a secondary sex ratio (at birth or hatching) close to 1:1, it notably shifts during development to a strong preponderance of females in mammals (as much as 77 out of every 100 adults in elk) and to a strong preponderance of males in birds (as much as 66 out of every 100 adults in starlings).

The explanation for these shifts in sex ratio is not at all clear, but probably centers on the genetic control of sex determination, the social system, and differences in physiological activity of the two sexes in the two classes. Mammals and birds have opposite types of sex determination; an XY sex chromosome constitution produces males in

Table 7–12 Some sex ratio changes exhibited by mammals.

SPECIES	JUVENILES (♂:♀)	ADULTS	SOURCE	
Brown rat	51:49 (1897)	41:59 (982)	Leslie et al. (1952)	
Muskrat	57:43 (18,832)	50:50 (5250)	Beer and Trux (1950)	
Cottontail	50:50 (2459)	46:54 (427)	Unpublished data, Wisconsin	
	PRENATAL	7–8 MONTHS	ADULT	SOURCE
Elk	53:47 (248)	52:48 (187)	23:77 (2232)	Cowan (1950)[a]
Mule deer			35:65 (661)	
Mountain goat			43:57 (272)	
Mountain sheep			44:56 (1450)	
	PRENATAL	1–14 DAYS	3–6 MONTHS	
Mule deer	53:47 (2299)	55:45 (808)	53:47 (13,046)	Robinette et al. (1957)

[a] Cowan's data refer to unhunted overpopulations where winter food was scarce.

Table 7–13 Some sex ratio changes in relation to the age of certain birds.[a]

SPECIES	SEX RATIOS (MALE:FEMALE) JUVENILE	ADULT	SOURCE
Hungarian partridge	50:50 (5)[b]	56:44 (7)	Keith (1958)
Bobwhite quail	51:49 (5)	62:38 (5)	Hickey (1955)
California quail	50:50	58:42	Emlen (1940)
Ruffed grouse	50:50	54:46	Dorney (1960)
Sharp-tailed grouse	49:51 (5)	55:45 (4)	Keith (1958)
STARLING	52:48	66:34	DAVIS (1959)
Mallard	51.2 (43,627)[c]	63.8 (24,526)	Bellrose et al. (1961)
Black duck	48.6 (3229)	61.3 (1183)	Bellrose et al. (1961)
Pintail	51.6 (10,325)	54.9 (9071)	Bellrose et al. (1961)
Canvasback	44.0 (5240)	56.8 (940)	Bellrose et al. (1961)
Scaup	49.7 (2924)	61.4 (1471)	Bellrose et al. (1961)

[a] The term juvenile refers to birds during their first fall or winter.
[b] Number of separate studies in parentheses.
[c] Stated as percent males; sample size in parentheses.

Table 7–14 California quail sex ratios from New Zealand.[a]

AGE (WEEKS)	SEX RATIO
6–13	34:66 (222)[b]
14–18+	55:45 (232)
Adults	66:34 (299)

[a] The author (Williams, 1957) believed that this trend was due to an initially higher mortality among males, which gradually shifted to females.
[b] Sample size shown in parentheses.

mammals and females in birds. This XY chromosomal constitution may be at least partly responsible for the biased loss of males in older cohorts of mammals and of females in older birds. Each gene on the X and Y chromosomes may be expressed, whereas in the XX-determined sex, heterozygosity can cover any deleterious effects of single recessive alleles. Thus, the embryo and later the adult with an XY chromosome dosage in its cells may be slightly more susceptible to physiological damage than the XX-carrying animal. We should note, however, that among mammals there is a tendency for a surplus of males to be born (Table 7–12), and among birds the ratio normally is either equal or there is a slight preponderance of females at hatching (Table 7–13, but contrast the data in Table 7–14). In terms of social systems common to these two vertebrate classes, it is possible with polygamy to have an excess of females without leaving females unmated; one male can mate with a harem of females, particularly in the ungulates listed in the bottom half of Table 7–12. Generally, birds are more paired in social behavior and less polygamous than mammals in the breeding season (see Chapter 9). Finally male mammals tend to be more active than the females, have larger home ranges, and generally have greater metabolic demands, perhaps leading to greater mortality rates. This is particularly true in territorial and other aggressive encounters. With birds, females tend to help the males defend territories and both sexes commonly participate in brooding the eggs and feeding the young. The females' extra metabolic demands of egg production may contribute to earlier mortality. It is noteworthy that in waterfowl species the sex ratios among the juveniles (the term here refers to birds during their first fall or winter) are about equal in the fall, whereas by the following fall the adults have striking sex-ratio differences, with up to 65 percent males (see Table 7–13). Keith (1958) found that between May 6 and July 15, the average male mortality rate in young waterfowl was 2 percent, whereas females experienced 8 percent mortality.

Two other trends to note in mammals are that young females with their first offspring tend to bear a preponderance of males at birth, and the younger the mother, the greater the chances that she will have predominantly male offspring. These trends are shown in white-tailed deer (Figure 7–6), where a 1½-year-old doe in the northeastern United States bears about 61 percent males but a 4½-year-old doe bears less than 50 percent males (McDowell, 1959). In laboratory colonies of brown rats, King (1939) found that 90-day-old females bear about 51 percent males while females over 550 days of age bear less than 47 percent males (Figure 7–7).

Sex-ratio phenomena are particularly important in population dynamics because where a reduction in the percentage of females occurs, there is likely to be a reduction in the rate of population growth. This effect is dependent upon the proportion that older cohorts represent in the population. Changes in older reproductively active cohorts in species with rapid turnover, affect the population greatly. Thus, if older cohorts show sex-ratio changes, these can be quite influential on population growth. A preponderance of males among older ungulates, for instance, would greatly depress population growth because the usual social pattern is one reproductive male with a harem of females. In waterfowl or small game species, the same effects of diminishing rates of population growth are noted when marked sex-ratio imbalance occurs in the juvenile stage.

For example, California quail have been introduced into New Zealand and breed there quite readily in the wild. Williams (1957, 1963) found that sex-ratio changes in these quail are correlated with age-ratio changes (Table 7–14). The greater the deviation in sex ratio, the greater the drop in the age ratio and the lower the rate of population gain. When the percentage of females in the fall populations drops in

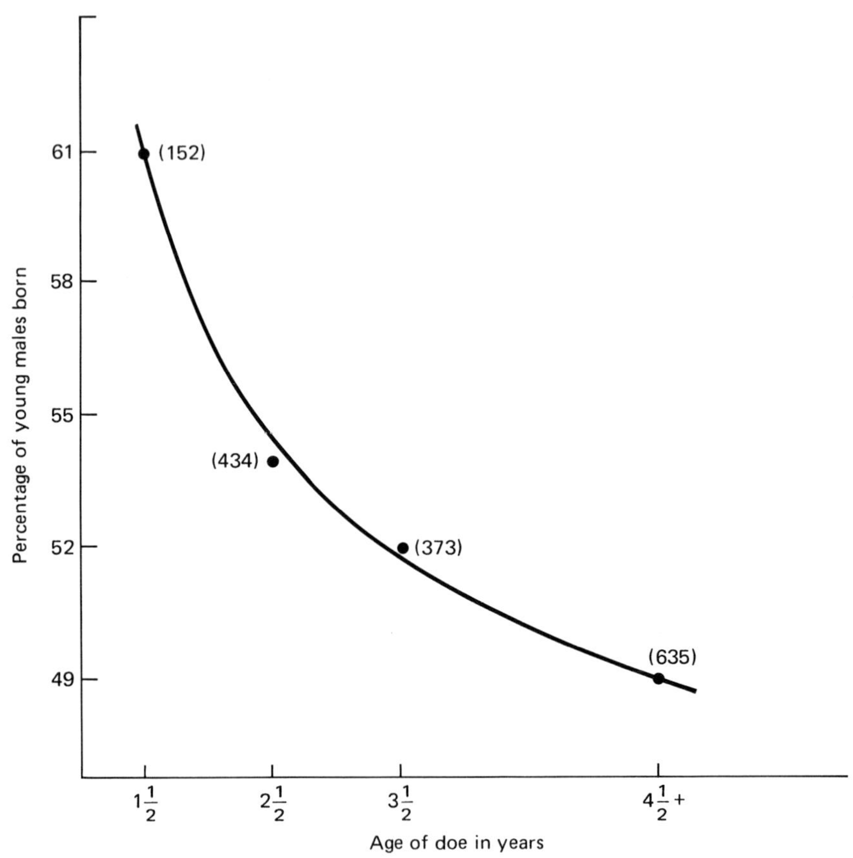

Figure 7–6 The effect of maternal age upon sex ratio of offspring in white-tailed deer in the northeastern United States. (Data from J. J. Dinsmore.)

any one year, then the juvenile/adult age ratio in the samples taken by hunters drops and more adults are present (Figure 7–8). When the percentage of females present in the fall populations increases, the juvenile/adult age ratio increases towards more juveniles.

The linkage of sex ratio and age ratio seems reasonably clear when one considers this example. An adult female quail produces approximately 2.5 young successfully. Therefore, 60 males and 40 females (at 2.5 young/female) give 100 juveniles. The juvenile/adult age ratio that fall is 100/100, or 1.0. But in a year when 60 males and 20 females (at 2.5 young/female) breed and produce 50 juveniles, the juvenile/adult age ratio will be 50/80, or 0.62 juveniles/adult. A change in age ratio indicates a loss of adult hens, which affects the population growth in the next breeding season and is also reflected in the adult sex ratio. Increased sex ratio of males may also lead to harrassment of the hens, causing nest abandonment and a reduced rate of population gain.

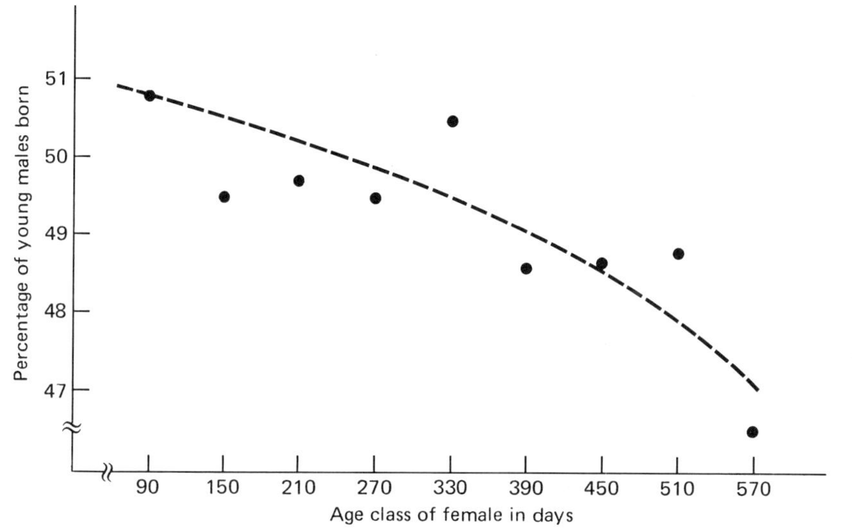

Figure 7–7 The effect of maternal age upon sex ratio of offspring in brown rats. (After H. D. King, 1939, *Amer. Anat. Mem., 17:* 1–72.)

Thus an imbalance in the sex ratio can have marked effects on population growth and density. The most productive populations tend to have a more balanced sex ratio and a high ratio of juveniles.

The preceding example is characteristic of the more general case in heterogametic species where selection appears to adjust even the primary sex ratio to favor maximum production of later offspring. Mendelian sex determination mechanisms predict one-to-one ratios, but as we have seen, observed sex ratios frequently differ from 1:1 by small, but significant, amounts. These modified sex ratios must involve genetic modifiers that alter expected Mendelian segregation. For instance, in the African butterfly *Acraea encedon* Owen (1970) postulates that the inheritance of the all-female broods is through a Y-linked gene (in butterflies and birds the female is XY) causing meiotic drive in the Y chromosome. The selective advantage favoring a modified sex ratio is believed to result from an optimization effect: selection favors a sex ratio which maximizes the product, $d \cdot s$, of the number of daughters in the population times the number of sons in the population (MacArthur, 1965). Analysis of the effects of differential viability of the two sexes shows that the stage in the life cycle to which MacArthur's maximization principle applies is that point at which male mortality and female mortality both cease to differ between families possessing different sex ratios (Spieth, 1974). Differences in viability that depend solely on an individual's sex will not affect the primary sex ratio at conception; instead, differential fecundity and viability leads to the production of an unequal primary sex ratio. For example,

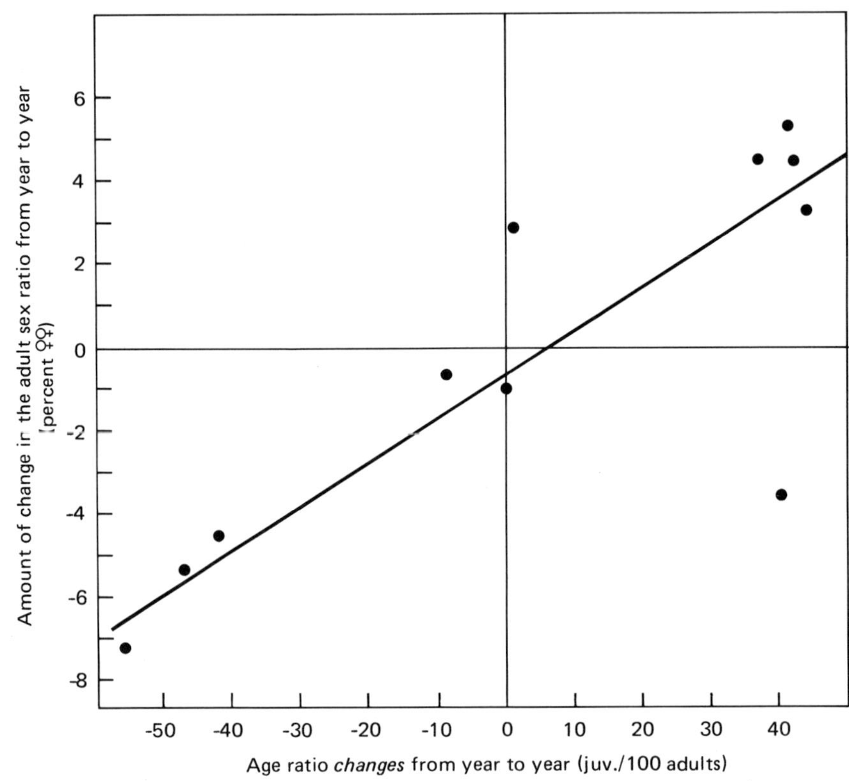

Figure 7–8 The relationship between adult sex ratio and age ratio of juveniles/adults in California quail populations in New Zealand. (After G. R. Williams, 1963, *J. Animal Ecol., 32:* 441–459.)

the average size of a successfully raised litter may depend in some way on the sex ratio of the litter. The parents may need to extend a greater amount of care to offspring of one sex to successfully rear it (e.g., the male being larger at each stage of growth than the female) and thus provide a selective mechanism for parental expenditure of energy to actually influence the sex ratio. This would fulfill Fisher's (1930) pioneering conclusion that selection favors a sex ratio that equalizes total energy expenditure upon the two sexes.

1 2 3 4 5 6 7 8 9 10 11

Life History Patterns and Selection in Populations

The interactions between organisms in populations and their secular environment determine probabilities of survival and reproductive activity as well as population size. Natural selection acts to adjust the parameters of the life history of an organism in such a way that the fitness of the individual is maximized, although like other phenotypic traits the life history involves a series of selective compromises to a suite of environmental variables (Wilbur et al., 1974). These life history attributes include fecundity, growth and development, age at maturity, longevity, and degree of parental care. One of the objectives of population ecologists is to explain the great diversity of observable life history patterns in terms of a minimum number of selective pressures.

K AND *R* STRATEGIES IN LIFE HISTORIES

MacArthur and Wilson (1967) coined the terms "*K* selection" and "*r* selection" to describe the two principal kinds of selection resulting from the effects of resources on the density of a species. These two opposing kinds of selection involve the carrying capacity (*K*) of the environment and the maximal intrinsic rate of natural increase (r_{max}). MacArthur and Wilson argued that populations are exposed to at least one predictable environmental change: the change caused by expansion of the population. Under *r* selection, the environment is not resource-limiting to the organism and the rate of change in population size will be relatively density-independent, as in the early stages of population growth in a new population. Genotypes will be favored which confer the highest possible intrinsic rate of natural increase. At the other extreme, *K* selection promotes increased ability to cope with physical or biotic factors under crowded conditions (as one approaches *K*, the carrying capacity of the environment) and genotypes of individuals which are relatively unaffected by high density will increase at the expense of other genotypes. This density-dependent selection presumably takes place as resources become limited.

As Pianka (1970) has emphasized, *r* and *K* selection represent the endpoints of a continuum and populations may exist at all possible densities in relation to resources. The extreme state for an *r*-selected organism would exist where there were no density effects and no competition. Its optimal life history strategy should involve maximal investment of energy in reproducing a great quantity of progeny with minimal contribution of matter and energy into each individual offspring. Hence, *r* selection ought to lead to high productivity; in other words, it is the quantitative extreme of an *r*–*K* continuum. The qualitative extreme is *K* selection, where density effects are maximal, with intense competition, and the environment is saturated with organisms. The optimal strategy becomes maintenance over a long lifespan and the production of a few exceptionally fit offspring that will more or less exactly replace senescent adults. *K* selection increases the efficiency of utilization of environmental resources. Hence, as an ecological vacuum is filled, selection pressures on a population will shift from primarily *r* selection to *K* selection. Correlates associated with the *r*- and *K*- selected extremes are summarized in Table 8–1.

The traditional characteristics of *r*-selected populations involve a large density-independent component of mortality. Populations emphasizing *r* selection typically inhabit regions with variable climates, such as the temperate zones, whereas populations with *K* selection are generally found in areas with relatively constant conditions, such as the tropics, where lower fecundity and slower development could act to increase competitive ability and hence overall individual fitness (Dobzhansky, 1950; Pianka, 1970). Of course, these kinds of selection are not restricted to the temperate zone and tropics. Populations under *r* selection frequently reside in marginal areas where conditions necessitate frequent recolonization (MacArthur and Wilson, 1967; Gadgil and Solbrig, 1972), and their population size fluctuates, frequently below *K*. Here selection favors early maturity with small size and

Table 8–1 Some of the correlates of *r* and *K* selection.

PARAMETER	*r* SELECTION	*K* SELECTION
Climate	Variable and/or unpredictable: uncertain	Fairly constant and/or predictable: more certain
Mortality	Often catastrophic, nondirected, density-independent	More directed, density-dependent
Survivorship	Often Type III (Deevey, 1947)	Usually Type I and II (Deevey, 1947)
Population size	Variable in time, nonequilibrium; usually well below carrying capacity of environment; unsaturated communities or portions thereof; ecologic vacuums; recolonization each year	Fairly constant in time, equilibrium; at or near carrying capacity of the environment; saturated communities; no recolonization necessary
Intra- and interspecific competition	Variable, often lax	Usually keen
Relative abundance	Often does not fit MacArthur's broken-stick model (King, 1964)	Frequently fits the MacArthur model (King, 1964)
Selection favors:	1. Rapid development 2. High r_{max} 3. Early reproduction 4. Small body size 5. Semelparity: single reproduction	1. Slower development, greater competitive ability 2. Lower resource thresholds 3. Delayed reproduction 4. Larger body size 5. Iteroparity: repeated reproductions
Length of life	Short, usually less than 1 year	Longer, usually more than 1 year
Leads to:	Productivity	Efficiency

SOURCE: Pianka (1970).

rapid development, with a large ratio of offspring mass to parental mass. After the first reproductive period, poor survivorship is characteristic; thus *r*-selected species are frequently *semelparous* (a single burst of reproduction and short-lived). Resources are allocated to reproductive structures in the organism at the expense of long-term maintenance.

The predictability and permanence of the habitat determines the type of selection, as has been most recently emphasized by Schaffer (1974), Southwood et al. (1974), and Wilbur et al. (1974). As a general rule among animal groups, many authors have pointed out, it is logical that in a relatively nonlimiting environment such life history components as early maturity, large clutches, and minimal parental care should maximize individual fitness thereby becoming integrated into the life history scheme. In more limiting environments, delayed maturity (which allows sufficient opportunity for learning to exploit resources), smaller clutches of offspring, and larger young or hatchlings capable of better handling competitive situations might be expected (Wilbur et al., 1974). Terrestrial organisms seem to fall by groups into these selection classes. Thus, most terrestrial vertebrates and perennial plants seem to be relatively *K* selected, whereas most annual plants and insects (and perhaps terrestrial invertebrates in general) are apparently *r* selected (Pianka, 1970). In fact, environmental conditions prevailing at different periods of geological time likely encouraged evolution of each of these groups (Southwood et al., 1974). Insect radiation occurred principally in the Permian and Triassic, when conditions are thought to have been very variable. The primary bursts of vertebrate radiations occurred during the warm, humid, and stable climatic regimes of the Jurassic, lower Cretaceous, Eocene, and Oligocene periods.

Pianka (1970) has shown evidence that increased body size is correlated with a shift to *K* selection among animals, perhaps because it reduces environmental resistance in many ways (e.g., fewer potential predators, better buffering against changes in the physical environment). The fossil record shows trends in increasing body size in many groups. Body sizes and generation times of insects and vertebrates are largely nonoverlapping. Pianka suggests that the attainment of a generation time exceeding a year may well be a threshold event in the evolutionary history of a population. In addition to requiring a larger body size and longer life-span adapted to cope with the full annual range of physical and biotic conditions that prevail at a given locality, perenniality leads to a rather drastic shift from *r* to *K* selection under a substantially lessened element of "surprise" in the environment.

Cole (1954) introduced the term *iteroparity* to refer to the life history phenomenon of repeated breeding in long-lived species where organisms with great longevity (e.g., redwood trees) reproduce over many years but only rarely reproduce successfully. Murphy (1968) argued that there is a dynamic interdependence of three parameters—longevity, growth rate, and age at first maturity—in determining life history patterns. He suggested that evolutionary pressure for *long life, late maturity, low fecundity*, and *repeated reproductions* may be generated wherever the mortality of young or prereproductives is high or variable. Such a situation could occur in a physical environment where density-independent factors kill the young, or in a biotic environment having intense competition with the reproductive individuals. On the other hand, evolutionary pressure toward *short life*, *early reproduction, high fecundity,* and a *few reproductions* (*semelparity,* or only one reproduction), will be generated by high or variable adult mortality. If there is uncertain survival from zygote stage to first maturity, Murphy (1968) shows from both competition and genetic models that selective pressure for iteroparity results, which may entrain pressure for reduced energy allocation for reproduction in order to ensure longer life. The stability of tropical faunas, in his view, might be simply a reflection of the relatively long lives of individual members of the community and their relatively low birth rates.

As Wilbur et al. (1974) have pointed out, data on life history parameters are often consistent with more than a single hypothesis. Clutch-size evolution in tropical birds, for instance, has been explained in three different ways by three different investigators. Cody (1966, 1971) postulated that the smaller clutch sizes of tropical birds compared with temperate relatives (Figure 8–1) largely related to density of birds in relation to resources. Under conditions of intense competition and presumably living near their carrying capacity in a smaller niche much of the time, tropical birds lay smaller clutches (Table 8–2). Temperate bird populations generally have less limited resources during the breeding season. Lack (1968) argued that larger clutches could be reared in the temperate zone because longer day lengths increased the time available for adult food gathering. Skutch (1961, 1967) emphasized the higher levels of predation in tropical areas and hypothesized that better parental care and hence nesting success resulted from smaller clutches.

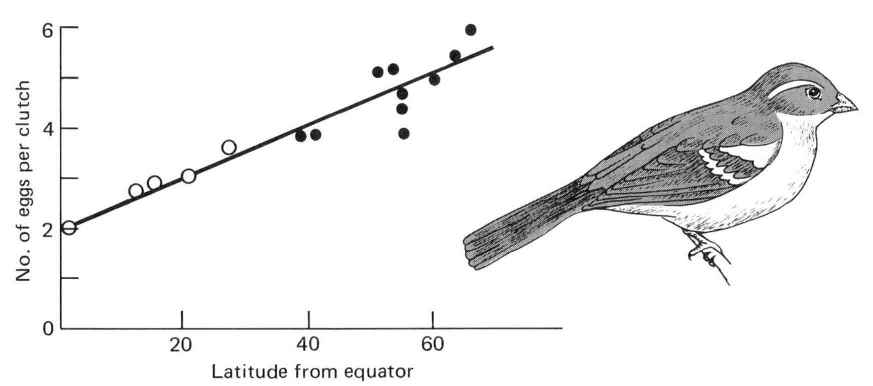

Figure 8–1 A graph of clutch size (ordinate) against latitude (abscissa) for buntings of the genus *Emberiza* (Source: M. L. Cody, 1966, *Evolution, 20:* 174–184.) The open circles indicate species nesting in Africa south of the equator, whereas the solid circles depict species nesting in Europe and Asia. The two sets of points for these ground-feeding, finchlike birds fit the same line. The illustrated *Emberiza* species is the three-streaked bunting (*Emberiza orientalis*) that occurs locally but not uncommonly in the light bush country of Tanzania. (Source: J. G. Williams, 1967, *A Field Guide to the National Parks of East Africa,* Houghlin Mifflin, Boston, p. 285.)

Table 8–2 Clutch size in passerine birds in relation to habitat.

HABITAT	NO. SPECIES	AVERAGE CLUTCH SIZE
Tropical forests	82	2.3
Tropical savanna, etc.	260	2.7
Tropical arid areas	21	3.9
Middle Europe	88	5.6

A further example of the problems involved in a simplistic acceptance and application of the r–K selection explanation may be seen in the green sea turtle *Chelonia mydas,* a transoceanic species that migrates long distances to reach a limited number of island or mainland beaches for reproduction (Figure 6–13). At the beaches, females deposit several clutches at intervals during a single breeding season, each containing about 100 eggs (Carr and Ogren, 1960). The matings, which occur offshore, provide sperm for the eggs to be laid during the female's next return journey to that beach two or three years in the future. The large herbivorous adults have among the longest life expectancies known in the vertebrates, yet the tiny young are subject to very high mortality at hatching. The extremely high fecundity and small young suggest r selection promoting expansion of the population, yet the long adult life is expected in K-selected species. In this instance, Wilbur et al. (1974) suggest that the long adult life and high fecundity might well have evolved in response to juvenile mortality which had a high mean and high annual variation. The green turtle cannot be called a K-selected or r-selected species without obscurring the interplay of selective factors responsible for these characteristics.

OTHER SELECTIVE FACTORS

The causal determinants of adaptive strategies in the evolution of life histories are therefore broader than mere population density. The density of the population in relation to resources, the trophic and successional position of the population, and predictability of survivorship patterns all appear to be important in selection for life history type (Wilbur et al., 1974). The female adults of the checkerspot butterfly *Euphydryas editha* oviposit on different foodplant species in different colonies in the western United States. The newly hatched first instar larvae can exercise no foodplant preference, as they lack sufficient powers of movement to leave the plant on which the eggs were laid (Singer, 1971). The instantaneous probability of oviposition on a particular plant species in a given insect population depends on many factors, including the chemical preference, behavior, and population structure of the insect, and on the distributions and densities of the various foodplants. A decrease in density of the preferred oviposition plant may instantly increase the probability of oviposition on several secondary foodplants. Consequently, the distribution of young (prediapause) larvae on the alternative foodplants is determined by the oviposition preference of the adult. Prior to the obligatory diapause, the larvae leave the foodplant in the third instar. The dry summer and fall are spent in diapause under rocks, and high mortality occurs. During the postdiapause instars in the following spring, the larvae are much more mobile and their distribution becomes influenced more by their own foodplant preferences and the new densitites of available host species than by the ovipositional preferences of their short-lived parents in the preceding year. When the butterfly laid its eggs in large batches within a few days and the resultant young larvae had no choice, one would say it was an *r*-selected response in the life history. The maturing larvae, however, show attributes of *K* selection in enhancing their own survival by responding selectively to resource factors in the biotic environment.

The trophic position of a species in the community affects its dependence upon resources and life history characteristics. Herbivores and lower-order predators would often be limited by predation to densities below the carrying capacity, whereas predators high in the food pyramid would often be limited primarily by food supply. Prey strategies that could reduce mortality from predation should evolve. Effective defense mechanisms such as poisonous chemicals, hardened exoskeleton, or large body size, ought to lead to such life history characteristics as long adult life, delayed maturity, and even large eggs and young. Conversely, if these defense mechanisms are ineffective, then we would expect short adult life expectancy, early maturity, and very high reproductive efforts soon after attaining adulthood (Wilbur et al., 1974). Predator satiation presents another adaptation to predation pressure. A population of periodical cicadas or forest legume trees will reproduce synchronously and flood the environment with large numbers of adults or young for a brief period (Lloyd and Dybas, 1966; Janzen, 1969). A predator has difficulty in consuming more than a small fraction of this suddenly available resource (Figure 8–2). Janzen (1971) has emphasized the trophic relations between *Cassia grandis* trees and their insect and mammalian predators as important life-history determinants. Two bruchid weevils and several moth species destroy almost all of the seed crop in the vicinity of the parental trees unless vertebrate dispersal agents scatter the seeds. Individual trees bloom about every two years in the dry season and carry the seeds on the tree for 11–12 months. Insect predators bore into the seed pods at this time. Sweet odoriferous casings attract deer and other herbivores. High fecundity, large seeds, widely spaced reproductive periods, and extended parental care, seem best explained by these trophic relationships with other organisms.

a

b

Figure 8–2 Mass 1973 emergence of a brood of 17-year cicadas in Pennsylvania. (a) Single adult after emergence and (b) adults shedding nymphal skins, and nymphs crawling up on vegetation in preparation for adult emergence. (Photos by Christine Simon.)

EVOLUTION OF COMPLEX LIFE CYCLES

A complex life cycle is one which involves passage through two or more ecologically distinct phases. Each phase has its own set of predator–prey, competitive, physical–environmental, and resource interactions; no overlap exists among the factors which limit abundance in each phase (Istock, 1967). Some indication of the complications involved in such life cycles could be seen in the checkerspot butterfly life history (Singer, 1971) cited earlier. Istock (1967) develops a pioneering theoretical treatment of the ecological and evolutionary exigencies imposed by a complex life cycle.

To a large degree in insects and other species with complex life cycles, the ecologically distinct phases evolve independently of one another, each having its own distinct morphological and physiological adaptations to its respective environment (Figure 8–3). The principal restraints are on those behavioral, developmental, and morphological characteristics necessary for making the transitions between phases. When the population is simultaneously saturating both larval and adult environments, the conditions for the required demographic balance may be defined.

Let N_l be the number of larvae alive, N_a the number of adults alive, R_g be a gross reproductive rate (mean number of offspring born to a female) and R_0 be a net reproductive rate (the fraction that survive). Upper limits for N_l and N_a would be K_l and K_a, the saturation values or maximum carrying capacities of the larval and adult environments. $R_{g,e}$ is the reproductive rate when the effects produced by both larval and adult phases are just sufficient to produce equilibrium for the population as a whole (Figure 8–3).

Thus we have $\Delta N/\Delta t = 0$ at demographic balance in this population, where Δt is the average time required for the death of K_a adults (Istock, 1967). Over this time interval, let N_m be the total number of

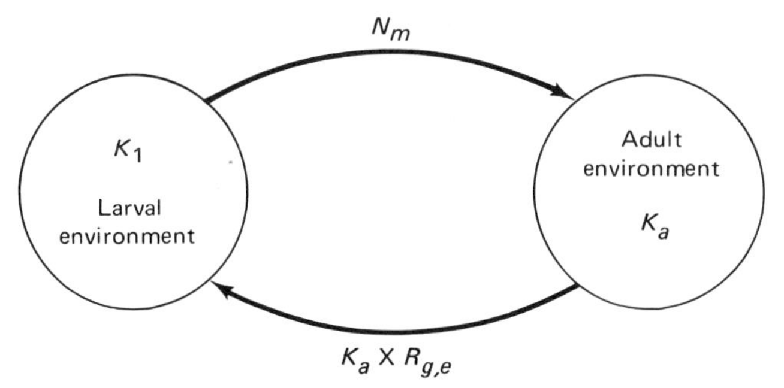

Figure 8–3 A schematic representation of the relationship between larval environment, adult environment, and the demographic functions that must unite the two segments of the population in these environments. See text for explanation of symbols. (Source: C. A. Istock, 1967, *Evolution, 21:* 592–605.)

Table 8–3 The four possible complex life-cycle states.

CASE NO.		ECOLOGICAL INTERPRETATION
1)	$N_m > K_a$ $N_a R_{g,e} > K_l$	Both environments saturated
2)	$N_m > K_a$ $N_a R_{g,e} < K_l$	Adult environment saturated Larval environment not saturated
3)	$N_m < K_a$ $N_a R_{g,e} > K_l$	Adult environment not saturated Larval environment saturated
4)	$N_m < K_a$ $N_a R_{g,e} < K_l$	Both environments not saturated

SOURCE: Istock (1967).

individuals passing from the larval phase to the adult portion of the life cycle. By definition, $N_m < N_l$. More significantly, $N_m > K_a$ and $N_a R_{g,e} > K_l$ must hold if the population is to maintain the highest possible equilibrium numbers in both environments (Figure 8–3). The units in these inequalities must be numbers: mass, volume, or energy units are not valid. Alterations of the latter two inequalities make it possible to define the four possible states for a population with a complex life cycle. These are listed in Table 8–3. Case 1 represents the optimal case, that is, the maximum exploitation of the ecological advantages of a complex life cycle, and it seems improbable that any species continuously realizes Case 1. The reversal of one of the inequalities in Case 1 will yield either Case 2 or Case 3. In these circumstances, either adult fertility and abundance become insufficient to saturate the available larval environment, or conversely the rate of metamorphic transition from larvae to adults will become too low to saturate the adult environment (Istock, 1967). This imbalance may occur in either changing or constant environments. Thus, tadpoles can survive to metamorphose in environments such as temporary or semipermanent rain pools in the tropics, saturating that larval environment's capacity and experiencing very rapid growth rates (e.g., Wassersug, 1974), but the adult environment is not saturated. Case 4 is undoubtedly the most common life cycle state in nature.

A hypothetical example where the inequalities of the optimal case (1) are almost satisfied (except for failure to saturate the adult environment in early spring) is shown in Figure 8–4. Specific times in the adult reproductive period are related to specific times in the larval growth period to obtain corresponding values of saturation and corresponding flow rates between phases. It should be emphasized that K_l and K_a are not likely to be simple constants for natural populations, and that through any single turn in the life cycle they are functions of time.

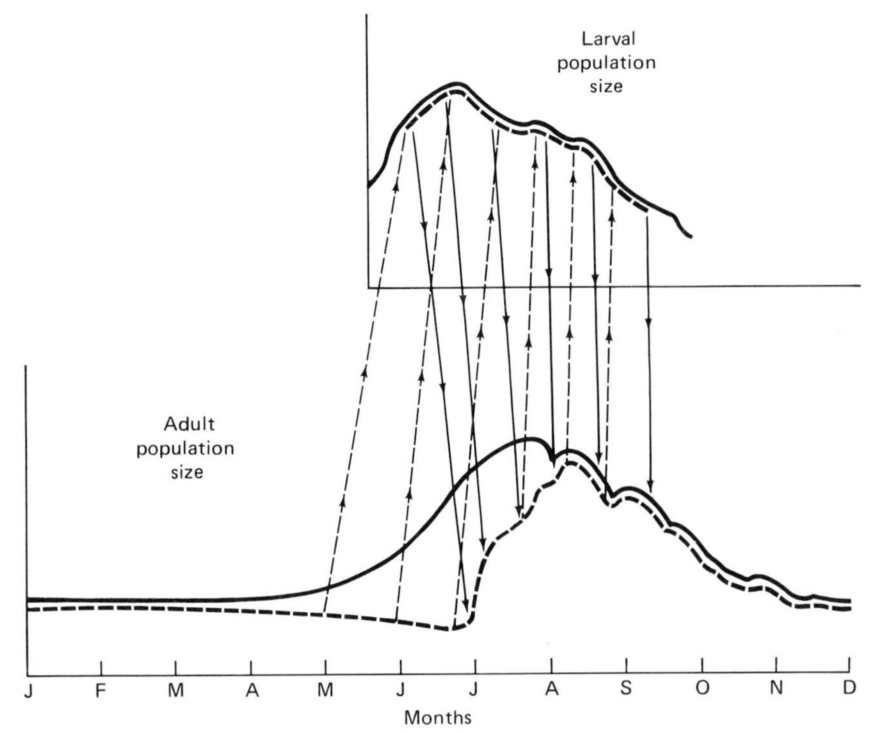

Figure 8–4 Representation of the dynamics of a population with a complex life cycle during one year. The solid lines represent the time-specific larval and adult saturation values, and the dashed lines show the actual sizes of larval and adult segments of the population. The lines with arrows symbolize transfer between phases, but are not meant to indicate a tight functional relationship. In fact, it is probably unlikely that such a relationship exists for most organisms with complex life cycles. Note that the rate of movement between phases also changes with the season. (Source: C. A. Istock, 1967, *Evolution, 21:* 592–605.)

REPRODUCTIVE STRATEGIES IN NONAVIAN VERTEBRATES

Aside from birds, the best-surveyed vertebrate groups for reproductive strategies are lizards (Tinkle et al., 1970) and mammals (Weir and Rowlands, 1973). Lizards, like the birds (Lack, 1966, 1968), may be divided into two groups on the basis of their age at first reproduction. Early-maturing lizards have shorter adult life expectancies, are almost always multiple-brooded, and produce relatively small clutches (in contrast, birds of this type produce large clutches); oviparity is almost the universal type of reproduction and these lizards tend to be tropical and temperate in distribution. Lizards with delayed maturity must have greater survivorship, are almost always single-brooded, and have larger clutches. Viviparity (bearing live offspring) is just one form of the latter strategy inasmuch as almost all viviparous lizard species produce one litter per year and have a significantly later age at first reproduction. The single-brooded, delayed maturity group of lizards are primarily temperate in distribution.

Late-maturing lizards, birds, and other vertebrates achieve stable populations through the adjustment of other life history parameters such as degree of parental care (important in many vertebrates; see Ricklefs, 1973, for general discussion), adult life expectancy, clutch size, and the frequency of clutches (Cole, 1954). In lizards, three adaptations counter the cost in fitness imposed by delayed maturity (Tinkle et al., 1970): the production of larger clutches (partly a result of growth to a larger size before maturity); increased parental care and viviparity; and a longer reproductive life expectancy. As Tinkle et al. (1970) pointed out, birds have not evolved viviparity but the large raptors and sea birds couple a long breeding life expectancy for the adult with increased parental care of their one or two chicks such that there is a high probability of the young surviving to maturity, thereby balancing the reproductive cost of delayed maturity.

In mammals, no obvious patterns of reproduction can be correlated with taxonomic group or environmental type (Weir and Rowlands, 1973). Even within a single genus (e.g., voles of the genus *Microtus*), reproductive patterns may be different. The family Chinchillidae is one of the few higher taxa for which members of all three known genera (two of them monotypic) have been studied. Each displays completely different characteristics, the most bizarre being that of polyovulation in the plains viscacha (with 200–800 eggs being produced at each ovulatory episode) and the use of only one ovary in the mountain viscacha. The marsupials appear to be more successful reproductively than the placental mammals (eutherians), contrary to most evolutionary dogma, which has marsupials on an intermediate stage towards the evolutionary excellence of eutherian mammals. The clearest trends comparable to those demonstrated in birds and lizards are the high rate of increase and large broods produced in the small myomorph rodents—particularly the murids and cricetids—which have short generation times. More intensive comparative studies remain to be done among mammals before broad generalizations about selective factors on reproduction may be safely made.

1 2 3 4 5 6 7 8 9 10 11

Mating Systems and Behavior in Populations

Sexual reproduction is a regular feature of diploid plant and animal species. The blue-green algae, certain fungi, and bacteria are haploid, and therefore lack meiosis and sexual differentiation, but these groups still have several means of achieving genetic recombination. In theory, sexual reproduction is unnecessary if an organism is particularly well suited to its environment; it could persist perfectly adequately by asexual reproduction (fission, budding, and so forth) of the same "winning" genotype. Under a constantly changing natural environment of physical factors and biotic competition, however, genetic recombination is logically considered to be advantageous and organisms possessing sex would be predicted to survive while asexual competitors will lose out. The precise reasons for the selective advantages of sexual reproduction are still an unresolved controversy in population biology (Williams, 1971). It is clear that the independent assortment and recombination of genes in meiotic processes allow new combinations of genes to arise at each generation in a population and hence maintain genetic variability. Several mutations arising in several individuals may be recombined into the same individual in sexually reproducing species. The effects of recessive deleterious alleles are suppressed in diploid organisms and may even prove advantageous in the heterozygous state (e.g., the protection conferred against malarial parasites by the sickle cell anemia allele in a heterozygous human).

Sexual reproduction and its contributing selective factors differ greatly in expression among the various divisions of organisms. The two sexes may be separate (*dioecious*), with male and female individuals usually distinguished by the relative size of the gametes. Dioecious organisms include most arthropods, most vertebrates, and some plants. Approximately equal numbers of males and females occur in most dioecious species populations. Bisexual organisms are *monecious* (hermaphrodites); gonads for the two sexes occur in the same individual. Most plants and nonarthropod invertebrates are hermaphroditic.

In both monoecious and dioecious organisms, the genes are normally recombined through outbreeding (*heterogamy*) in which separate and unlike individuals mate with each other. This may take place directly through actual physical contact of the two sexes (e.g., insects and vertebrate animals) or it may necessitate an intermediary vector (e.g., wind, water, or animal pollinators in plants, and water currents in many aquatic invertebrates). At the other extreme is inbreeding (*homogamy*), in which genetically similar organisms (or the sexual products in the single hermaphroditic parent) mate with one another. These types of breeding structures in populations have distinct genetic outcomes, with inbreeding promoting genetic uniformity at a local level and outbreeding permiting genetic variability.

The types of sexual reproductive behavior in animal populations commonly involve close association between a single pair. The egg and sperm may meet inside the female (internal insemination, as in man and other mammals, birds, reptiles, and insects), or the male may fertilize the eggs as soon as they pass out of the female (external fertilization, as in fish and amphibians), or the male and female may shed their gametes simultaneously and in close proximity, with fertilization being purely a matter of chance (simultaneous spawning, as in oysters). In some species, sexually mature males and females simply mate in one of the above three ways as soon as they meet. In mammals and birds and less often in reptiles, amphibians, fish, and invertebrates, a courtship sequence involving a short or extended display of one sex to the other is a usual prelude to mating. Various types of pair bonds have evolved in these groups. Hence, in considering sexual reproduction and the breeding activities of animal populations, we can speak of *mating systems* and the selective factors that have molded them. In the remainder of this chapter we shall examine these population phenomena in animals from the viewpoints of courtship displays and mating behavior, mate selection, and the types and evolution of mating systems in the well-studied higher vertebrates.

COURTSHIP DISPLAYS AND MATING BEHAVIOR

Courtship displays involve usually conspicuous behavior which attracts attention and shows one or more particular features to the opposite sex. The communicative function of these visual, auditory, tactile or olfactory stimuli from the displayer involves a reaction in the audience, and unless a particular behavior has a sexual effect on the recipient it is not strictly "courtship" (Bastock, 1967). Nest-site displays, nest-building displays and greeting ceremonies may serve other functions than mating preparation. Morris (1956) proposes perhaps the most generally useful definition: "Courtship is the heterosexual reproductive communication system leading up to the consummatory sexual act."

Courtship functions in several significant ways to prepare animals for mating. Because each species generally has a particular set of courtship activities, it provides a means of species recognition such that males and females do not waste time, energy and gametes in interspecific hybridization attempts. Courtship displays reduce the normal aggressiveness of the males (particularly territory-holding species) and allow the female to be accepted into the male's proximity without hostility, conflict, or physical repulsion. Complex and long-lasting courtship patterns often serve to synchronize the internal physiological processes of the prospective mates and thus insure successful insemination when actual mating occurs. The ontogeny of the sexual cycle and mating behavior is therefore closely correlated with courtship activity.

The fruitfly genus *Drosophila* is well known for its utility in genetic and evolutionary studies, and courtship behavior among its more than 1000 described species shows a remarkable range of complexity (Bastock and Manning, 1955; Spieth, 1968). Drosophilids are gregarious in the morning and evening around their feeding and oviposition sites, which are generally rotting fruit, slime flux on a tree, a decomposing fungus, or a rotting mass of leaves. Many species may be in close proximity at the feeding sites; when the flies are not at the feeding–ovipositing–mating site, they scatter individually into the surrounding area where courtships, if they occur, must be accidental and infrequent. Spieth's (1968) observations show that the females spend practically the entire period of each diurnal visit in feeding, whereas the males spend only short intervals in feeding and devote the rest of their time to sexual investigation of other individuals. The males are promiscuous and initiate the courtships with a series of ritualized movements, whereas the females display both repelling and accepting actions (Figure 9–1). The Hawaiian species have superimposed upon the basic drosophilid mating behavior a true *lek* type of behavior in which each male selects and defends a particular courting arena near the food source and prospective ovipositional site. Here the

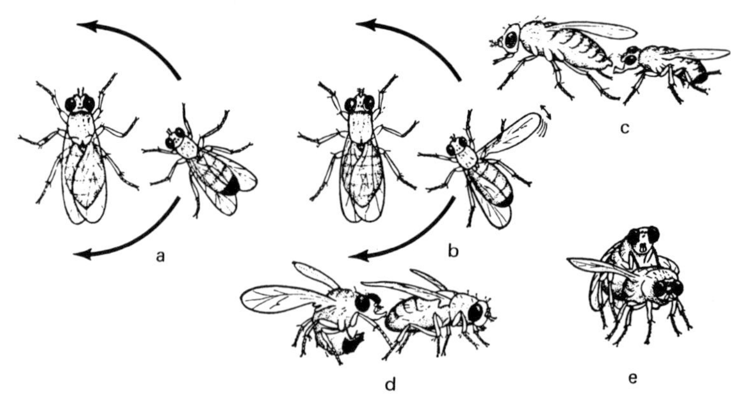

Figure 9–1 Courtship movements in the fruit fly *Drosophila melanogaster.* Preceding courtship, a male taps another insect with his forelegs to verify with his chemoreceptors the species identity of the prospective mate. The three main componenets of the courtship are (a) *orientation,* in which the male stands close to the female and faces her; (b) *vibration,* which consists of a vertical vibration of the wing nearest the female's head as he is orienting; and (c) *licking,* in which the male comes behind the female during orientation and vibration and licks her genitalia with a quick movement of his proboscis. Licking is usually followed by the male mounting the female and attempting copulation (d) by curling his abdomen downward and forward to touch her genital region with his. A receptive female cooperates by spreading her wings horizontally and exposing her external genitalia, permitting the copulation to occur (e). (Source: M. Bastock, 1967, *Courtship: An Ethological Study,* Aldine, Chicago, p. 54. Redrawn from A. Manning, 1965, *Viewpoints in Biology,* Vol. 4, Butterworth, London, pp. 123–167.)

males advertise their presence and the females are attracted to their lek territory.

True lek behavior, whether in Hawaiian drosophilids or in various birds and mammals, involves a number of promiscuous males assembling in close proximity and engaging in strikingly hypertrophied displays which attract sexually mature and receptive females to communal breeding grounds (Spieth, 1968). This includes the spatial separation of courtship from feeding and oviposition. The intense sexual selection has resulted in considerable sexual dimorphism such that the male exhibits clearly distinct structural characters and conspicuous display behavior (Figure 9–2). In lek behavior the resulting pair bond exists only for the period of courtship and copulation.

The diversity and complexity of courtship is nowhere greater than among the birds. Three phases are usually found in bird courtship display (Bastock, 1967). During *pairing* the prospective mates come together. In territorial species, the male's song and exhibitive displays provide both advertisement and threat. The approaching female will use appeasement signals or possess sufficiently distinctive plumage to be recognized as a female of the correct species. Nonterritorial birds

Figure 9–2 Lek display behavior as part of courtship in Hawaiian Drosophilidae flies. (a) *Drosophila grimshawi* male exhibiting lek display with tip of abdomen being dragged against plant stem; (b) *D. comatifemora* male exhibiting lek display on a fern frond. Note the elevated tip of the abdomen in this species. (Source: H. T. Spieth, 1968, *Evolutionary Biology*, Vol. 2, Prentice-Hall (Appleton), Englewood Cliffs, N.J., p. 180.)

also pair, often using communal pairing ceremonies. Greeting ceremonies, including the conspicuous mock-preening display (Figure 9–3), initiate communal displays in drakes of many duck species. After pairing time the flock lives peaceably together in large groups until coitus, when pairs will withdraw a little from the others. As just indicated in ducks, the second phase of courtship, *engagement*, tends to be an extended, generally nonaggressive period in which the male is dominant over the female but little fighting occurs. As the time for the third phase, *coitus*, approaches, the behavioral relationship of the pair is often reversed. The female becomes aggressive and the male timid. Soliciting signals are given in mutual or single-sex precoital displays (Figure 9–4). Ceremonial feeding or presentation of nest-building material results in the male losing its timidity and the female its aggressive tendencies.

The culminating step of courtship is actual copulation. Coital positions vary considerably among animal groups (Figure 9–5). Many male

Figure 9–3 Greeting ceremonies are often performed when drakes begin to gather for communal displays. The nonterritorial ducks, geese, and swans of the family Anatidae court on the water in the midst of a waterfowl community. Pairing and even the engagement period are communal, and pairs separate only for coitus. Here the drinking display (a) is followed by mock-preening in the mandarin duck *Aix galericulata*. When mates meet, very often both will drink; then the drake mock-preens (b), touching the wing feathers on the side nearest the female duck. The hood plumes are raised for drinking and stay up for mock-preening. (Source: M. Bastock, 1967, *Courtship: An Ethological Study*, Aldine, Chicago, p. 112. Redrawn from K. Lorenz, 1941, *J. Ornithol. Lpz.*, *3:* 194–294.)

mammals mount the female from the rear and dorsally (Figure 9–5a), whereas reptiles may have to contort themselves considerably to oppose their ventral genital openings (Figure 9–5c). The crouched copulatory position exhibited by felid cats is an aberrant form among mammals, and appears to be derived from ancestral arboreality in which most mating occurred in trees. Ewer (in Eaton, 1974) suggests that it has been retained in terrestrial cats because it minimizes risk to the female. This copulatory position reduces the chance of the male responding to the female as prey and killing her with one effective bite. Birds mount very quickly from the rear, whether in a terrestrial situation or on the water (Figure 9–5b). Butterflies and other insects generally copulate with both individuals facing in the same direction initially (the male being on top of the female or standing parallel to her, and curving his abdomen at an acute angle to "lock" his genitalia into hers), but after "locking," the pair will frequently straighten out and the male will face in the opposite direction from the female (Figure 9–5d). With many groups, repeated copulations or intromissions are the general rule. In the subterranean mole rat *Spalax ehrenbergi* of

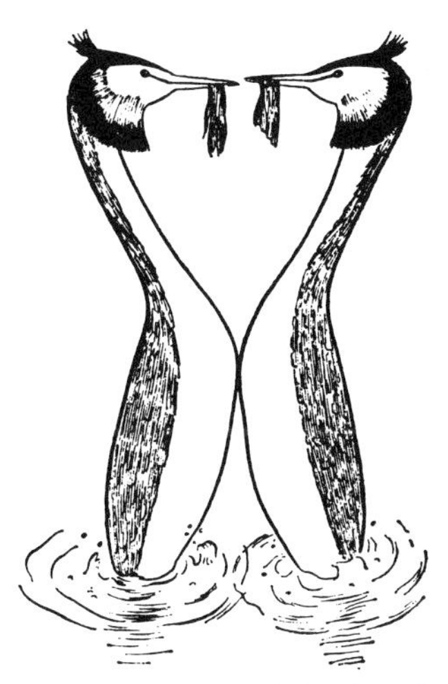

Figure 9–4 The "penguin dance," a mutual display during the precoital phase of courtship of the great crested grebe (*Podiceps cristatus*) in Europe. The male and female rise up breast to breast and sway together after each dives for aquatic weeds. (Source: M. Bastock, 1967, *Courtship: An Ethological Study*, Aldine, Chicago, p. 48. Redrawn from J. S. Huxley, 1914, *Proc. Zool. Soc. London, 1914* (2): 253–292.)

the Middle East, copulation may last more than 90 min and involve more than 60 separate mountings (Nevo, 1969). In this species, copulation is effected within elaborate breeding mounds built by females during the *Spalax* breeding season. At other times of the year, the sexes occupy separate territories and are quite hostile.

Iguanid lizards are an example of a group with highly species-specific display-action patterns in territorial behavior and courtship which serve to declare territories, regulate social structure, and increase reproductive isolation between species. The lizards of this family are found throughout North and South America, the Caribbean islands, the Galapagos Islands, Fiji and Tonga islands in the Pacific, and Madagascar. A characteristic display found in all members involves peculiar movements of the body commonly called nods, bobs, and push-ups. Carpenter (1966a) has analyzed the well-developed aggressive behavior in both sexes of each of the seven species of larva lizards (genus *Tropidurus*) on the Galapagos Islands (Figure 9–6). In both males and females, challenge displays are used to establish dominance and territory. Courtship behavior begins when a male approaches a female, stops, performs an nonchallenging assertion display, then continues to approach the female. The male begins head-nodding when within 6–12 inches of the female. The males of some island populations nod their heads similarly to the display–action pattern, while in other populations the courtship nodding is different from territorial

a

c

Figure 9–5 Variation in copulatory positions among animal groups. (a) African lions (*Panthera leo*) in Ngorongoro Crater, Tanzania, as male prepares to mount female who will shortly crouch in a typical felid position. (b) Flightless cormorants (*Nannopterum harrisi*) mating in the surf (male on top of female with his short stubby wings raised) off Narborough Island in the Galapagos. Most of the precopulatory activities of this pair took place on shore. (c) Land iguanas (*Conolophus subcristatus*) copulating (orange male on top and grasping the dark gray female) on South Plaza Island in the Galapagos. (d) Ithomiine butterflies (*Oleria kena*) copulating (male at left) on a mossy leaf in the Amazonian rain forest of eastern Ecuador. (Photos: a, T. C. Emmel; b,d, Boyce A. Drummond; c, Wayne Fitch.)

b

d

Figure 9–6 A female larva lizard, *Tropidurus delanonis,* from Hood Island in the Galapagos, in a challenge display on her territory. (Photo by Boyce A. Drummond.)

Figure 9–7 Both sexes of the Galapagos marine iguana (*Amblyrhynchus cristatus*) have enlarged conical scales atop their heads, which are used in head-butting contacts during territorial fights by males for females and by females for nesting space. (Photo by Boyce A. Drummond.)

nodding. If the female shows no rejection, tactile contacts ensue and finally the male gets a biting hold on the skin of the shoulder region. He then attempts to straddle her and swing his pelvic region into contact with the female's to effect intromission, which lasts from 15 sec to 3 min.

In another Galapagos iguanid, the marine iguana *Amblyrhynchus cristatus,* the males are seasonally territorial and have aggressive displays involving a species-specific display–action pattern. The rapid nodding of the head during courtship, the hold on the female taken by the male during copulation, and the digging of a burrow in which to lay its eggs, are behavioral characteristics shared with other iguanids (Carpenter, 1966b). Males will corral harems of females within their coastal territories. After insemination of the females, male territorial aggressiveness wanes and the females complete the process of reproduction with an increase in their own interfemale aggressiveness. As they aggregate on sandy beaches to dig nests, considerable displaying, fighting, biting, and head-butting take place. Head-butting among iguanids is known only in the marine iguana, and the enlarged conical scales capping the cranium of both sexes of this lizard are probably an adaptation for such encounters (Figure 9–7).

In ruminants such as sheep, deer, elk, gazelles, and antelope, the rams, stags, and bucks are not physically equipped to grab and hold on to the female for copulation if she is struggling. Thus copulation can come about only with the female's cooperation and this requires a strikingly cautious courtship by the ram when he is alone with an estrous female (Geist, 1971). If subordinate males are around, he will be stimulated to frequent breeding, at least during the first week of the rut. Under the rare circumstances when an estrous female is alone with a ram who shows little interest in breeding, the courtship behavior of estrous females may be observed (Figure 9–8).

MATE SELECTION

Selective factors molding mating systems in animal populations have been under intensive study since Darwin's theory of natural selection was first published in detail in *The Origin of Species* (1859). Charles Darwin himself considered sexual selection such an important topic that he devoted most of a later book to the subject (Darwin, 1871). He proposed two major selective forces in the evolution of sexual differences: (1) fighting and display among males for possession of females was the basis for the evolution of secondary sexual characteristics (e.g., size, horns) useful in aggressive behavior; and (2) sexual preference expressed by females at the time of mating accounted for the extreme development of plumage characters and other features of sexual dimorphism among males of birds and other organisms. The first aspect of sexual selection has been generally accepted, but the notion of female choice has been more controversial (Orians, 1969).

In general, one expects that mate selection is of greater importance to the female of a species than it is to the male. The male produces great quantities of readily replaced gametes (sperm) and is capable of mating repeatedly during any one breeding season. A female bird or mammal will usually produce not more than a few dozen to several

Figure 9–8 Courtship by the estrous female of the ram in mountain sheep. (a) The female performs a sudden coquette jump, pulling the startled ram after her; (b) the female turns and in a head-high horn threat prances at the ram; (c) she horns and forcefully butts the ram on the shoulder while the latter presents stiffly; (d) the female rubs the full length of her body along the ram's chest while horning his chin. (Source: V. Geist, 1971, *Mountain Sheep: A Study in Behavior and Evolution*, Univ. of Chicago Press, p. 218.)

hundred eggs in a lifetime, and an erroneous interspecific mating would claim an entire season's gamete production for a female. Males have strong sexual drives and tend to court rather indiscriminately in many species. As Orians (1969) aptly summarizes: "The inescapable conclusion is that mate selection will be practiced wherever sensory capabilities and locomotor abilities permit it and that females will, in the vast majority of cases, exercise a stronger preference."

In their choice of variable mates, females may show positive or negative assortative mating for particular characters, or they may exhibit no consistent tendencies for phenotypic assortative mating as in certain cerambycid and cantharid beetles (Mason, 1962). In the English tiger moth *Panaxia dominula,* negative assortative mating (where unlike moths mate more frequently than moths of the same phenotype) has been invoked as the selective mechanism maintaining the heterozygote form *medionigra* in the Cothill population in Berkshire (Sheppard, 1952). Sheppard demonstrated that nonrandom mating of heterozygotes occurs in which unlike genotypes are preferred by the females though the males are promiscuous.

Occasionally, directional selection against males of certain genotypes may occur on mechanical (i.e., structural) rather than behavioral grounds. Emmel and co-workers (unpublished) studied a population of Japanese scarab beetles, *Popilla japonica* Newman, in Spruce Run Valley, Virginia, near the Mountain Lake Biological Station in 1970. Since this species was introduced into North America at Riverton, New Jersy, in 1916 and spread across the continent within 40 years, this particular Virginia population had very likely not gone through more than 30–40 generations since its founding. Comparisons of elytral (wing-cover) size among males and females in mating-pair and nonmating samples of beetles indicated strong directional selection on the males for mating fitness (Figure 9–9). As measured by the metrical character of elytral length, males smaller than 5.4 mm were absent from the mating-pair subsample. On several successive sampling dates, mating males always averaged about 0.60 mm greater in elytral length than nonmating males. Mating and nonmating females were about equal in mean elytral length and variance. When elytral length of the male partner is plotted against elytral length of the female partner, a curve representing the minimum male length for successful pairing is obtained (Figure 9–10). The minimum mating size for males increased with increasing female size. For females up to about 6.5 mm in elytral length, successful pairing is possible providing the male is at least 5.4 mm in length. At the maximum observed female elytral length of 7.3 mm, the male apparently has to be about 6.3 mm in length. To this degree, then, mating is assortative for size. However, it is readily apparent from the graph that such large males are equally likely to choose very small females. The large cluster of points in the median range of female length suggests a possible stabilizing component perhaps mediated by female choice. Once the directional component is determined in the male sex (only males over approximately 5.4 mm will normally be able to successfully mount a female), the females that have average elytral length are most likely to mate successfully (stabilizing selection). Observation in both field and laboratory revealed the basis of the selection against mating success by

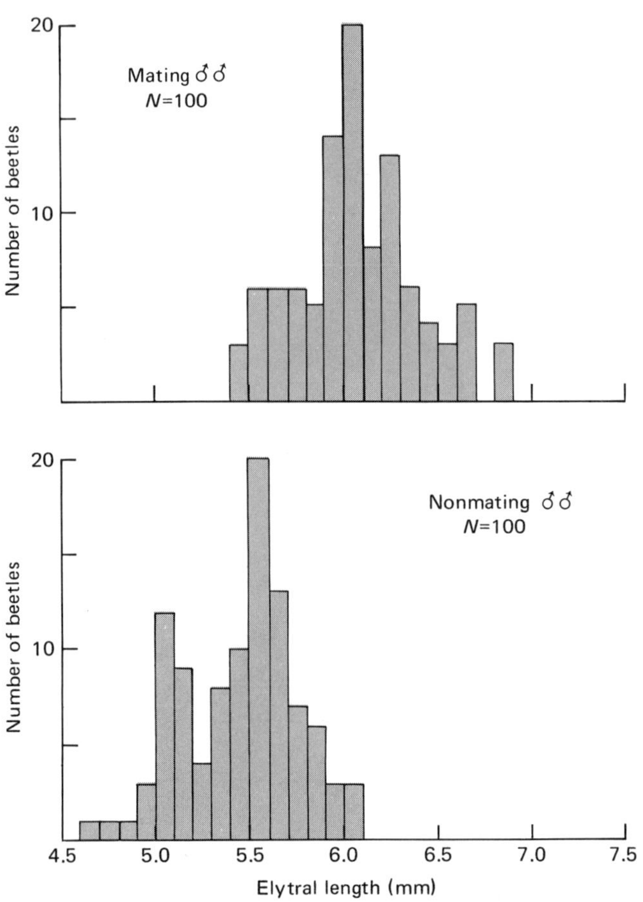

Figure 9–9 Histograms showing the number of male Japanese beetles of each size class present in mating and nonmating subsamples of a Spruce Run Valley, Virginia, population on July 30, 1970. It should be noted that some perfectly fit individuals are in the nonmating group, thereby diluting the differences between "mating" and "nonmating" subsamples.

small males. The male always climbs onto the dorsum of the female and positions himself by grasping and hooking onto the ventral edges of the female's thorax with his foreleg tarsi. If his forelegs are not sufficiently long, the male cannot maintain his position to insert his phallus and complete mating.

Negative assortative mating is of particular evolutionary interest because it is one of several systems that can promote polymorphism and a high level of heterozygosity within a population without the heterozygote necessarily being superior in fitness to either homozygote. However, preference for mating with unlike partners has only rarely been encountered in nature. Besides its occurrence between adult morphs of the tiger moth *Panaxia dominula* (Sheppard, 1952, 1953), dis-

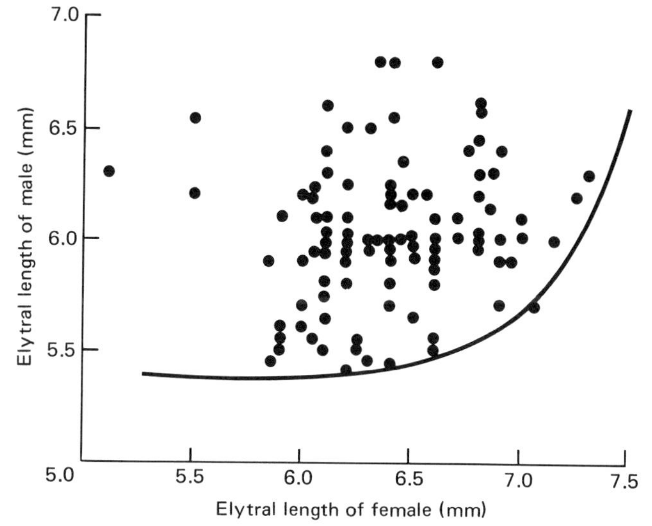

Figure 9–10 Elytral length of male partner (vertical axis) plotted against that of female partner, for 100 mating pairs of Japanese beetles in the Spruce Run Valley, Virginia, population sample collected on the same date as those shown in Figure 9–9. The curved line shows the apparent minimum length of the male required for successful copulation with females of various elytral lengths.

assortative mating is reported between certain mutant lines of *Drosophila melangaster* (Rendel, 1951), strains of *D. melanogaster* derived from natural populations (Parsons, 1962), and in wild populations of the polymorphic white-throated sparrow *Zonotrichia albicollis* (Lowther, 1961; Thorneycroft, 1966). Frequency-dependent disassortative mating has also been discovered in laboratory colonies of six *Drosophila* species (see Ehrman, 1969 for review). All of these reported situations involve a positive contribution of this type of mate selection to the maintenance of a balanced polymorphism in the population.

In the particularly interesting case of the polymorphic butterfly *Anartia fatima,* however, assortative and disassortative mating both work *against* the maintenance of a balanced (stable) polymorphism (Emmel, 1972, 1973). This widespread neotropical nymphalid butterfly shows a striking color dimorphism in both adult sexes (Figure 9–11). In Costa Rica the yellow-banded form occurs at higher frequencies in warm lowland environments while the white-banded form predominates in cooler lowland and highland localities (Figure 9–12). Within these two sets of areas there is considerable stability of morph frequency both over time (dry season to wet season and year to year) and among geographically separated populations (Table 9–1). Thus, the polymorphism in *Anartia fatima* appears relatively balanced and within any one population, seasonal and annual adjustments are slight.

At one locality, near San Vito de Java in southern Costa Rica, in-

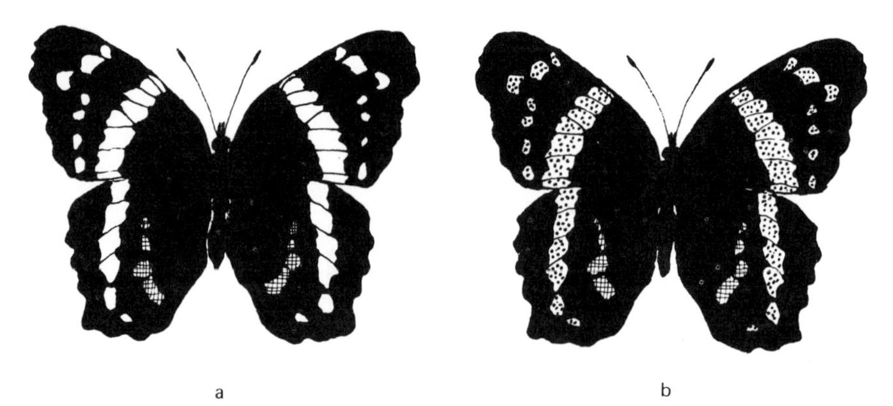

Figure 9–11 The white-banded (a) and yellow-banded (b) morphs of *Anartia fatima* (Nymphalidae : Nymphalinae), dorsal surfaces. Cross-hatched basal hindwing spots are red and are common to both forms. (Source: T. C. Emmel, 1972, *Evolution, 26:* 96–107.)

tensive mating-behavior studies were carried out in the dry season (March) and wet season (August) of 1968. The peak of flight activity occurred in the morning hours and mating activity reached its climax between 1030 and 1100 hr. Males would fly in a slow zigzag pattern between 0.3 and 0.6 meter above the ground or above the vegetation, searching for newly emerged virgin females. When a sitting *Anartia fatima* is seen, the male dives down and if the resting butterfly does not fly up, the male will flutter above it in "approach behavior" for several to 30 or more seconds before alighting and attempting to copulate.

Emmel (1972, 1973) and later Taylor (1973) used dead model specimens pinned out in grassy areas to elicit normal approach behavior from male *Anartia.* The color patterns of some models were altered with colored inks to assess their role in influencing early courtship behavior. Eight categories of butterflies were used as models: (1) normal white-banded females; (2) normal yellow-banded females; (3) white-banded females with the red patch obscured to match the brown ground color of the wings; (4) yellow-banded females with the red spots obscured; (5) white-banded females with white band obscured with brown ink, red patch showing; (6) yellow-banded females with yellow band obscured with brown ink, red patch showing; (7) white-banded females with white band colored red ("super-red" model); (8) white-banded females with white band colored blue, and red patch colored blue (used in dry-season experiments; in the wet-season experiments, this eighth category utilized black color instead of blue).

The mate-selection experiments with these eight classes of models in the dry season and wet season indicated that both white and yellow males prefer white females as compared to yellow females in about a

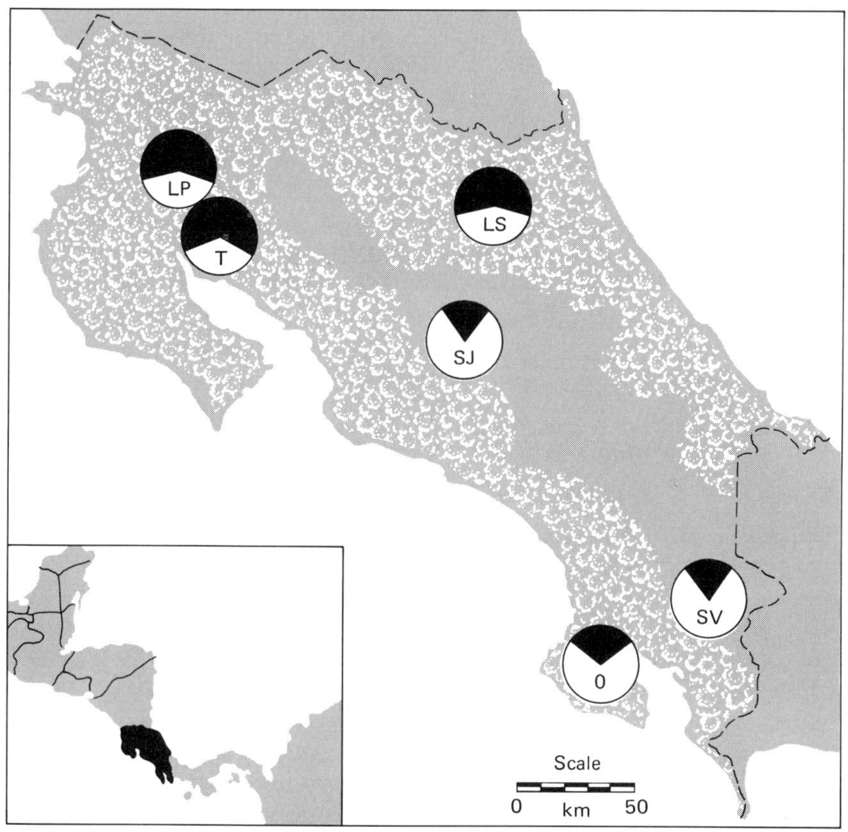

Figure 9–12 The relative frequency of the yellow (dark shading in circles) and white morphs of *Anartia fatima* in six populations in Costa Rica. The tree shading indicates lowland tropical forest (below approximately 1000-m elevation). The location of Costa Rica in Central America is shown in the inset. LP = La Pacifica, T = Toboga, LS = La Selva, SJ = San Jose, O = Osa, SV = San Vito. See text and Table 9–1 for details. (Source: T. C. Emmel, 1972, *Evolution, 26:* 96–107.)

2:1 ratio (Tables 9–2 and 9–3). In the dry season, for instance, white-banded females attract both white (32 percent) and yellow (29 percent) males about twice as often as yellow-banded females attract either morph (16 and 19 percent, respectively). This relative proportion of mate selection remains constant even when the red basal patch is marked out (columns 3 and 4 of Table 9–2). Female band color was shown to be the most significant sign stimulus to the males; when this was eliminated, models were approached in about equal frequency by white-banded and yellow-banded males (columns 5 and 6 in Table 9–2). This behavior may be related to the high ultraviolet reflectivity of the white pigment and the absorbent quality of the yellow pigment.

Table 9–1 Morph frequencies in six populations of *Anartia fatima* in Costa Rica. January–March samples in the tropical dry season, July–August samples in the wet season.

LOCATION OF POPULATION	DATE OF SAMPLE	YELLOW MORPH (%)	WHITE MORPH (%)	TOTAL BUTTERFLIES
San Vito	15 March 1968	21.3	78.7	80
	9 August 1968	22.9	77.1	205
	15–17 August 1969	22.2	77.8	117
San Jose	26 January 1967	25	75	36
	17 August 1969	21	79	19
Osa	July–August 1968	29	71	14
	August 1969	37	63	16
La Selva	1–8 August 1969	55.8	44.2	539
	18 August 1969	64.3	35.7	84
Taboga	6–12 July 1968	56.9	43.1	109
La Pacifica	12–16 July 1969	57.5	42.5	214

SOURCE: Emmel (1972).

The combination of normal band color and the red basal patch (which is present on both morphs) is important for the total attractiveness of a model.

Visual selection by assortative (white males) and disassortative (yellow males) mating, then, is exerting very high selection pressures against the yellow morph in these *Anartia* populations. Yellow band color would be expected to be eliminated through the intense mate selection against yellow females. Furthermore, yellow males may be at a slight disadvantage from tending to mate relatively later in the morning than white males (Figure 9–13). Virgin females tend to mate almost immediately upon emergence (which is usually within several hours of sunrise) and are protected by several behavioral and physical (spermatophore "plug") mechanisms against multiple matings. The earlier-flying white males seem to accomplish most of their courtship activity before adverse weather changes in the afternoon in these tropical localities, and thus they would apparently have the advantage in finding the virgin females and leaving the majority of the offspring for the next generation. The reasons why yellow band color does not disappear from populations will be considered below.

The evolutionary consequences of a mating system such as described for *Anartia fatima* are of particular interest since the males exercise the principal (if not the entire) choice in mate selection. In the vast majority of animal species, females exercise a stronger preference (Orians, 1969). Because males in butterflies may mate repeatedly throughout most of their lives while females of many species mate

Table 9–2 Mate selection by *Anartia fatima* during the tropical dry season (March 1968) experiments at San Vito, in terms of approach flights by cruising males (white-banded and yellow-banded morphs) to eight different models.

WILD MALES OF EACH MORPH MAKING APPROACH	FEMALE MODEL CATEGORY								APPROACH FLIGHTS (EXCLUDING MODELS 7 AND 8)
	(1) WHITE ♀	(2) YELLOW ♀	(3) WHITE ♀ NO RED	(4) YELLOW ♀ NO RED	(5) "WHITE" ♀ (BAND MARKED OUT)	(6) "YELLOW" ♀ (BAND MARKED OUT)	(7)[a] ALL-RED ♀	(8)[b] ALL-BLUE ♀	
	NUMBER OF APPROACH FLIGHTS TO MODEL								
White ♂♂	188	95	97	39	74	87	(36)	(17)	580
Yellow ♂♂	35	23	13	5	24	20	(11)	(13)	120
Total ♂♂	223	118	110	44	98	107	(47)	(30)	700
	PERCENTAGE OF TOTAL MALE APPROACH FLIGHTS								
% of white ♂♂ responding	*32.4*	*16.4*	*16.7*	*6.7*	*12.7*	*15.0*	—	—	99.9%
% of yellow ♂♂ responding	*29.2*	*19.2*	*10.8*	*4.1*	*20.0*	*16.7*	—	—	100.0%
Total white and yellow ♂♂ responding	*31.8*	*16.9*	*15.7*	*6.3*	*14.0*	*15.3*	—	—	100.0%

[a] Only used in 5 experimental set-ups.
[b] Only used in 2 experimental set-ups.
All model categories except (7) and (8) were replicated in seven simultaneous experiments.

Table 9–3 Mate selection by *Anartia fatima* males during the tropical wet season (August 1968) experiments at San Vito. See Table 9–2 and text for explanation.

WILD MALES OF EACH MORPH MAKING APPROACH	FEMALE MODEL CATEGORY								TOTAL APPROACH FLIGHTS (ALL MODELS)
	(1) WHITE ♀	(2) YELLOW ♀	(3) WHITE ♀ NO RED	(4) YELLOW ♀ NO RED	(5) "WHITE" ♀ (BAND MARKED OUT)	(6) "YELLOW" ♀ (BAND MARKED OUT)	(7) ALL-RED ♀	(8) ALL-BROWN ♀	
NUMBER OF APPROACH FLIGHTS TO MODEL									
White ♂ ♂	44	23	16	16	10	23	0	3	135
Yellow ♂ ♂	7	7	7	4	0	4	1	2	32
Total ♂ ♂	51	30	23	20	10	27	1	5	167
PERCENTAGE OF TOTAL MALE APPROACH FLIGHTS									
% of white ♂ ♂ responding	*32.6*	*17.1*	*11.8*	*11.8*	*7.4*	*17.1*	*0*	*2.2*	100%
% of yellow ♂ ♂ responding	*22.0*	*22.0*	*22.0*	*12.5*	*0*	*12.5*	*3.1*	*6.3*	100%
Total white and yellow ♂ ♂ responding	*30.5*	*17.9*	*13.7*	*12.0*	*6.0*	*16.2*	*0.6*	*3.0*	100%

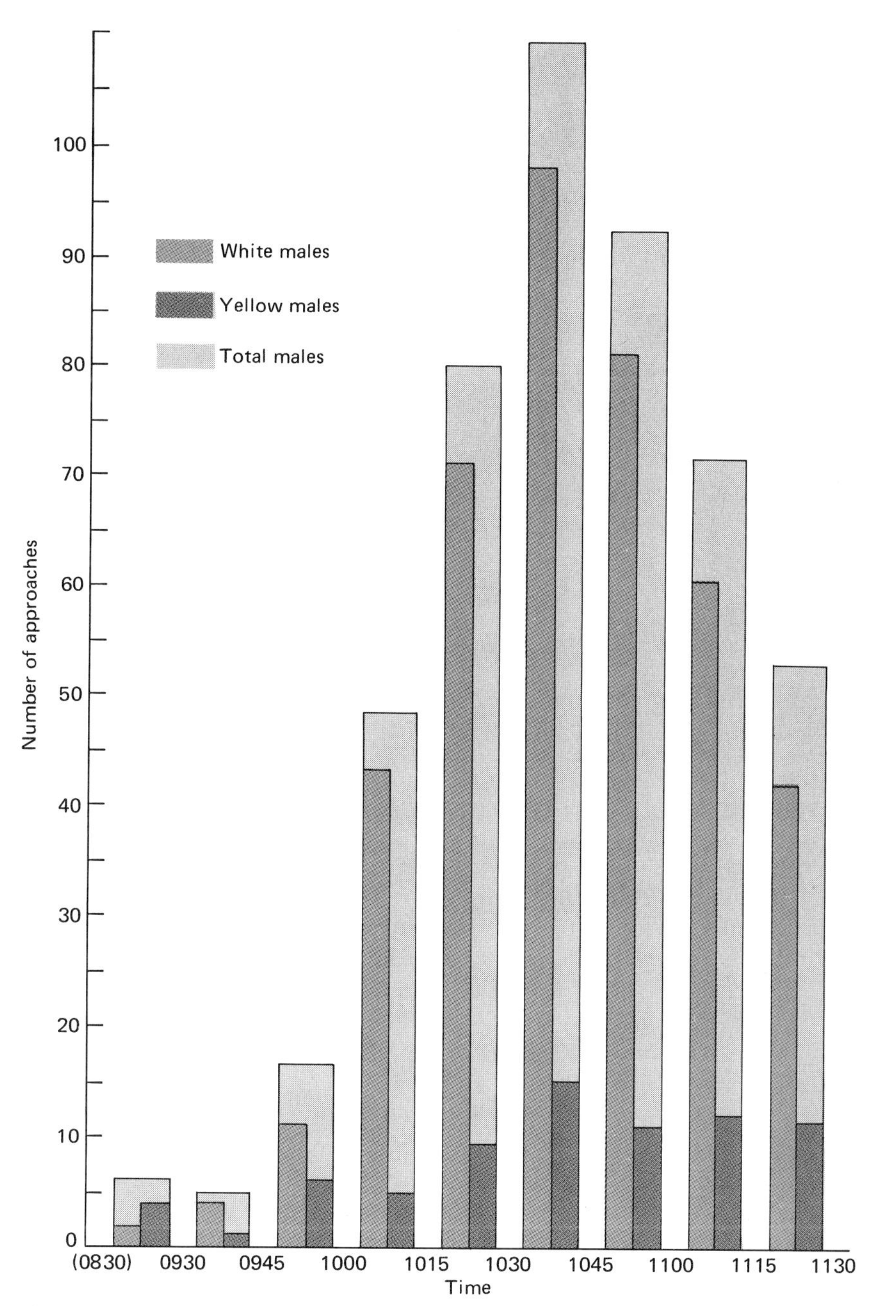

Figure 9–13 Histograms showing the number of white and yellow male *Anartia fatima* approaches to all models during 15-min intervals through a single morning at the San Vito, Costa Rica, experimental area. (The first time interval on graph is a 1-hr total because of low flight activity from 0830 to 0930.) Yellow males tend to peak in mating activity relatively later in the morning than do white-banded males. See text for details. (Source: T. C. Emmel, 1972, *Evolution, 26:* 96–107.)

only once or twice, one would expect female choice to be the key factor, causing sexual selection between the male morphs. (Sexual selection implies a female choice between different male morphs, in the original usage of Darwin (1859) and Wynne-Edwards (1962), among others; mate selection is used here to distinguish male discrimination between female morphs.) However, this is clearly not the case since live females did not exhibit preference and since both morphs of the male sex did show significant discrimination in favor of the white female morph. Thus, the mating system is working against the maintenance of variability in the population, and the equilibrium points of the observed balanced polymorphism must be determined by equally strong factors counterbalancing the negative effects of both assortative and disassortative mate selection during courtship. These counterbalancing trends for the yellow morph may include (1) physiological advantage in the adult stage for the yellow form in warm climates, as previously found in certain yellow and white dimorphic *Colias* butterfly species (Hovanitz, 1948, 1950); (2) mimetic selection favoring the slightly mimetic yellow morph; (3) mating advantage of rare males; (4) mate selection relative to total population density; (5) a cryptic advantage; and (6) a hypothetical differential fecundity or survival advantage in the immature stages (Emmel, 1972, 1973).

TYPES AND EVOLUTION OF MATING SYSTEMS

Mating requires individuals of the two sexes to become synchronized in activities and join in physical proximity. The types of behavioral mechanisms which bring about mating in animal populations may be collectively referred to as *mating* systems.

The loosest structural form of mating systems is found in *promiscuity,* in which males and females have complete freedom to mate with any individual they encounter. Although promiscuous mating systems are not infrequent and may be found even among bands of chimpanzees, there is often some form of discrimination by the females which prevents a truly panmictic population. The purest form of promiscuity is probably reached in wind-pollinated terrestrial flowering plants and certain marine invertebrates such as barnacles, which release their gametes to be dispersed randomly by wind or ocean currents.

Longer pair bonds involve two or more individuals in a close relationship. In *monogamy,* one male and one female share a pair bond and both parents typically care for the young. Orians (1969) emphasizes that monogamy is relatively rare among mammals, as the physiology of mammalian reproduction dictates a relatively minor role of the male in the case of offspring, but should be the predominant mating pattern among birds, where the only activity for which males are not equally adept as females is egg-laying. Almost all insectivorous and carnivorous birds [in fact, 91 percent of all birds species, according to Lack (1968)] are monogamous, as are many terrestrial carnivores where the male can aid in prey capture.

In *polygamy* the mating system involves simultaneous pair bonds of one individual with more than one member of the other sex. *Polygyny* refers to one male having pair bonds with two or more females at the same time, while *polyandry* includes mating systems where one female mates regularly with several males. Probably because of the relative reproductive roles of males and females, the latter state is quite rare among all animal groups,although it appears to occur in some of the button quails, painted snipe, jacanas, tinamous, and rails (Lack, 1968). Males in other groups such as rheas, emus, and kiwis, normally incubate the eggs and care for the young, but most of these species are apparently monogamous (Orians, 1959).

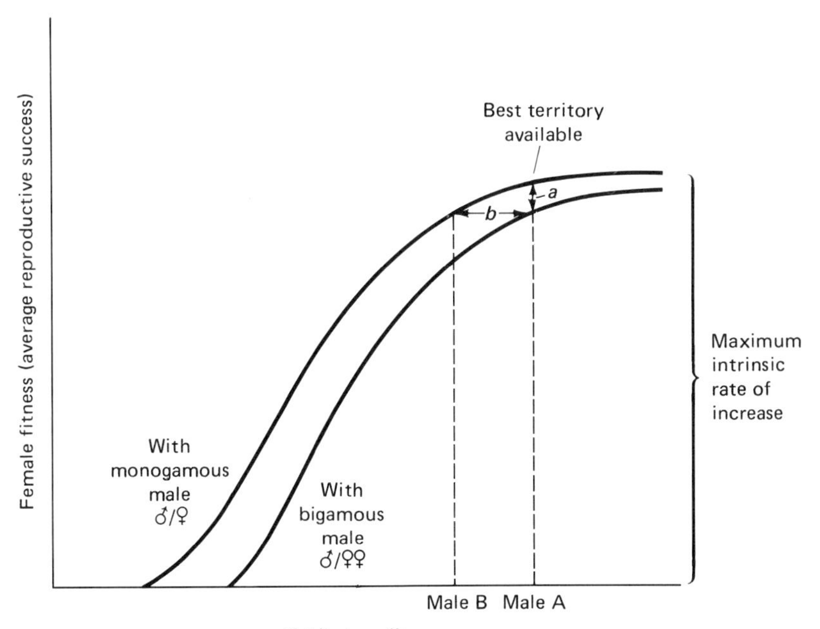

Figure 9–14 Graphic model of the conditions necessary for the evolution of polygynous mating systems. Reproductive success of females is correlated with environmental quality, and females select mates that give them the highest individual fitness. Distance *a* is the difference in fitness between a female mated monogamously and a female mated bigamously in the same environment. Distance *b* is the *polygyny threshold,* or the minimum difference in territory quality held by males in the same general region that is sufficient to favor bigamous matings by females. (After G. H. Orians, 1969, *Amer. Naturalist, 103:* 589–603. Modified by E. R. Pianka, 1974, *Evolutionary Biology,* Harper & Row, p. 121.)

The role of monogomous males in providing food and protecting the brood against predators or inclement weather may lead to the evolution of *polygyny* if the quality of habitat is such that the female's reproductive success will be higher when she mates with an already-mated male holding a superior habitat, rather than mating with an unmated mate on a poorer habitat. The difference in quality of habitats held by mated and unmated males which is required to make bigamous mating *advantageous* for a newly arriving female has been designated the *polygyny threshold* (Verner and Willson, 1966; Orians, 1969) and is depicted in Figure 9–14. Polygyny would be expected to evolve only when the females of a species are regularly presented with this situation.

Finally, as Jolly (1972) has noted, even the most highly evolved vertebrates such as the primates and ungulates exhibit every possible social structure, from solitariness through very casual association through tightly knit troops; mating systems vary from promiscuity to regular polygamy to strict monogamy. In the primates, males of many species are incorporated within a group throughout their lives and do play a significant role in the upbringing of the young. Eisenberg (1966), who has carried out the broadest survey of mammalian societies and mating systems, prefers to emphasize the similarities among different orders of mammals and the fact that it is impossible to draw hard and fast phylogenetic distinctions in behavior. The one general conclusion that can be drawn is the apparent absence of any true polyandrous mating system among mammals.

1 2 3 4 5 6 7 8 9 10 11

Seasonality and Populations

The population biology of plants and animals is affected as much by the seasonality of the climate as by any other physical or biotic factor in the environment. The aspects of population structure that we have been looking at in earlier chapters, namely, dispersal behavior, genetic composition, life history patterns, mating systems, population growth and dynamics, regulatory systems, and the effect of age and sex, are intimately adjusted to seasonal changes in climate and resources. Here we consider responses of populations to these seasonal changes in tropical and temperate regions (Figure 10–1). Let us first look at some general principles concerning the factors that determine climate.

TEMPERATE AND TROPICAL CLIMATES

Rather surprisingly, the first writings that hypothesized a link between climate and terrestrial location came from Greek philosophers of the Pythagorean and Eleatic schools of the sixth century B.C. On the theoretical basis of perfect form rather than personal observation, they postulated a spherical earth. They observed that the seasonal travel of

Figure 10–1 The strongly seasonal appearance and reproductive activities of desert annuals makes a flower-strewn carpet of the Mojave Desert in California in the early spring. (Photo by T. C. Emmel.)

the sun across the latitudes caused changes in climate and thus they set up the five-zone climatic classification system used by us even today: a central tropical zone around the equator, two mid-latitude temperate zones, and two frigid polar caps. Thus the relationship of the angle of the sun in determining climates was one of the great early insights of the mind of man regarding his environment (Bailey, 1964). Quantitative data on the relationship between solar height and heat from the sun was actually not collected until the Nineteenth Century. Alexander Von Humboldt created the first isothermal map in 1817 to show mean annual temperatures. Within a few decades, major outlines of mean, maximal, and minimal thermal distributions along latitudinal parallels were established.

At this point, it might be well to mention that the concept of *temperateness* was and still is a biological rather than a physical description. In other words, the temperate zones were defined historically on the basis of the temperatures that were most comfortable to man. Actually, it is clear that the most equable temperatures are found in the tropics and not in the temperate zone, where dramatic daily and seasonal temperature changes frequently occur. In contrast, the temperature in the tropics remains nearly constant all year on the equator at any particular altitude.

As should be obvious from the early observations on latitudinal position and climate, the driving force of climatic change is solar energy. The distribution of the input of this energy over the earth's surface varies with latitude for two reasons. First, when radiant energy hits the rounded surface of the earth, the same amount of energy is scattered over a larger surface area in the far northern or southern hemispheric regions than in an equatorial area, simply because of the incident angle. In addition, if a beam of light hits the atmosphere at an angle north or south of the vertical position of the sun, in order to penetrate to the earth's surface it must penetrate a deeper layer of air. The result is that a considerable amount of light is scattered before it ever reaches the surface. Thus, average annual temperature tends to decrease relatively uniformally towards either pole.

To understand how climate is controlled across the earth, we should recall three basic principles about atmospheric gases that affect processes of energy transfer: (1) warm air rises, (2) rising air cools, and (3) cooling air reduces its moisture-holding capacity. In the atmosphere, energy transfers occur through air currents, which are brought about in three principal ways: (1) *convection,* which is simply an upwelling of warm air (Air becomes lighter as it becomes warmed by the sun, and therefore it rises); (2) *orographic lifting,* where a mass of moving air is being pushed over a rise in the earth's terrain (Frequently, then, this means that air is moving from a heated high-pressure area near sea level to a cooler low-pressure area at a higher elevation); (3) *frontal* movements, in which a warm air mass is pushed over a cold air mass, the latter being heavier and lying next to the ground.

The basic major air pattern on the surface of the earth is thermally caused. The angle of the earth causes the rays of the sun to fall more directly on the equatorial than on the polar regions. As a result, air at the equator is warmed more rapidly than elsewhere, and according to the well-known behavior of warm air, it rises. This expanding warm air flows poleward at high altitudes, resulting in an increased weight of air over the poles. Hence, at the equator we find predominant low-pressure areas and at the poles, high-pressure areas. The air masses moving this northward or southward direction from the equator rise above their equatorial source to the tropopause, the boundary occurring at about 15 km above the earth's surface between the troposphere, a region of much air movement and most of our clouds and "weather," and the higher stratosphere, a region of stable air. By the time these air masses reach the poles, the level of the tropopause has dropped to about 10 km above the earth's surface. Thus, as a consequence of the rising air at the equator, we have low pressure there and with air settling at the north and south poles, we have high pressure at those points on the earth's surface.

The earth is constantly rotating on its axis from west to east, of course, and this sets up deflections and eddies in the returning air near the surface, as the air masses accumulating at the poles flow southward from the North Pole and northward from the South Pole toward the equator to equalize the distribution of air in a finite space, the earth's atmosphere. The general result is a secondary high-pressure area at about 30° latitude North and South, where air descends. The falling air creates high-pressure areas there, and the deserts of the world are found in this subtropical high-pressure area: the cold dry air masses warm as they fall, taking up water. The secondary low-pressure area occurs at about 60° latitude North and South, with an upwelling of moisture-laden air from the earth's surface. This area is the subpolar low. Figure 10–2 shows this idealized pattern of air circulation.

These idealized vertical movements of atmospheric currents are deflected owing to the rotation of the earth about its axis. With the planet rotating from west to east, an object in one of the polar cap regions will travel a relatively short distance during a 24-hr period compared to an object on the equator, which will travel about 25,000 miles as the earth makes a complete rotation around its axis. If the object such as an air mass is changing direction and moving either north or south, the rotational velocity will cause the mass to veer off to one

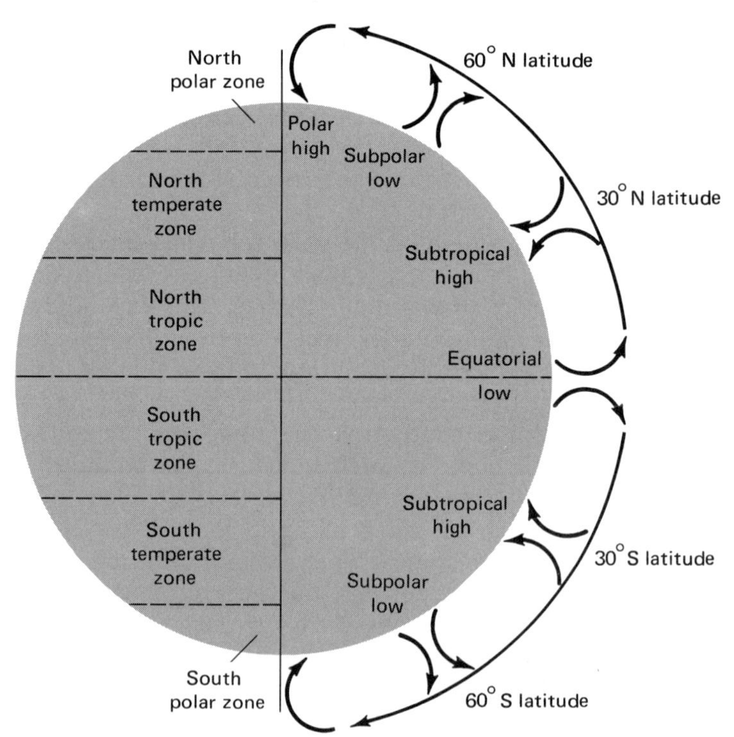

Figure 10–2 The idealized primary and secondary air movements in the earth's atmosphere, ignoring deflections caused by axial rotation.

side. Thus, the basic patterns of the westerly winds are set up in the temperate latitudes between 60° and 30° North or South, while the northeast tradewinds form between 30° North and the equator and the southeast tradewinds circulate between 30° South and the equator. The idealized picture of these atmospheric circulation patterns is depicted in Figure 10–3, but the presence of oceans, large lakes, or major topographic relief such as the high mountains on the western edge of North and South America may exercise considerable influence over regional air circulation and weather.

Climatic conditions are additionally affected by the eliptical annual orbit of the earth around the sun and by the inclination of this planet's

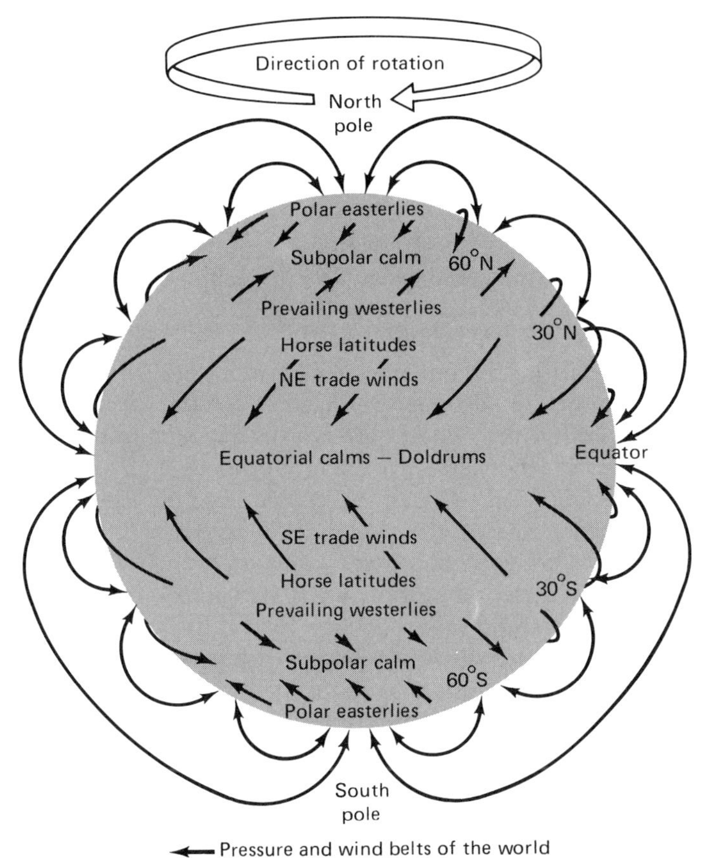

Figure 10–3 The general circulation patterns of the earth's atmosphere. The vertical profile of air movements is shown against latitude on the edge of the globe, and prevailing wind currents on the earth's surface appear across the face. These belts of moving air move north and south with the seasons. (Source: B. D. Collier, G. W. Cox, A. W. Johnston, and P. C. Miller, 1973, *Dynamic Ecology*, Prentice-Hall, Englewood Cliffs, page 73. After T. A. Blair and R. C. Fite, 1965, *Weather Elements*, Prentice-Hall, Englewood Cliffs.)

equator at an angle of 23°27′ relative to this orbital plane. The annual shift in the position of the sun in the sky, with the corresponding changes in the length of day, is caused by the angle that the spinning earth holds in this journey around the sun. If the polar axis of the earth were exactly perpendicular to the plane of the earth's orbit around the sun, the sun would *always* be directly overhead at noon at the equator and the angle of its rays would increase evenly toward either pole. But the polar axis is tilted so that the earth's equator forms an angle with the plane of the earth's orbit around the sun. Whereas this angle currently has a value of about 23°27′, it has varied somewhat during the geological history of the earth.

In the course of the earth's annual journey around the sun, then, the direct rays falling on the earth's surface gradually move upward to a limit of about 23½ degrees north of the equator, then downward to the corresponding southern limit. If we examine this situation diagrammatically (Figure 10–4), we see that the sun is directly over the equator at two times during the course of the year: on March 21, when spring begins in the northern hemisphere, and on September 23, when autumn begins in the northern hemisphere. The sun moves north from the equator between March and June, and by June 21 it is directly over the Tropic of Cancer, 23°27′ north of the equator. Now the sun begins its annual journey south and six months later on December 22, it has reached a point directly over the Tropic of Capricorn located 23°27′ south of the equator. This annual movement of the vertical orientation of the sun has profound effects on local climate. The tropical areas around the equator are always exposed to warm temperatures and a relatively constant day length of around 12 hr, which does not vary more than about 30 min at any time of the year. In the northern temperate areas of the globe, day length and temperature will vary dramatically between summer, when the sun is at its northernmost point, and winter, when the sun is far south of the equator. Because the air masses and land surfaces heat rather slowly as the sun moves north or south, seasonal changes occur in even the tropical areas. Thus, we have tropical rainfall changes coinciding the intertropical convergence, the junction between falling and rising air masses which are created as the sun moves north or south. The intertropical convergence lags behind the sun by about one month and does not move quite as far (the maximum extent being about 15° North to about 15° South). The rainy season in the tropics moves with this intertropical convergence boundary.

From the point of view of worldwide climate, then, we can extract three main points from this discussion. First, heat energy from the sun creates movement of air or winds (i.e., air currents). Second, the earth's rotation causes deflections from the idealized north–south movement of these air currents. Third, the seasonal shifting of the ori-

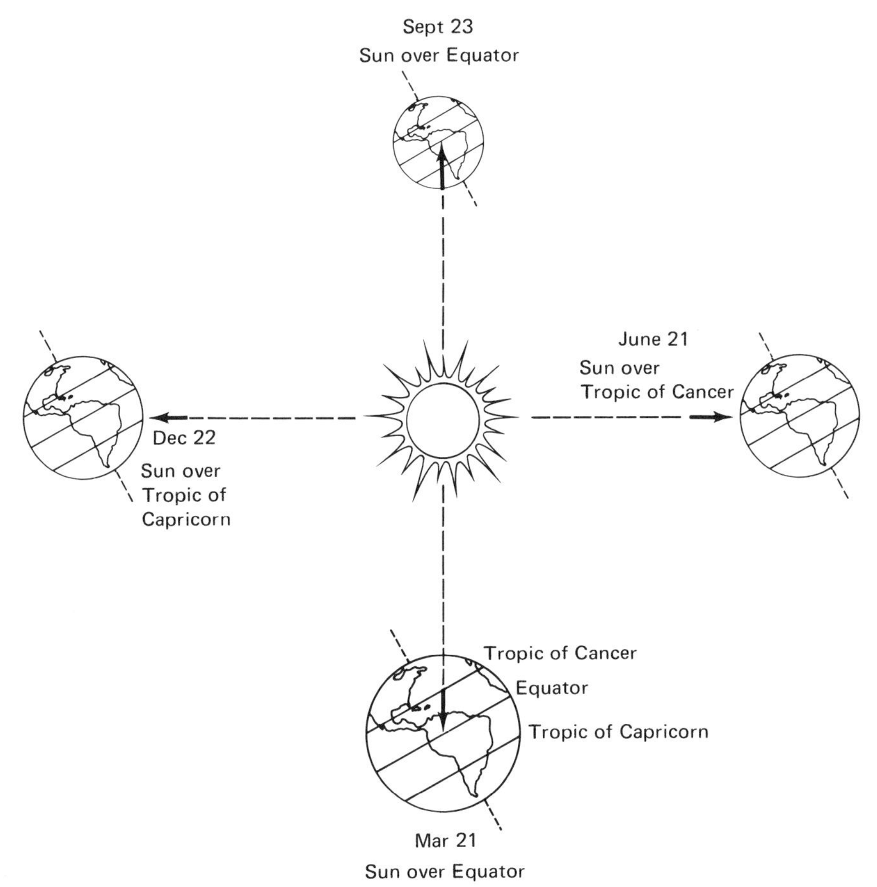

Figure 10–4 In the course of the earth's annual journey around the sun, the direct rays of light gradually move upward to a limit around 23½ degrees north of the equator (on June 21), then downward to the corresponding southern limit (reached on December 22). The equator is crossed twice (September 23 and March 21).

entation of the sun to the earth's axis causes the intertropical convergence to move about, moving the accompanying rain belts at the boundary of the equatorial low pressure areas with it (Figure 10–5).

SEASONALITY AND TEMPERATE ZONES

In temperate portions of the northern and southern hemispheres, the cold season or winter regularly alternates with a warm season or summer during the annual cycle. Depending on latitude, there will be intermediate spring and fall climatic periods of variable length between these two climatic extremes. Cold winters have led to major biological adaptations in organisms, such as the deciduous habit in

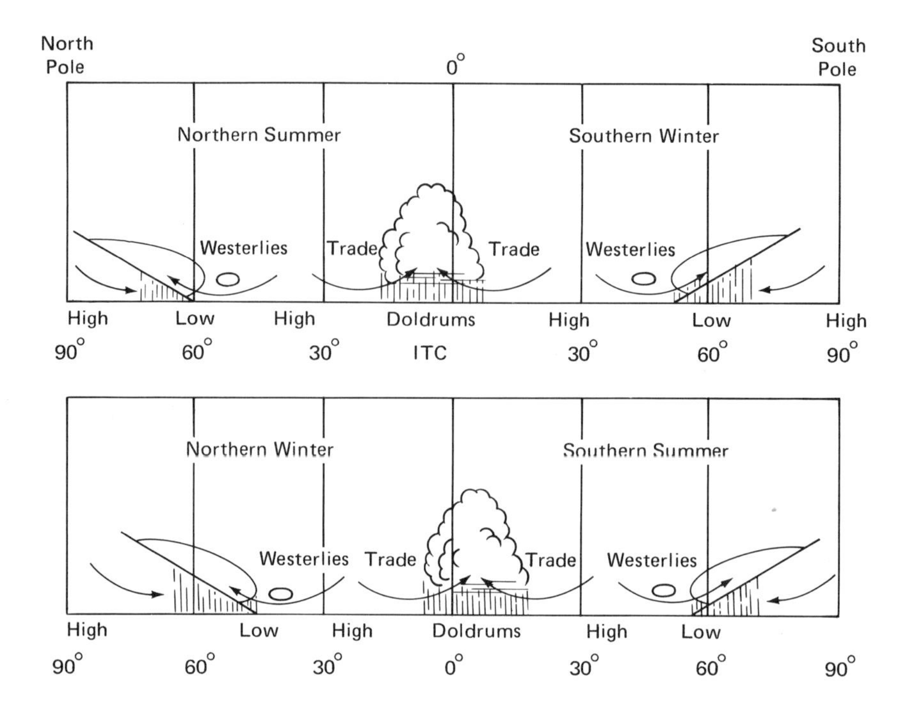

8	7	6	5	4	3	2	1	2	3	4	5	6	7	8
Sparse precipitation all seasons	Precipitation in all seasons; more in summer	Winter rain, summer dryness	Slight winter rain	Dry all seasons	Slight summer rain	Summer rain, Winter dryness	Rain all seasons	Summer rain, Winter dryness	Slight summer rain	Dry all seasons	Slight winter rain	Winter rain, Summer dryness	Precipitation in all seasons, more in summer	Sparse precipitation all seasons

a

the flowering plants and hibernation behavior in many warm-blooded and cold-blooded animals. The period of extended low temperatures requires some physiological ability to adapt one's enzyme systems and cellular fluids to survive freezing temperatures. That this has not been a particularly difficult problem is shown by the large number of species from all major animal groups that employ the strategy of hibernation to survive the temperate zone winter. Plants generally respond by withdrawing circulation of their fluids from their outer extremities and by shedding thin leaves and other exposed living tissues that might suffer damage if frozen during the winter cold.

The seasonal distribution of food in temperate zone forest communities is of course dramatically affected by the loss of leaves from the trees and shrubs. Unless an animal such as a predatory fox is capable of feeding on other organisms which it may encounter during the

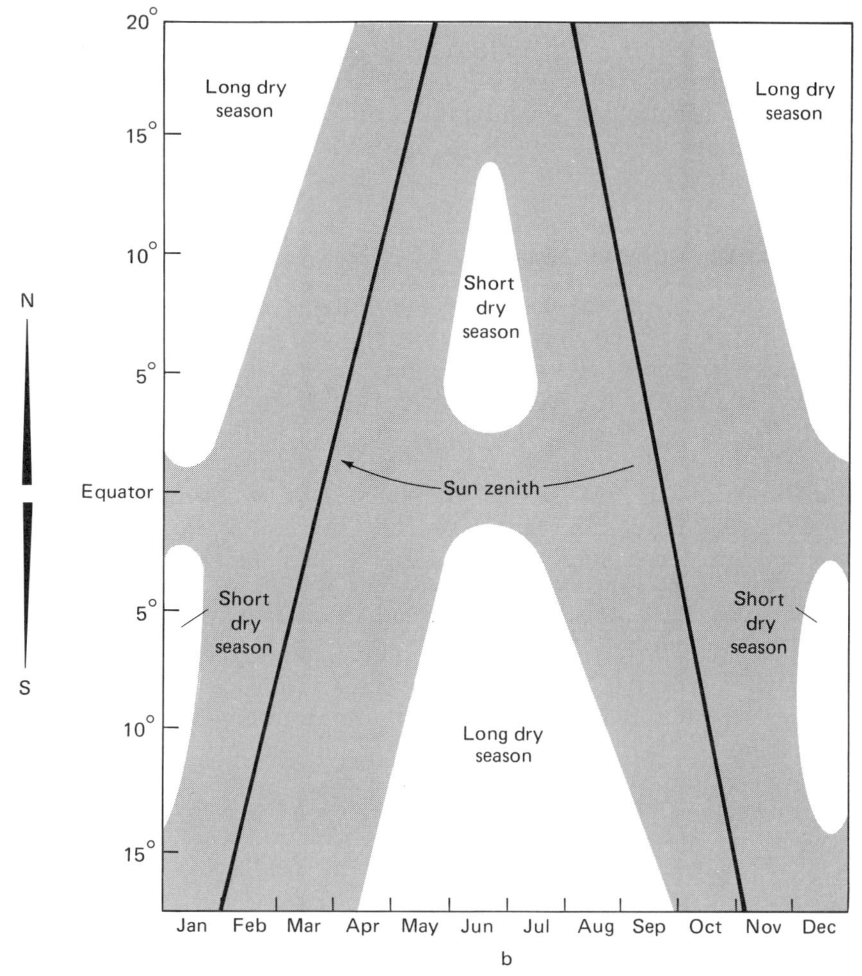

Figure 10–5 (a) The seasonal characteristics of weather at latitudes from the equator to the north and south poles, emphasizing the shifting of the inter-tropical convergence (ITC). (b) The tropical wet and dry seasons in relation to latitude between the Tropic of Cancer and the Tropic of Capricorn.

winter, the species must hibernate or die. Several general strategies for circumventing the problem of food supply during the long winter hibernation are employed. Subcutaneous deposits of fat may be built up by eating large amounts of food during the late summer months. Many small mammals store seeds, nuts, or other commodities that can be eaten during occasional short periods of activity during the winter. Another alternative solution to hibernation is for the bird, mammal, or insect to migrate out of the temperate zone for the onset of winter, and organisms as diverse as monarch butterflies and warblers carry on migrations involving thousands of miles of movement to winter in the warm southern or tropical climates. Reproduction in this temperate area takes place principally in the spring and early summer and hence is strongly seasonal. The severe winter season, in sum, must be adjusted to by hibernation, by migration, or by physiological adaptation in the form of enzymes and cellular fluids that can withstand stringent conditions.

SEASONS IN TROPICAL ZONES

In contrast to the warm–cold alternation of seasons in the temperate zones the tropics have an alternation of wet and dry seasons. The definition of what constitutes a dry season is more or less arbitrary, but in the wetter parts of the tropics, dry season is defined as the period when less than 100 millimeters (4 inches) of rain falls per month. The extreme condition in the tropical dry season would be when no rainfall occurs. The wet season is defined as a period during which more than 100 mm of rain falls per month, but usually this amounts to over 400 mm. Some localities may receive over 1000 mm (40 inches) per month in the rainy season. The onset of the rainy season or wet season varies according to the movement of the intertropical convergence (recall the one-month lag in rainfall moving along behind the change in vertical orientation of the sun). Thus, whereas the wet season starts in May in Costa Rica (with the sun being vertically overhead there in April), it is early or mid-July in the more northern region of central Mexico before the rains begin, even though the sun has reached the Tropic of Cancer in northern Mexico by June 21st. In general, the start of these wet seasons is a highly predictable event; hence, the ending of the dry season is highly predictable, also, and in the causational interpretation of biological events seemingly related to seasonality, it is important to keep this distinction in mind. The dry season and wet season are quite different environments to most organisms in seasonally dry areas, and as we shall see later, they present different requirements for optional adaptation. Among the most important are the cues for their temporal frame of reference; that is, which season is beginning and which is ending.

A tropical dry season is analagous in many ways to the cold winter of the northern or southern temperate zones. During the dry season, plants stop producing new foliage and in many areas may largely drop their leaves as an early indication of water stress. Some generalizations of interest to population biologists about the tropical dry season are the following (mainly after Janzen and Emmel, unpublished):

The tropical dry season involves a period of clear weather. During the rainy season, extensive cloud cover cuts down the amount of insolation that reaches the earth's surface. This can result in the loss of opportunity for insects and reptiles, for instance, to sit in the sun for thermoregulatory purposes. At a site in the tropical deciduous forest in Costa Rica, a day during the dry season typically has 9–10 hr of unoccluded sun, while in the wet season, about 6–7 hr of each day have unobstructed sunlight. Late-afternoon clouds build up during the wet season and rain usually falls by 2:00 or 3:00 P.M. In the rainforest of the Atlantic lowlands in Costa Rica, approximately 8 hr of daylight are present during the typical dry-season day, whereas only 2 to 3 hr of sunlight occur during the typical day in the wet season.

The daily temperature variation in the dry season is considerably greater than the wet-season variation in temperature. During the dry season, the temperature may range from a low of around 20°C at night to as much as 35° or 38° during midday. In the wet season, these temperature extremes are considerably less so that there is less difference between the nocturnal and diurnal temperatures.

There is no cold winter available to lower poikilotherm body temperatures and thus slow the rate of biochemical processes in the cells. Organisms such as insects that depend on external sources of radiation to maintain their body temperature must develop ways to survive the dry season without the aid of cold to slow enzymatic processes. Thus, if the species passes the dry season in a state of estivation, adaptive changes have to occur in the enzymes to prevent the animal from literally digesting itself in metabolic processes during its period of suspended animation.

The seasonal changes in tropical populations are keyed to seasonal changes in rainfall, whereas in the temperate zone these changes are more typically cued to changes in day length and in air temperature. The beginning of the rainy season in the tropics is very abrupt and therefore provides a good point of synchrony for plant and animal populations. Leaf production, reproductive activities, and so forth can start in synchrony among members of the whole population. In the temperate zone, only daylength provides an absolutely regular and predictable timing device to begin these activities, for although air temperature may regularly increase or decrease at specific times of the years, it is not absolutely predictable. Thus from an environmental viewpoint the hazards of starting a new cycle of reproduction in the tropics are less here than in the temperate zone, which is a less predictable environment. Once the rainy season starts in the tropics, it rains every day for six months or more depending, on the length of the cycle and thus selection for synchronization on the start of the wet season means that the organism is assured of a particular set of environmental conditions after this date.

One important result of adding a dry season in the tropics is that a new insect community may be added. In a tropical equatorial rainforest where rainfall is more or less evenly distributed all year round, there are relatively few species of ground-nesting bees. However, in the tropical deciduous forest areas of Costa Rica (8 to 11° latitude North of the equator) and the northern Neotropics, a large number of ground-nesting bees are specialized to emerge and carry on all their life cycle activities during the dry season, when nesting conditions are ideal and the large flux of flowering deciduous trees provide a rich source of nectar and pollen for the bee's own metabolic needs. Thus, one can effectively double the number of species in a geographic area by having a dry-season fauna in addition to a specialized wet-season fauna.

HOW ANIMAL POPULATIONS SURVIVE THE DRY SEASON

There are at least eight major ways in which animal populations circumvent the effects of the dry season.

Estivation or at least greatly reduced activity in burrows deep under the ground: Lungfishes (*Protopterus*) in Africa survive the dry season by burrowing deep into the mud of their drying ponds and living in an encased cavity until the next rainy season arrives. The water percolating through the soil then releases them from their concrete prison. Many rattlesnakes and other Central American reptiles estivate in underground burrows to escape the intense heat and desiccating effect of high winds during the dry season.

Diapause in one stage of the life cycle: Many insects utilize this strategy in order to escape the effects of the dry season. Hemimetabolous insects, those that go through a series of nymphal stages in development, may diapause (go into an inactive state) in either the egg stage or the adult stage. In the case of holometabolous insects—those with a complete metamorphosis—diapause may occur in the egg, larval, pupal, or adult stage. Lycaenid butterflies such as hairstreaks (*Thecla* species) that feed on particular vines or trees which lack green foliage during the dry season will lay their eggs on the stems of these plants at the end of the rainy season and then die. The eggs pass the dry season in a diapausing stage without hatching. A number of moths, such as those in the family Arctiidae (tiger moths), diapause in the third or fourth larval instar, when their food plant dries up; with their covering of protective hair and the ability to sustain some slight shrinkage upon partial desiccation, these larvae survive quite well under stems and logs and other partially sheltered places. Many saturnid silk moths and hymenopterous insects survive the dry season as pupae. In seasonally dry forests such as the tropical deciduous forests of Costa Rica, adults of a number of insect species such as pierid butterflies and nymphalid butterflies go into reproductive diapause at the start of the dry season. They rest quietly in low undergrowth along river beds most of the time during the next three or four dry months, until the first rains come again. At that point, adults initiate reproductive activities; new eggs and sperm are produced for the next generation, and after mating or remating, females begin laying eggs. Within a week or two after the arrival of the first rains, these adults die and for a month or so, there are few adult butterflies of these species to be seen in the forests. Larvae of the first wet-season generation develop during this period following the death of their parents.

Local movement of individuals and populations to moister areas: In the seasonally dry Central American tropical deciduous forest, individual animals such as tree iguanas (*Iguana iguana*), *Ctenosaura* lizards, and Felidae mammals, as well as entire populations of other species (e.g., howler monkeys and capuchin monkeys), move to the wetter gallery forest areas along permanent streams, where the vegetation remains green and provides a food source during the dry season for the herbivores, and where the concentration of herbivores provides a better food supply than the surrounding dry forest for the carnivores. Some butterflies such as *Eurema daira,* a small yellow pierid species, go through a local migratory movement combined with adult diapause. At the start of the dry season, *Eurema* butterflies move down from hills in drier areas of Costa Rica to these riverine forest areas. Within several weeks, large concentrations of adults form on the low undergrowth under protective shade trees. Few individuals move off from these dry-season roosts to consume nectar at nearby flowers; most of the time they hang motionless from twigs and branches of this undergrowth. Examination of testes and ovaries of these adults reveals that they are in reproductive diapause. When the first rains come, usually at the first of May, diapausing populations of *Eurema daira* immediately move out of the gallery forest areas and disperse widely through the hills, mating and laying eggs as they go. The adults die within a week and then during the remainder of the wet season, several generations of nondiapausing adults and their younger stages pass rapidly in succession until late December, when the last wet-season adult generation begins to go into reproductive diapause and starts its movement back to moist stream bottoms.

A switch of food preferences to sources that are available in the dry season: The Tamandua anteater (*Tamandua tetradactyla*) switches from eating mainly ants during the wet season to eating mainly termites during the dry season, gaining a greater source of water from the moister food to face the desiccating climate of the dry season. Howler monkeys (*Alouatta villosa*) graze on canopy tree leaves during the wet season, but during the dry season they change opportunistically to available flower buds of the many blooming species, eating not only nectar but obtaining a rich source of food in the flower petals and stamens themselves.

Subsist on stored food: Animals that employ this strategy simply stop eating during the dry season, having presumably stored up sufficient food in the form of subcutaneous fat during the preceding wet season to allow for survival. *Ctenosaura* lizards and land tortoises located far back in dry forest areas away from any permanent streams seem to employ this strategy in Central America.

Reduce population size: In the seasonally dry tropical deciduous forest of the New World tropics, *Heliconius* and ithomiine butterflies and the spectacular iridescent blue butterfly *Morpho peleides* (Young and Thomason, 1974) pass the dry season in continued but greatly lowered reproduction. The few individuals that survive into the next wet season lay eggs, of course, under optimal conditions and the population builds up rapidly again to its pre-dry-season levels.

Avoid the dry season by migrating out of lowland areas or seasonally dry areas: The tremendous large-scale migrations of ungulates in East Africa (discussed in Chapter 1) are a classic example of this strategy for avoiding seasonally dry areas in the tropics. Organisms capable of migrating scores or hundreds of miles on land or in the air commonly migrate to escape effects of a seasonal dry period.

Thus, animal populations respond to the dry season to survive periods of severe environmental stress in a number of ways. The climatic changes and responsive biological changes occurring in animal populations profoundly affect the seasonal activities of plant populations in the tropical forest, not only from the point of view of sexual reproduction but also because of the impact of herbivorous animals. Likewise, seasonality of climate causes fluctuations in animal population sizes, and a number of seasonally related environmental factors influence the seasonal distribution of breeding in the tropics, despite the nearly constant annual temperature regimes there. We examine these problems below.

REPRODUCTION AND FRUITING IN TROPICAL PLANT POPULATIONS

Janzen (1967) proposed that many tree species in lowland Central America have evolved the timing of their flowering and fruiting period to coincide with the dry season. The adaptive significance of this synchrony lies in maximizing vegetative competetive ability of individuals and in maximizing the use of pollinating and dispersal agents. What *are* the advantages and disadvantages of sexual reproduction during the dry season? That is, we have already talked about mechanisms synchronizing various activities of many species to seasonal changes, such as physiological stimuli or the sudden start of the rainy season. These are proximate causes of dry-season seasonality. Now we want to shift our focus to the ultimate causes, namely the advantages and disadvantages of seasonality in sexual reproduction in plant populations. There is little argument with the evidence for frequent synchronization of flowering and fruiting in the dry season (Figure 10–6), as it occurs even among tree populations in tropical rain forest communities which experience a slight dry season.

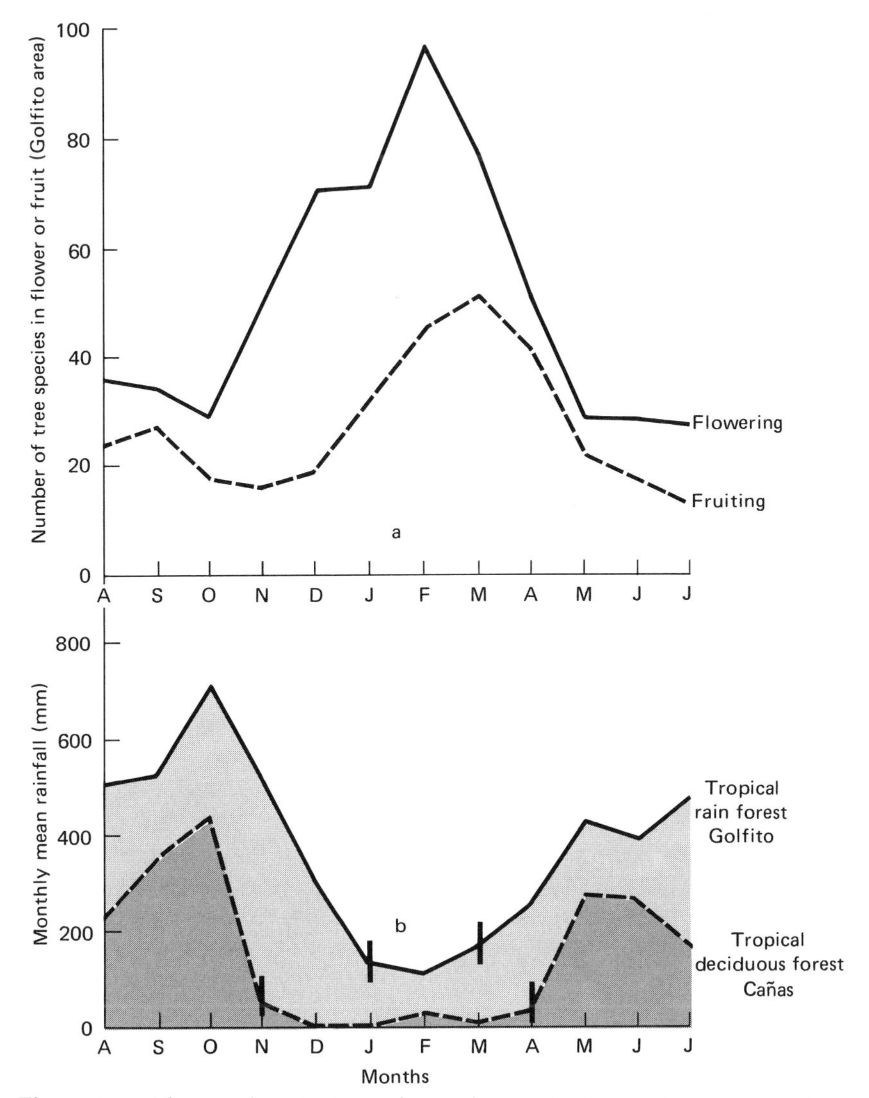

Figure 10–6 The synchronization of sexual reproduction of trees within the dry season in Central America. (a) Periodicity of flowering and fruiting in rain forest tree species in Golfito, Costa Rica. (b) Periodicity of mean monthly rainfall in the evergreen rain forest area of Golfito (1941–1960) and the drier tropical deciduous forest area of Cañas (1951–1960) in Costa Rica. The dry season in each site, as indicated by reduced vegetative growth and phytophagous insect densities, occurs between the vertical bars. (Adapted from D. H. Janzen, 1967, *Evolution, 21:* 620–637.)

Selective forces promoting dry season flowering and fruiting involve (1) competition, (2) pollinators and dispersal agents, and (3) lack of available water. The dry season is characterized by a lack of vegetative competition, for trees are generally inactive and lack leaves. At the start of the rainy season, there is a major flush of vegetative growth and at this time, maximum stem elongation and leaf production occurs. To flower at this particular time would place the individual at a severe disadvantage compared to other actively growing trees and vines in the community. The lack of vegetative competition during the dry season and its intense character during the wet season should cause flowering and fruiting to shift toward the dry season. Reserves stored during the wet season growing period could then be used in flowering and fruiting. Also, to maximize use of moisture still in the surface soil from the wet season, there should be a tendency to flower and fruit as early as possible in the dry season. Early flowering and fruiting also permits fruit maturation before the beginning of the rainy season and ensures that seedlings have a maximum chance of receiving sunlight in early successional stages, as well as having the entire rainy season to grow a root system before the next dry season ensues.

Activities of pollinators and dispersal agents during the dry season are also favored. With the paucity of insects and free-standing water, coupled with the almost complete absence of foliage to act as shade, birds face a severe deficit in their water budget during the dry season. Thus, nectar-bearing flowers and succulent fruits form an important source of water as well as nutrition at this time of the year. Additionally, the lack of leaves in the canopy provides ready visibility of the fruit and doubtless facilitates location of flowers and fruits by animals. Many bats and birds are specialized to feed on flowers and succulent fruits, and the activities of such animals should be important in promoting flowering and fruiting throughout the entire year, including the dry season as well as the wet season. Bird and bat activity at individual plants which flower or fruit at a season when many of the other plants are not in reproductive activity should result in increased survival of those plants that reproduce at odd times. Thus there would be considerable selection pressure by these vertebrate pollinators and dispersal agents towards dry-season flowering and fruiting. A large number of other common tropical mammals are flower, nectar, or succulent fruit eaters, including monkeys, opossums, squirrels, deer, peccary, kinkajou, and coati. With young leaves, free water, and insects in relative scarcity, nectar and fruits become increasingly important in dry-season diet with such species. Adaptations on the part of the flowering or fruiting tree to insure a larger and more succulent flower and fruit crop during the dry season should insure that those trees attract a larger visitor population during this period of food shortage.

Thus, trees that meet this need of their vertebrate visitors will tend to spread at the expense of others that do not flower or fruit in the dry season.

Dry-season weather conditions favor diurnal and nocturnal insects in their primary role as pollinators because of the increased amount of sunshine per day during the dry season and the lack of rain to fill flowers (which are knocked off stems or have their nectar diluted by precipitation). These insects find maximum availability of this food during the dry season. For ground-nesting bees and other insects, nesting conditions are improved during the dry season, when rain water does not flood entrances and bare surface areas are exposed by the loss of leaves and falling water levels in ponds and rivers. Problems with fungal growth in underground nest cells are reduced by the drier air. Also, air temperatures during the dry season reach higher levels earlier in the morning than during the rainy season and thus there are more total hours when tropical insects have the necessary temperature range for maximum activity.

Besides biological selection for dry-season activity, the dry air and strong winds characteristic of the Central American dry season provide physical environmental selection conducive to flowering and fruiting in the dry season. Dry air aids in the operation of exploding seed pods (characteristic of several families of plants) due to differential shrinkage of different parts of the valve as it dries. Heavy winds are common during the dry season in Central American lowland deciduous forests and in the absence of foliage, they can reach the forest floor. This is an excellent time for wide dispersal of airborne seeds, as Janzen (1967) has emphasized. Finally, pollen damage by rainfall would tend to be one more selective force favoring flowering during the dry season.

Of course there is substantial selection *against* flowering in the dry season for many species because of the lack of water in the soil. Plants with shallow root systems or poor water-storage ability and those that lack drought-resistance in the aerial shoot rarely flower then, or at least during the last two-thirds of the dry season. Advantages to flowering and fruiting during the rainy season are that many animals do not eat merely green vegetation but require nectar and fruits for food and those needs will selectively promote sexual reproduction of plants at that time if these animals are pollination or dispersal agents. Trees and herbaceous plants that lack water-extraction or nutrient- and water-storage abilities necessary to reproduce sexually during the dry season should also find an advantage to reproduction in the wet season.

In conclusion, Janzen (1967) proposed with outstanding evidence that "The peak in flowering and fruiting of tree species in the lowlands of Central America is a result of selection for sexual reproduction at the most opportune time in the year, rather than a result of immutable physiological processes which can only occur at that time of the year." Other investigators have found similar types of selective factors influencing the evolution of fruiting seasons in tropical forest plants. Snow (1965) was studying the ecology of forest birds in the Arima Valley of Trinidad, an island off the northeast coast of the South American country of Venezuela, when he started looking at the seasonality of fruiting in trees and shrubs of the genus *Miconia* (family Melastomaceae). These shrubs and trees have fruits which form a major part of the diet of the smaller frugivorous birds such as manakins and tanagers. This observation suggested a possible selective mechanism for the evolution of flowering and fruiting seasons in this genus. The fruiting seasons of 18 species of *Miconia* were remarkably well dispersed throughout the twelve months of the year, providing a continuous supply of ripe *Miconia* fruits (Figure 10–7). Every month, at least two species are in fruit, and in most months there are three or more. A reasonable postulate is that if one started with a number of species growing together in the same area and all fruiting at the same time in response to climatic changes, any species that started to shift its flowering and fruiting season would obtain a strong selective advantage. In Snow's (1965) view, the advantage would be twofold: First, if fewer other species were in fruit at the same time, the fruits of the species which we are supposing to be shifting in season would be more likely to be eaten by birds and its seeds thus dispersed; and second, its seedlings would be less exposed to competition from seedlings of other species. Selection would therefore almost certainly favor the breakdown of the original synchrony and lead to a state in

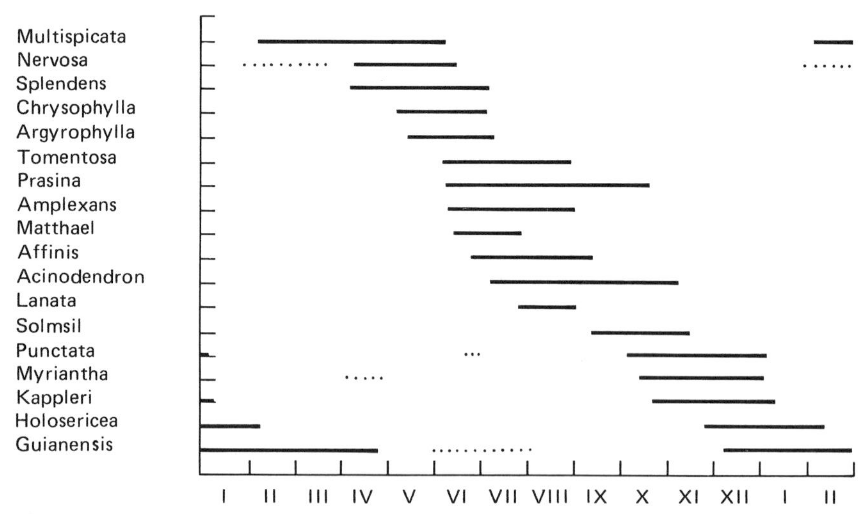

Figure 10–7 Fruiting seasons of 18 species of *Miconia* in the Arima Valley, Trinidad. Dots indicate occasional out-of-season fruiting. (Source: D. W. Snow, 1966, *Oikos, 15:* 274–281.)

which the fruiting seasons of the various species were as far as possible staggered throughout the year, such as in fact is found in the Arima Valley of Trinidad. An important consequence of such a staggering of fruiting seasons would be that a more constant fruit supply would be available throughout the year, thus providing conditions suitable for the maintenance of sedentary populations of the fruit-eating birds and other animals. This would in turn favor the more efficient dispersal of their seeds and generate the selective pressure necessary for the maintenance of the staggered fruiting season (Snow, 1965).

REPRODUCTION AND FRUITING IN TEMPERATE PLANT POPULATIONS

As in the case of tropical forest species, the seasonal reproductive activities of plants in temperate zones likely arise from the same sorts of ultimate evolutionary causes. A great many trees and shrubs flower in the early spring, often even before vegetative growth occurs or else immediately after leaves are produced. In agriculturally important trees such as peaches, apricots, and other members of the genus *Prunus,* or common Eastern American forest species such as *Ulmus* (elms) and *Acer* (maples), there is an obvious ecological and evolutionary advantage to placing the fruiting and flowering periods early in the growing season, reducing conflict with the energetic demands of the later vegetative growth phase. There is also an obvious advantage in placing one's seeds on the ground early in the growing season, giving the seedlings the maximum amount of time for rooting and their first year of growth. Trees such as *Quercus* (oaks), *Prunus,* and *Juglans* (walnut) have slowly maturing fruit but also have early flowering in the spring. An evolutionary advantage of slowly maturing large fruit can be seen in that the energy necessary for the development of these fruits is not inimical to, that is, does not cause a drain on, the energy required for vegetative growth and competition with other species in the forest. Anderson and Hubricht (1940) suggest that early blooming plants in the understory, such as rhododendrons, may have their reproductive activity concentrated in the period before the canopy trees leaf out, so as to take advantage of the time of maximum light on the forest floor. The vegetative activity of rhododendron and other understory plants at the same time as flowering occurs provides additional evidence in support of this hypothesis.

Activities of new pollinators and predators of temperate zone plants may also influence the flowering, fruiting, and general reproductive biology of the species. The monument plant, *Frasera speciosa* (Gentianaceae), is a conspicuous and abundant perennial in the Rocky Mountains where it forms discrete colonies with groups of spectacular inflorescences (up to two meters high), which appear after four to seven years of vegetative growth from a large fleshy root. Inflorescences are almost completely absent in some years and in great abundance in others. This sporadic yet synchronous flowering is the most conspicuous feature of the reproductive biology of the species and its adaptive value was investigated by Beattie et al. (1973) in a mountain valley in Colorado. The blooming *Frasera* plants attract and maintain a remarkable number and variety of insect visitors despite the low percentage of plants with inflorescences and the occasionally wide geographic separation of individuals. This host of floral visitors insured a seed set which is never less than 52 percent. Seed set in *Frasera* is not significantly affected by predispersal herbivory, that is, insects or other animals destroying seeds before dispersal. Beattie and his coworkers viewed this local synchrony of flowering coupled with the almost total absence of flowering in some years as a strategy for predator avoidance. These features of the reproductive biology also reduce pressure from the sympatric plant species competing for pollinators. Reproductive advantages are therefore conferred to a minority species which might otherwise rapidly decline to extinction in the presence of intense competition from other plants. The authors suggest that

> the flowering regime of *Frasera* combines a predator avoidance system which yields widely dispersed colonies in space and time with a pollination system which successfully exploits the maximum diversity of floral visitors and maintains excellent seed set whenever and wherever the colonies appear. The systems are clearly complementary in preventing the build-up of predator populations while maintaining an attractive forage source for potential pollen vectors. A combined effect is to maintain the abundance of the species in a variety of stress environments, in return resulting in a remarkably wide geographic success. It is considered likely that similar systems will turn out to be very common among entomophilous plants in both temperate and tropical regions.

The impact of herbivore animal populations on seasonality in plant populations is also beginning to be investigated. An interesting temperate zone situation has been described by Breedlove and Ehrlich (1968), whereby an herbivorous insect has apparently caused a directed selectional change in the seasonality of flower and seed production in a plant. This system involves a perennial lupine, *Lupinus amplus,* growing in populations near Gothic and Crested Butte, Colorado. The herbivore is a small lycaenid butterfly, *Glaucopsyche lygdamus.* Females of this small blue butterfly oviposit only on the immature pubescent portions of inflorescences (Figure 10–8). Breedlove and Ehrlich found that if some flowers in the inflorescences were already open, the butterfly would not oviposit (Table 10–1). When larvae hatch, they feed on the wing and keel of the corolla and the developing stamens contained within the keel. Because of this herbivory,

Figure 10–8 The lycaenid butterfly *Glaucopsyche lygdamus* (ventral surface) and the perennial lupine *Lupinus amplus* near Crested Butte, Colorado. (left) Inflorescence of *Lupinus* without open flowers; (right) inflorescence with open flowers at base and unopened flowers at apex. (Source: D. E. Breedlove and P. R. Ehrlich, 1968, *Science, 162:* 671–672.)

flowers attacked by lycaenids usually do not reach maturity, thus preventing pollen release and seed set, and subsequently they fall off. This lycaenid butterfly destroys nearly 50 percent of the potential seed production in the populations of the lupine at Gothic. This intense selective pressure suggested that the lupines are exposed to a long-term

Table 10–1 Distribution of *Glaucopsyche* butterfly eggs on the two types of *Lupinus* inflorescences shown in Figure 10–8.

	NUMBER OF INFLORESCENCES	
NUMBER OF EGGS	NO OPEN FLOWERS	OPEN FLOWERS
0	43	120
1	53	8
2	14	2
3	9	0
4	3	0
5	1	0
6	1	0
7	0	0
8	1	0

SOURCE: Breedlove and Ehrlich (1968).

attrition of their seed production. Lupines depend on scattering a large number of seeds over a broad area because they germinate only upon disturbance and scarification of the seed coat. With such an immense selective effect from just one predator, we might expect to see the evolution of behavioral as well as morphological adaptations to repel predator attack. One selective response of the plant to lycaenid attack has apparently been the advancement of flowering time. Breedlove and Ehrlich note that

> The Gothic population of *L. amplus* seems to have been pushed to its earliest limits, as many examples of frost-killed and damaged inflorescences were observed. The butterflies oviposit strictly on the immature inflorescences, indicating that the plants on which flowers mature before the adult butterflies emerge, or early in the flight season, would be least subject to damage. There is no other obvious reason for the early flowering, as seed production is completed with more than a month of growing season remaining.

REPRODUCTION IN TEMPERATE AND TROPICAL ANIMAL POPULATIONS

Let us now look at some of the factors affecting seasonality of breeding in animal populations in the tropics as compared to the temperate zones. Among the environmental factors causing seasonality in breeding of animal populations are air temperature, water temperature, the period of maximum availability of food to the young of a breeding population, and the general availability of resources to the adult population. We shall look at ultimate factors again as we did for plant populations, and will ignore proximate factors such as photoperiod which may trigger seasonality of breeding.

Air temperature

In California, Dobzhansky and Epling (1944) studied the seasonal cycles of *Drosophila pseudoobscura* at different elevations on Mount San Jacinto in California. The occurrence of this fruitfly in different months of the typical annual cycle is shown in Table 10–2. At the lowest elevation in Andreas Canyon (800 ft above sea level) the population reaches its maximum peak in March and April. Higher in the transect at Pinyon Flat (4000 ft) the population peaks are reached in April and May. At Idyllwild (5300 ft) the maximum population levels shift to June and July. These seasonal maxima are relatively constant at each site with reference to the time of year, but population density itself fluctuates from year to year, and this seems to be partly a function of rainfall. Dobzhansky and Epling showed that flies breed throughout the period of maximum abundance by releasing large numbers of

Table 10–2 Occurrence of *Drosophila pseudoobscura* in different months along a climatic and altitudinal gradient from the Colorado Desert near Palm Springs up Mount San Jacinto, California.

LOCALITY	J	F	M	A	M	J	J	A	S	O	N	D
Andreas Canyon	—	x	X	X	x	·	·	·	—	—	—	—
Piñon Flat	—	x	x	X	X	x	·	·	—	—	—	—
Keen Camp	o	o	o	—	x	X	X	x	—	—	—	o
Idyllwild	. . .	. . .	. . .	. . .	x	X	X	x	. . .	. . .	. . .	. . .
Aldrich	172	131	512	1113	562	2	6	0	1	62	144	231

SOURCE: Dobzhansky and Epling (1944).
Key: (·) few flies or none; (—) fairly abundant; (X) maximum; (x) increasing to or decreasing from the maximum; and (o) unknown, but probably none or only occasional.

flies (e.g., 3297 at Idyllwild on one day) which were homozygous for the recessive mutant gene *orange eye*. Genetically marked flies of this type were found throughout the summer after their release in mid-June. These were not survivors of June releases but were young individuals which had since hatched in the wild (as determined by examination of reproductive organs). The degree to which the population of a given area contracts and expands is apparently a function partly of rainfall. However, the climatic factor of *air temperature* clearly determines the time of year that seasonal maxima are reached. The temperature must be above 50° (10°C) for *Drosophila pseudoobscura* to be active. Once the daily temperature range includes diurnal temperatures above this minimum level, seasonal maxima begin to develop. Changes in combinations of air temperature, humidity, and light levels influence the daily activity of these flies, but on a seasonal basis, air temperature seems to be the principal factor influencing seasonal maxima of emergence.

In tropical lowlands, temperature varies little and has no significant influence on seasonality of emergence or births. Instead, rainfall and plant production of foliage and flowers or fruits influence breeding in both vertebrates and invertebrates, as we shall explore in the sections on maximum availability of food for young of breeding population, and in the availability of resources for the adult population (*vide infra*).

Water temperature

For organisms living in water during at least part of their life cycle, the temperature of the aquatic environment will influence breeding activity. Tropical organisms do not face these kinds of temperature problems because water temperature remains nearly constant throughout the year, just as does air temperature at any particular elevation. In the temperate zone, though, seasonal changes in water temperature may assume considerable importance in regulating seasonal activity of breeding. Johnson (1968) has examined the seasonal ecology of the dragonfly, *Oplonaeschna armata,* whose larvae inhabit rocky canyon streams in the arid southwestern mountains of the United States. In Arizona and New Mexico, Johnson found that habitats supporting *O. armata* have the following attributes: the water fluctuates widely from flood stages during summer thundershowers to near desiccation at other seasons, and is moderated only by springs along the drainages; winter snowfall varies from year to year and contributes little to the level of surface water and then only for short periods in the early spring; the aridity of the general terrain produces near complete isolation between streams and in the lower portions, the water disappears into the ground before reaching the desert basins. The summer floods make the stream bed a harsh habitat and the regularity of flooding each year with its potential accompanying damage constitutes a strong selective agent. The surprising survival ability of these dragonflies suggest that there is some compensating adaptation to flooding, and Johnson's investigation of a population located in the Magdalena Mountains of New Mexico demonstrated the existence of some unique life history adaptations.

Dragonflies have an incomplete metamorphosis. The egg of *O. armata* is laid in shallow water-filled depressions and the larva develops there, going through 10 or 12 instars and taking three years to mature. At this point the larvae crawls out on an exposed rock surface less than 60 centimeters above the water and the larval skin splits, allowing the adult to emerge. The new adults go through a 12 to 20 day maturation period in the open forest on mountain slopes above the streams. Following maturation they fly in patrol-like flight back and forth over the stream. One can obtain an exact record of the number of adults emerging in a population by counting the number of fresh untenanted exuviae (cast exoskeletons) on the rocks and tree trunks by the stream each morning. Table 10–3 shows the numbers of dragonflies emerging and their sex ratio at emergence for the Water Canyon population in the Magdalena Mountains. The population size of adults in Water Canyon was estimated at close to 50 insects. The remarkable emergence pattern found by Johnson indicated that all adults emerged in nine days and that 40 percent of them emerged on the first day, 20

Table 10–3 *Oplonaeschna armata:* Numbers emerging and sex ratio at emergence for the Water Canyon population in New Mexico.

YEAR	NUMBER OF EXUVIAE	PERCENTAGE OF MALES
1963	42	40
1964	48	42
1965	40	30

SOURCE: Johnson (1968).

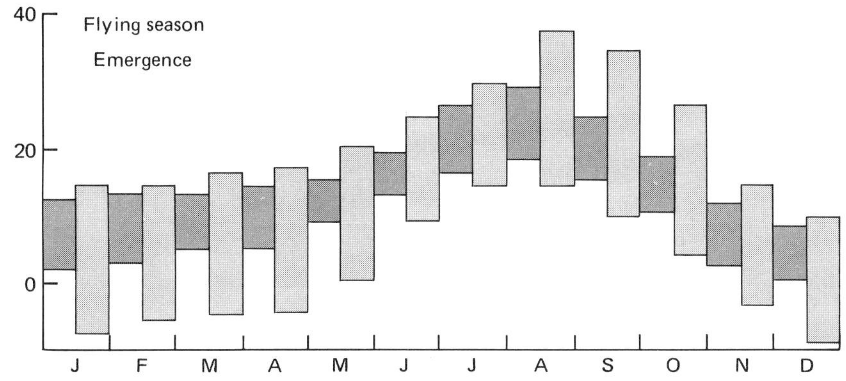

Figure 10–9 The emergence time and flying season of adults of the dragonfly *Oplonaeschna armata*, and the monthly ranges for maximum and minimum temperatures for water and air (15 cm aboveground) in Water Canyon, Magdalena Mountains, New Mexico, during 1965. Temperature ranges for water are indicated by checkered bars, and values for air by clear bars. Months are shown on the abscissa and degrees centigrade on the ordinate. (Source: C. Johnson, 1968, *Amer. Midland Naturalist, 80:* 449–457.)

percent more on the second day, and 10 percent more on the third day. The flying season lasted about 31 days, ending on August 5; thus, the flying season terminated before maximum air temperatures were reached in the areas. However, oviposition by the new females was timed to occur during maximal water temperatures (see Figure 10–9). Johnson found that approximately three weeks are required for completion of egg development, which occurs faster at warmer water temperatures. Very rapid growth rates characterize the early instars so that the larvae reach a mean length at 4.0 mm by October of their first year of growth. The maximum speeding of development may be important in a selective sense to get the first-year larvae to a size where they can successfully cling to the rocks during periods of flash flooding from late summer rains. Thus, the selective factor of water temperature proves more important to timing of breeding in these aquatic insects than do air temperatures, both in terms of activity of the aerial adults and aquatic larvae.

Maximum availablity of food for young of breeding population

The importance of this factor in influencing seasonality of breeding is easily recognized in temperate zone populations. Herbivorous insects time their emergences to coincide with the period of maximum availability of their food, whether this is pollen, nectar, or foliage. Many herbivores must time their seasonal activity carefully to avoid plant defenses. Paul Feeny (in Ehrlich, 1970) reports that larvae of the winter moth *Operophtera brumata* will not mature satisfactorily on oak leaves two weeks older than those on which the larvae normally feed. In nature, larval development is completed rapidly very early in the growing season, before the protective tannins are laid down in the young oak leaves.

In grazing animals such as deer or antelope, in which breeding takes place in the fall and gestation during the cold winter season, the births take place in the spring when the plant productivity of the community increases dramatically. Likewise, Owen (1966) and others have mentioned many examples of tropical mammal species in areas with a major dry season, where births are timed to occur at the beginning of the wet season. Tropical terrestrial insect populations also erupt at those times even though the air temperature range may be approximately the same all year.

Pipkin (1953) looked at the interesting question of whether cyclical population densities that have been demonstrated in temperate-zone populations of *Drosophila* would occur in tropical *Drosophila* populations inhabiting areas with a constant wet warm climate. Using almost daily field collections, she examined monthly population densities in *Drosophila* species on Moen Island in the Eastern Caroline Islands of the South Pacific. On this wet tropical island, annual rainfall amounted to about 3198 mm (125.9 inches) and was essentially evenly distributed all year round (5–18 inches per month). The only climatic change was a windy season occurring from January through April, when winds of up to 10–24 miles per hour blew across the islands. Mean monthly temperature varied from 79.8°F in August to 81.5°F in December. Daily temperature variation was somewhat greater: 10–12°F. Relative humidity stayed remarkably constant around 80–86 percent, dropping as low as 75–78 percent during the windy season.

Large fluctuations in population size of two *Drosophila* species were apparently due almost entirely to the presence or absence of fallen ripe breadfruit near the traps used for censusing. These trees have maximum fruiting in the spring and autumn. Pipkin found that *Drosphila ananassae* (with a 16-day life cycle) increased its population much more rapidly than its competitors in the brief spring fruiting season of the breadfruit tree. The population of *D. hypocausta* (with a 26-day life cycle) could not take sufficient advantage of that brief fruiting period to expand its population beyond its normal low level.

Thus, in an area in which temperature and humidity were practically constant all year, cyclical population expansions of *Drosophila* species could be expected, depending upon the fruiting seasons of certain tropical trees, the fruit of which serves as food for *Drosophila* larvae and as breeding grounds for adults. Small month-to-month fluctuations of *Drosophila* populations may occur between fruiting seasons of the major tree sources of food (e.g., breadfruit), and these are due to irregular local variations in the alternate food supplies.

Seasonal influence on breeding activities in tropical bird populations may also be observed, even in wet forest areas with no major climatic variations through the annual cycle. Miller (1963) studied breeding cycles of the avifauna in an equatorial cloud forest at San Antonio in the western Andes of Colombia at about 6500 ft elevation. Here climatic conditions are relatively uniform, with fog almost continually sweeping over the ridges. Many epiphytes and mosses hang on the trees, and there is a dense tangled understory. Miller observed that in this region the trees and shrubs followed no seasonal pattern of flowering. In fact, within a species individual trees came into reproductive activity asynchronously, and different individuals always could be found in flower and fruit the year round.

Total rainfall at this tropical montane site is about 53 inches per year. Lying on the equator, the area experiences two somewhat wetter seasons of three months each, alternating with two dry seasons; the major dry season lasts from June to September and the shorter one from January through February. During dry seasons, about 1 to 3 inches of rain fall per month, while in the wet seasons 5 to 8 inches of rain may be expected. The pattern of heavy clouds hanging over the ridges creates moister conditions than might be expected in other lowland forests having 53 inches of rain per year. The monthly minimum temperatures are from 56.7° to 58.6°F and the monthly maximum temperatures are 68.4° to 73.8°F, showing remarkably little range. The greatest difference in daylength throughout the year is only 24 min; that is, 12 hr ± 12 min.

The seasonality of breeding data is summarized in Figure 10–10 for 111 species found breeding in this area. The maximum number of species breeding varies from a high of 56 in March to lows of 9 and 13 species for November and December, respectively. The curve indicating number of breeding species per month in Figure 10–10 demonstrates several important facts. First, the entire year is utilized for breeding by the bird community, thus spreading out the general impact on the environment of the heavy food requirements connected with raising young. Second, the amplitude of variation in breeding by the species present in the community is moderate, for the high levels are roughly three times as great as the low. Finally, in the course of the year there is only one distinct peak, in the month of March. This is correlated with the departure of northern migrants.

Miller examined a number of factors to determine correlations of these breeding peaks with seasonality in environmental conditions. Variations in photoperiod and temperature were slight, and there is little likelihood of a significant effect. Rainfall is distributed across two wet and two dry periods; yet inside the forest, the contrast in vegetative conditions between seasons is very slight, whether one looks at the amount of leaf production or presence of fruit. In fact, when one

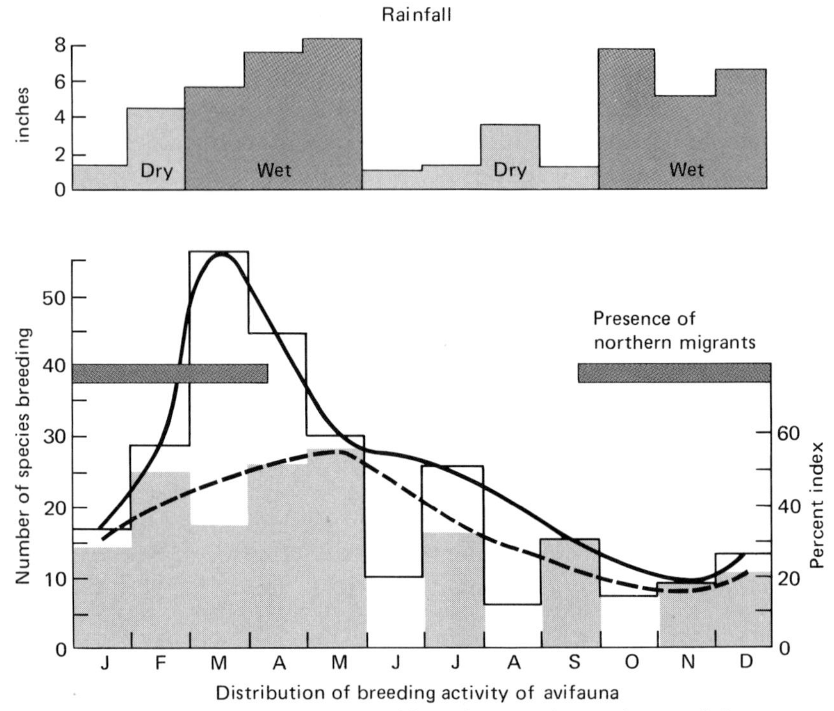

Figure 10–10 Seasonal distribution of breeding in the avifauna of the western Andes in relation to wet and dry periods and the presence of North American migrant species. Open columns and curve with solid line represent monthly totals of numbers of species found breeding. Shaded columns and broken line indicate a breeding index in which number of breeding species is expressed as percent of the total of specimens collected. Black bar marks period of presence of southern migrants. (Source: A. H. Miller, 1963, *Univ. Calif. Publ. Zool., 66:* 1–78.)

examines the breeding index values shown in Figure 10–10, there is no correspondence with the wet and dry seasons. Individual species, as opposed to the entire community, however, may show correspondence with the wet and dry season. The peak for the avifauna as a whole falls in a wet period but the low points fall in the other, equally wet period. Both dry periods have intermediate breeding indices, though one is longer and more pronounced. So Miller concluded that there was no significant overall correlation between breeding of bird populations in the avifauna and rainfall or other physical environmental factors in this cloud forest area.

By far the most important correlation was found with respect to the influx and exit of migrant bird species. The bars in Figure 10–10 show the periods in which migrant bird species from North America are present in the cloud forest or on its borders. These migrants are chiefly insect-eating members of the warbler, thrush, and flycatcher groups.

They represent about a 10–15 percent increment in the cloud forest bird fauna. The peak of the cycle of resident breeding is from March to late April or early May, and thus the following period, when the young resulting from this peak are making the greatest demands on food resources, would be from late in April on into June and July. At this time, the forests are relieved of the competitive impact of the migrant species. The greatest number of returning migrants comes from late September through November, and breeding of resident species is low then on into January. As Miller (1963) states, "Whatever may be the causal mechanisms lying back of these correlations, the fact remains that the load on the environmental resources is evened out around the year to a considerable degree by the occurrence of the lowest levels of breeding at the time when the number of North American visitants is greatest."

In Africa, Moreau (1950) examined breeding seasons of birds at Amani, an African equatorial counterpart of cloud forest in the Neotropics. The heavy long rainy period from March to May there is avoided for breeding and in this interval, about half of the annual total rainfall of 80 inches is received (14 inches per month). In fact, there is no breeding at all in April and May and low points in breeding occur in March, June and July. The peak of breeding of African forest birds in this area is in the period from October to December, which represents 10 to 20 times the amount of breeding occurring in the low points of the annual cycle. Thus, in the South American Andean station studied by Miller, we see that the annual weather cycles are much less pronounced than at the montane African cloud forest site, and the breeding of the avifauna may be much more evenly spread around the year. In an evolutionary and selective sense, then, other biological influences such as the competition of northern migrant species have selective influence in shaping the seasonality of breeding among birds of that community.

Snow and Snow (1964) looked at breeding seasons and annual cycles of lowland South American land birds. Their observations over a 4½-year period were carried out in rain forest receiving nearly 100 inches of rain per year, in the central part of the Northern Range of Trinidad. In this region of the Neotropics, there is a single dry season (January to May). Vegetation flowers throughout the year, though peaking somewhat in April and November. The fruiting peak is reached in April to June, with a minor peak in November. Insect life is most abundant at the beginning of the wet season in June. Snow and Snow found that the breeding season of most bird species in Trinidad is related to food availability. Species with long breeding seasons tend to have a main peak of breeding in April to June, and then adults molt their feathers during the period from July to November. A minor peak of breeding occurs in October and November, following the molt.

These authors consider food supply to be the principal evolutionary factor ultimately determining the season at which each species breeds. Proximate factors influencing breeding are much more obscure, and initiation of rainfall at the start of the wet season may be the most important one for at least some species. Likewise, in a general survey of the nesting seasons of Central American birds in relation to climate and food supply, Skutch (1950) found a well-marked breeding season for many species, with a peak in April to June. During the dry season from January through March, breeding activity gradually increases and reaches its maximum in April, just before the initiation of the wet season. Breeding and nesting then fall off from June to August and in the last four months of the year there is little or no breeding. Birds with specialized eating habits such as the nectar feeders breed at a different time from the majority, but in each case Skutch could show a correlation of breeding with the season when the species' usual food was at its greatest abundance. He also gives convincing evidence that the time when most of the less specialized species were feeding their young (May to June) coincided with the time when the supply of insects and ripe fruit is at its height in Central America.

The availability of resources for the adult population

Of course in many species of animals, a combination of several main ultimate factors influence seasonality of breeding. Asplund (1967) examined the ecology of 13 species of lizards in the Relictual Cape Region of Bohau, California. In August and September, tropical oceanic cyclones bring rain to this region, increasing the amount of available food, water, and shade from the growing vegetation. Some nine species of lizards out of the total fauna of 13 begin breeding, ovipositing, and/or hatching at this time, a period when arthropods and plants are at their maximum abundance but when other factors of ecological importance to these lizards, namely food, water, and shade for temperature regulation, are also available.

Like the cloud forest birds we considered in the preceding section, crustaceans and other arthropods such as spiders may have to time their molting cycles to seasonal changes in the environment. The ultimate evolutionary reasons for such seasonality are selective, of course: to pick the period during the annual cycle when the individual will be at minimum risk as regards predation and desiccation and food is maximally available. Birds molt feathers when they do not have to breed or compete with migrants. Arthropods often molt in seasons when predators are confronted with an abundance of prey species, risks of desiccation are low, and food is abundant for their rapid growth. The wet season fills these requirements in tropical regions while the spring or early summer is best in most temperate areas. Proximate cues for seasonal molting can be changes in daily photoperiod (Aiken, 1969), oscillations in endocrine level, rainfall patterns, or other internal and external triggers.

TROPICAL SEASONALITY AND POPULATION FLUCTUATIONS

As a result of seasonality in breeding of tropical insect populations and other animals whose life-spans last less than a year, there are considerable seasonal fluctuations in population sizes of adults and earlier life history stages. We conclude this chapter on seasonality and populations by looking at some examples of fluctuations in animal population sizes with respect to seasonality and the evolutionary or ecological factors that influence this change.

The seasonal effects on the population dynamics of insect communities in the tropics are relatively little studied. Janzen and Schoener (1967) found that in the understory of primary forest areas the total numbers of individuals of insects of all species were remarkably similar between the wet and dry season; however, the number of *species* during the dry season is reduced by about 25 percent. They suggest that some species may not be able to withstand the decrease in moisture level. In second-growth vegetation exposed to the sun on the edges of the forest, there is a threefold increase in the number of individuals during the dry season. This is accompanied by about an 18 percent increase in number of species and a definite increase in the number of individuals per species (5.9 to 7.9). These data may also indicate that to a point, reduced water levels may be beneficial to insect density. As we shall be noting, butterflies seem to survive and function better in the adult stage if they are active and fly principally during a drier period. In a very wet environment, such as Costa Rican rainforests studied by Janzen and Schoener, increased dryness and increased direct insolation (resulting from less cloud cover) may actually promote insect activity.

In the tropical rain forest on Barro Colorado Island in Panama, Emmel and Leck (1970) found considerable fluctuations in adult population size from month to month for most of the 92 species of butterflies being censused (e.g., Figure 10–11). Certain species flew mainly in the dry season while others flew in the wet season; however, many species reached their population peaks in the transition period between wet and dry seasons. This effect is probably due both to an overlapping of dry and wet season faunas and to the favorable junction of environmental factors for adult tropical butterfly activity at that particular time. The principal environment factors of selective importance for developing the seasonality of activity seem to be the amount of available sunlight for thermoregulation and the condition of appropriate larval host plants.

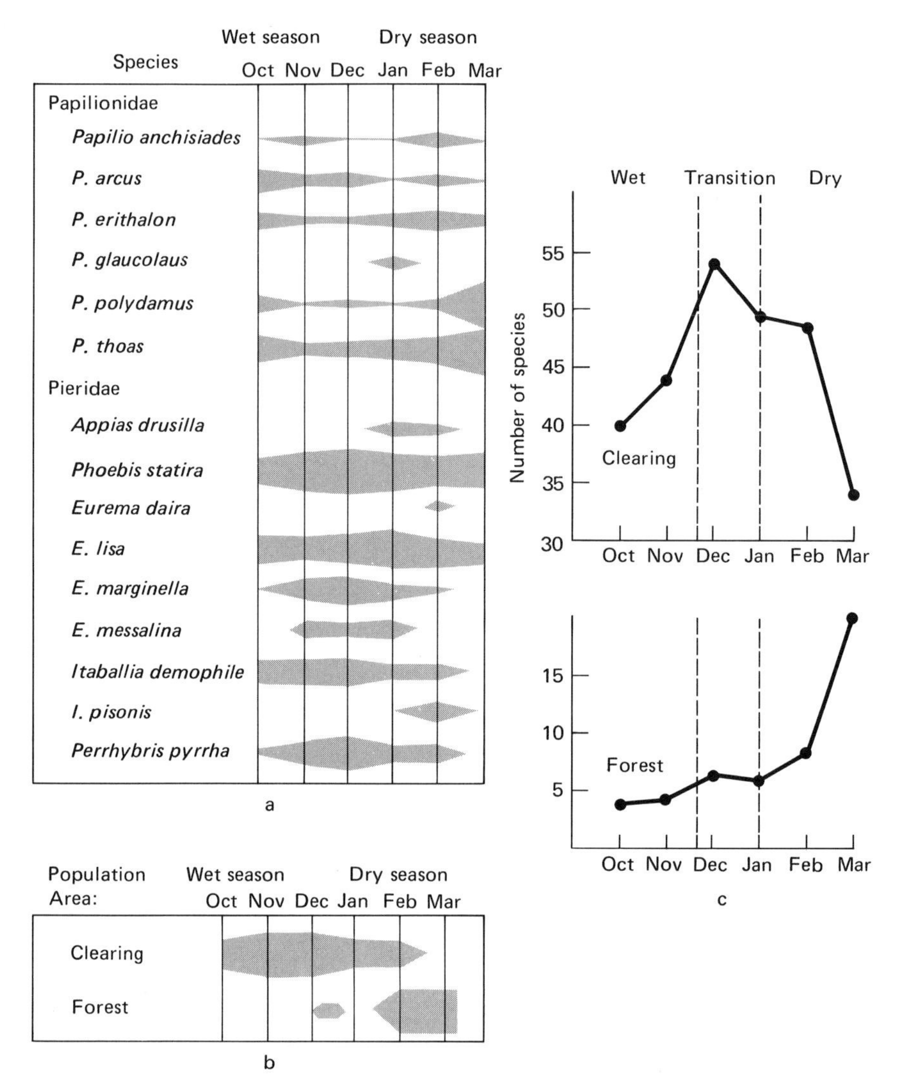

Figure 10–11 Seasonal changes in organization of tropical rain forest butterfly populations in Panama. (a) Patterns of flight activity of adult populations of butterflies in the families Papilionidae and Pieridae through the second half of the wet season and the first half of the dry season. The width of the bar indicates relative adult population density for a given species during the course of each month. (b) The pattern of flight activity for the pierid *Itaballia demophile*, showing an apparent seasonal movement of adults from an open second-growth clearing to the cooler and more humid forest as the dry season intensifies. (c) Number of butterfly species observed in study areas in a second-growth clearing (top) and inside the rain forest (bottom) per month from wet to dry season, 1968–1969. Maximum diversity is reached at the seasonal ecotone between the end of the wet season and the intensification of the dry season. (Source: T. C. Emmel and C. F. Leck, 1970, *J. Res. Lepidoptera, 8:* 133–154.)

An *ecotone* is a transition area between two adjacent communities. Treating the wet-season butterfly fauna and the dry-season butterfly fauna as separate communities, the transition period between the wet and dry seasons may be called a *"seasonal ecotone"* and is a temporal analogy of the spatial concept of the ecotone. Emmel and Leck (1970) suggest that this seasonal ecotone might be a general phenomenon influencing tropical species diversity, in that one could find the greatest number of active species (of short-life-cycle animals) between two distinct seasons partly because both wet and dry season communities may be represented. With diurnally active insects such as the butterflies, the most reproductively favorable conditions also may exist at this time and hence the seasonal ecotone fauna does not merely represent an overlapping of communities but one which has responded in an evolutionary sense to the most satisfactory breeding period during the annual cycle, a cycle which exists even in a tropical evergreen forest. Data from Costa Rican sites and elsewhere (Emmel, unpublished) indicate that diversity increases only at the gradual wet-season—dry-season seasonal ecotone (December in the northern Neotropics), not at the sharp point of dry season—wet-season transition (April or May in the northern Neotropics). This presumably is a result of adult intolerance in dry-season species of the rainy conditions suddenly initiated by the start of the wet season, and the maximum suitability and availability of larval host plants for development of immatures.

In several montane temperate zone localities, a somewhat analogous overlap of spring and summer butterfly communities has been shown (Emmel and Emmel, 1962, 1963; Emmel, 1964) near Florissant, Colorado (8600 ft elevation) and Donner Pass, California (7200 ft). The point of maximum species diversity occurs when the newly emerging adults of summer-flying species overlap with the older surviving adults of spring-flying species. The periods of prime conditions of larval host plants for the two groups are seasonally separated and this overlap of adult flight is not as significant as in the tropical situation, although some competition among adults for certain nectar sources may occur in temperate areas. Rainfall is minimal during the spring and early summer flight seasons in these localities, and has no significant influence on timing of adult emergence of either spring or summer species (Emmel and Emmel, 1963; Emmel, 1964).

In Liberia, Fox et al. (1965) found that most butterflies have adjusted their annual cycles to the rainfall cycle and few adults are found during the heavy rainy period (June 1 to October 1) in this West African country. The heavy, nearly continuous rains of the rainy season probably destroy most adults and hence exercise heavy selection against flight during that time period.

Marked seasonal changes in vertebrate population density have also been noted in complex tropical forest communities. Sexton (1967) estimated population sizes of the lizard *Anolis limifrons* in the rainforest on Barro Colorado Island during the 1963 and 1965 dry seasons. In 1963 a mean number of 7.5 lizards were found per 64-square-meter quadrat and in the dry season of 1965, only 2.5 lizards occurred per quadrat. The dry season during the year 1962–1963 was relatively wet while the dry season in 1964 to 1965 had considerably less precipitation in Central Panama. Thus, even within the tropical rain forest, vertebrate populations were affected by the intensity of the dry season.

In contrast, Hirth (1963) found that neither *Ameiva quadrilineata* nor *Basiliscus bittatus* lizard populations changed in density over a two-year period on the beach at Tortuguero in northeastern Costa Rica. These vertebrates bred all year round at Tortuguero and yet had their main egg-laying season in April to June, whereas the largest populations of juveniles were present in August and September. In seasonally dry areas such as the tropical deciduous forest of northwestern Costa Rica, the large iguanid lizards *Iguana* and *Ctenosaura* restrict their reproductive activities to just before and during the first part of the rainy season, prior to the major flush of new vegetative growth. The wet season then brings a large increase in population size with the hatching of new juveniles. In a Bornean rain forest studied by Inger and Greenberg (1966), males and females of four species of lizards appear to breed continuously throughout the year, in contrast to the usual restricted breeding seasons of temperate lizards (e.g., see Tinkle, 1967). The total annual rainfall at this site is around 5700 mm, and no month has less than 100 mm of rain. The temperature, relative humidity, and day length varied only slightly at this equatorial site. The stability of food availability was not assessed in this study, but may be presumed to be essentially uniformly abundant throughout the year.

In Panama, the phyllostomid bat *Artibeus jamaicensis* is capable of reproducing at several times during the year. Young are born in March or April and July or August, coinciding with the end of the dry season and the first half of the rainy season. Embryos from conceptions in the rainy season become dormant from September to mid-November during the height of the rainy season, and then resume normal development at the end of the rainy season (Fleming, 1971). This fruit-eating bat apparently has adjusted its natality periods to the times of maximum food availability (Figure 10–12).

The reproductive period of the clay-colored robin *Turdus grayi* seems to be affected more by nest predation than by food availability in the Panamanian rain forest. This insectivorous song bird breeds in the dry season when food resources are comparatively low. Although the rainy season brings more food, breeding stops at this time, when

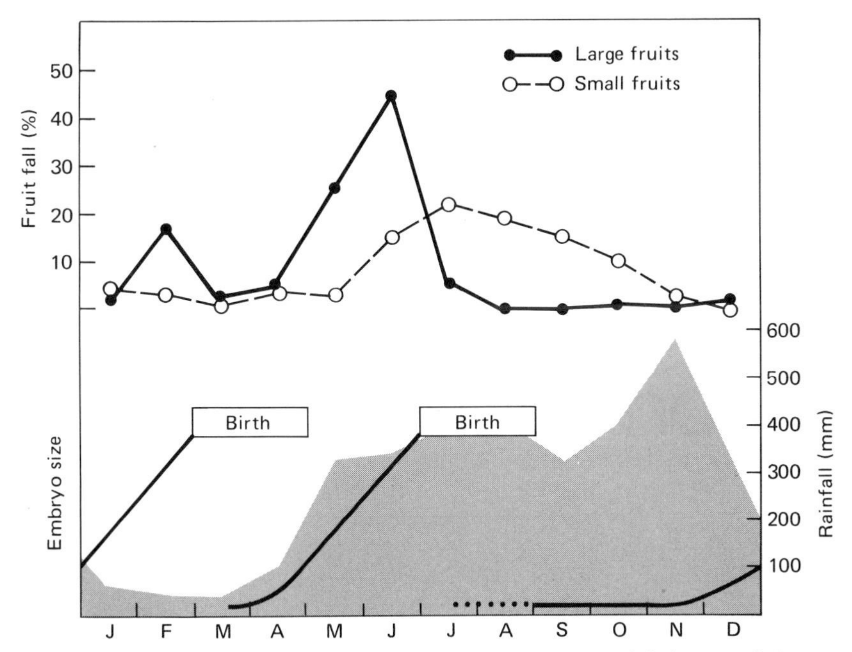

Figure 10–12 The relationship between rainfall, fruit availability, and the natality periods of the neotropical phyllostomid bat *Artibeus jamaicensis* in Panama. Embryos are postulated as being conceived in July or August (dotted line at bottom). (Source: T. H. Fleming, 1971, *Science, 171:* 402–404.)

predators destroy as many as 85 percent of the nests once the rains begin (Morton, 1971). Birds breeding in the dry season have, in contrast, about a 42 percent chance of fledging young.

Hence, we see that seasonality in reproductive behavior and population density occurs in both vertebrates and invertebrates in the tropics as well as temperate areas, but that a variety of ultimate and proximate explanations are involved. Effects of this seasonality on life history patterns have been explored in Chapter 8. In Chapter 11, we consider in more detail a subject that affects seasonality as well as other aspects of population structure: namely, interactions of unrelated populations in biological communities.

1 2 3 4 5 6 7 8 9 10 **11**

Interactions of Unrelated Populations in Communities

The stability and complexity of biological communities in an ecosystem are directly dependent on the interactions of populations of different species living in that ecosystem. Organisms of one species affect the growth, health, behavior, and general population biology of organisms of another species through many channels. Most of these can be grouped under the headings of competition, predation, and various ecoevolutionary strategies employed to reduce competition such as cooperative symbiosis and species differentiation. Evolutionary solutions to the problem of the reduction of competition in the functioning of natural communities also include competitive exclusion and character displacement. Interactions of attack, defense, and behavioral response often involve chemical agents as well as physical force. The interactions of unrelated populations which promote or restrict the number of species present in an area contribute to the overall evolutionary and ecological problem of species diversity (MacArthur, 1972). Let us begin by looking at the problems entailed in competitive interactions of unrelated populations in communities.

COMPETITION

Competitive interactions between individuals involves common requirements for a resource in limited supply. The usual result of this mutual striving is a reduction in fitness or population density in the short term. Although no definition of competition is universally used by ecologists, a broadly acceptable definition would include the emphasis that competition is the biological interaction between two or more individuals that occurs when (1) a necessary resource is in limited supply, or (2) resource quality varies and demand is quality-dependent (McNaughton and Wolf, 1973). Two or more individuals may be competing for different qualities of a resource type—for instance, the best nesting space or the meadow area with grasses richest in protein content.

Competition may occur between individuals of the same species, and this is usually intensified by the fact that the individuals are genetically similar and occupy the same niche. Competition between members of the same species is intraspecific competition. If competition occurs between members of two different species, interspecific competition is said to be occurring. When competition occurs by direct interactions between competitors, such as eating available foliage or occupying some of the same habitat space, the species are said to be involved in *interference competition*. Competitors that reduce the amounts of a scarce resource available for other organisms are exerting rather indirect inhibitory effects, and this situation is known as *exploitation competition*. When competitors directly reduce the amount of food or space available to other species through individual interactions such as aggressive encounters, interference competition is taking place. Rather than suffer the disadvantages of competitive interactions, species usually experience selection that minimizes the effects of competition by increased specialization and diversification of resource-exploitation behavior.

Although competition theory has had a tremendous impact on ecological and evolutionary thinking, the problems involved in interspecific competition remain one of the most controversial fields of population biology and ecology. We shall look at competitive situations in nature and then relate them to the general model of competitive interactions and their theoretical outcomes.

The problem of competitive interaction in interspecific competition is intimately related to population density, as we have seen in Chapter 5. If there is no interspecific competition in a region, the individuals of the subject population should more or less evenly distribute themselves through the habitat. The first individuals in the area will find an abundance of resources, but with an expanding population, eventually the balance of demand and supply must equalize. Likewise, the level of intraspecific competition cannot increase infinitely.

At higher population densities, deviant individuals that use less optimal but also less hotly contested resources or habitats will be favored by selection; thus, the population may spread its utilization demands over a variety of resources and habitats. In interspecific competition, it should be obvious that if the species involved are too similar to one another in their requirements, they cannot indefinitely coexist. This outcome is usually generalized by saying two species cannot occupy the same niche, at least indefinitely, and is commonly called *Gause's principle,* after G. F. Gause, a pioneer Russian investigator of competition. This principle, the competitive exclusion principle, has been redefined by numerous ecologists since Gause's 1935 formulation, primarily because of the difficulty in defining the term *niche* (Chapter 6). A niche encompasses virtually all facets of the biology of a species. Since two species, by definition, are genetically distinct, no two species would have the same niche. With different niches, they should not be competing but ought to be able to coexist. A broader definition of the principle of competitive exclusion, offered by Wilson and Bos-

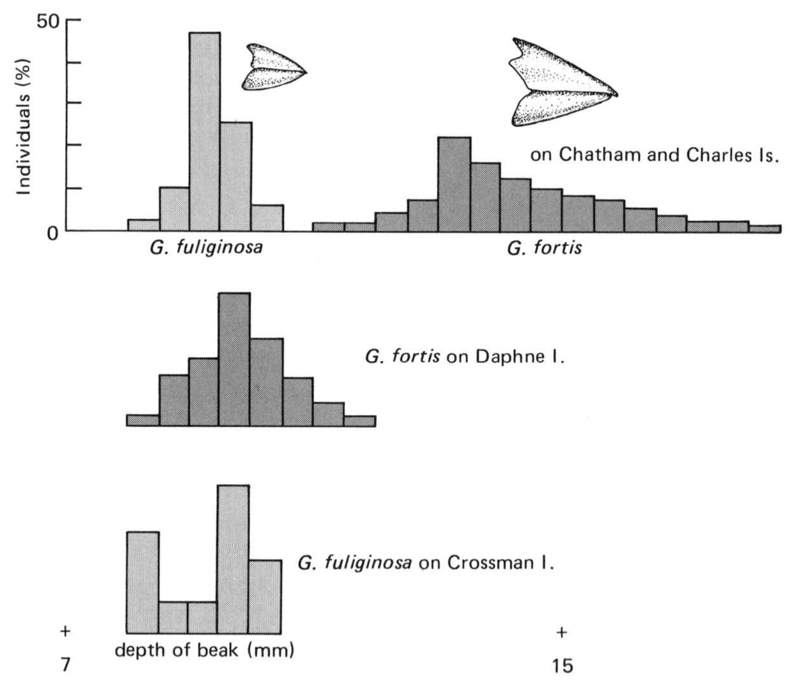

Figure 11–1 Character displacement in beak size when two species of Darwin's finches (Geospizinae) occur on the same island. Both *Geospiza fortis* and *G. fuliginosa* have medium-sized beaks on islands on which they occur alone. On Chatham and Charles Islands, their competition for similar food (mainly seeds) in sympatric situations results in selection for notably different beak sizes capable of specialization on a limited range of food items. (Source: S. Carlquist, 1965, *Island Life,* Natural History Press, New York, p. 374.)

sert (1971), is that no two species that are ecologically identical can long coexist. A pair of species can coexist only if parts of their niches are broadened or redefined such that new resources are utilized separately by each species. Some overlapping utilization of a resource is a common situation and a primary question becomes, how much overlap in utilization of a resource is necessary before competition becomes severe enough to produce exclusion of one of the species (MacArthur and Wilson, 1967). A number of population biologists have attempted to answer this question.

Three usual outcomes of competition are coexistence through character displacement or niche partitioning (Figures 11–1 and 11–2), survival in adjacent habitats through competitive exclusion, or extinction.

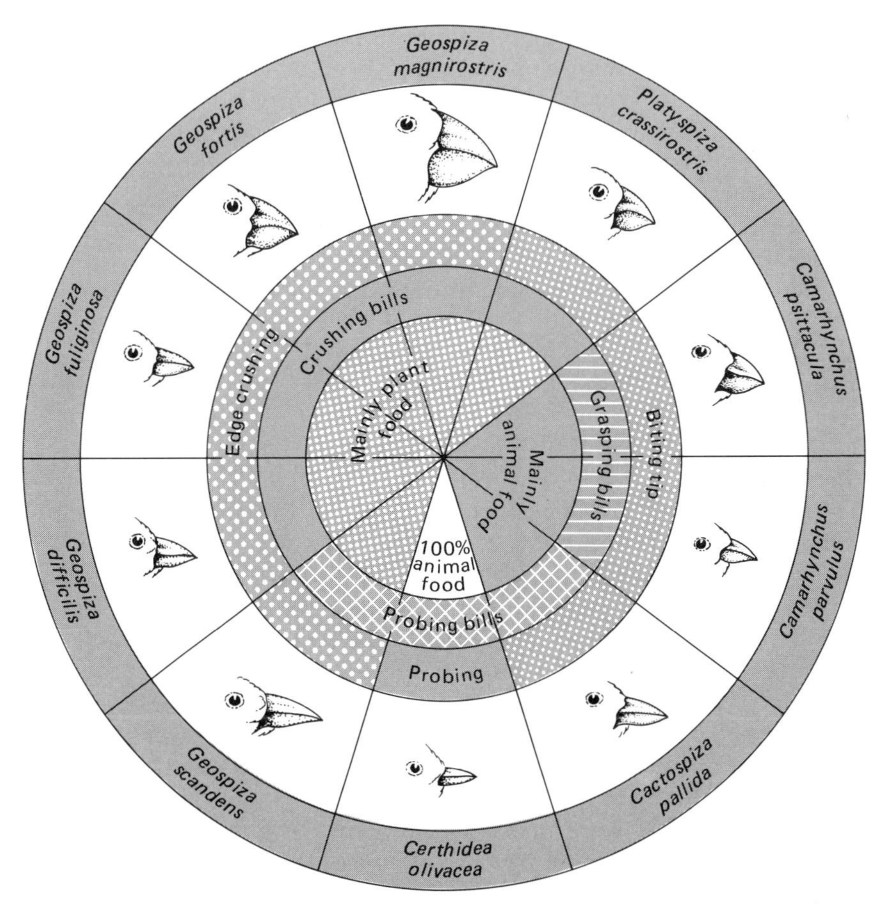

Figure 11–2 Niche partitioning: schematic representation of the relationship between bill structure and feeding habits in 10 species of Darwin's finches (Geospizinae) from Indefatigable Island in the Galapagos Archipelago. (Source: R. I. Bowman, 1961, *Univ. Calif. Publ. Zool., 58:* 1–302.)

The distinction between niche partitioning and competitive exclusion is fine, and examples may be interpreted either way. Both represent the more general case of resource partitioning to be discussed in the next section. Thus, Cody (1973) reported that several species of sea birds coexist geographically and on the same breeding rocks despite a possible 80 percent overlap in overall diet. The birds apparently partition the feeding niche so that each species forages at a different distance from shore. This situation perhaps resulted from competitive exclusion, as we shall see below. In sympartric species of earthworms in Wales, England, specialization occurs on organic matter at different stages of decomposition. This small overlap in resource requirements is not sufficient to cause displacement or extinction of one of the species (Piearce, 1972). When niche partitioning is not possible, competitive exclusion commonly occurs.

As long as two species are able to utilize different portions of the environment such that each is not suppressed by excessive numbers of its competitors, coexistence is the easiest solution to the problem of competition. But if one species produces enough individuals to prevent the population of the other species from increasing, competitive exclusion results in time or space. The two species may begin utilizing different seasons in which to live and breed, or they may move into adjacent but distinct parts of the habitat.

One of the best documented examples of the influence of interspecific competition in causing exclusion is Connell's study (1961) on the distribution of several species of barnacles. On the intertidal rocky shore of Scotland, the adults of two species of barnacles occupy two separate horizontal zones with a small area of overlap (Figure 11–3). The young of the species from the upper zone are found in much of the lower zone, but seldom survive since few adults are found there. The upper species, *Chthamalus stellatus,* has its lower limit of distribution determined by competition with the lower species, *Balanus valanoides,* for attachment space during growth. Connell found that *Balanus* settled in greater population densities and grew faster than *Chthamalus.* The former species smothered, undercut or crushed the *Chthamalus* even if adult *Chthamalus* were transplanted to low levels of the shoreline by the experimenter. The few surviving *Chthamalus* which had settled at various sites in the *Balanus* zone were smaller than the uncrowded ones in the upper intertidal zone. Since smaller barnacles produce fewer offspring, competition tended to reduce reproductive efficiency in addition to increasing mortality. The upper limit for the *Balanus* barnacles was mainly determined by physical factors such as desiccation. Predation by *Thais* snails on these barnacles tended to decrease the severity of this interspecific competition in the lower intertidal zone. Competitive exclusion was demonstrated to have caused the observed absence of adults of *Chthamalus* barnacles from the lower zone.

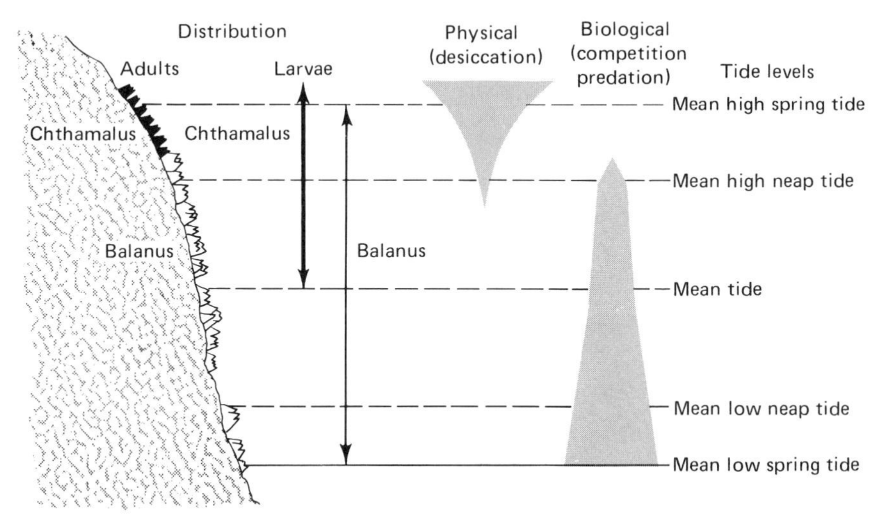

Figure 11–3 Competition between two barnacle species populations in an intertidal zone in Scotland. The young of each species settle over a wide range, but those of *Chthamalus stellatus* are subsequently overgrown or pried off by the faster-growing *Balanus balanoides* in areas in which physical factors do not limit the occurrence of individuals of the *Balanus* population. The zonation of these barnacles, then, is caused by competitive exclusion. (Source: J. Connell, 1961, *Ecology, 42:* 710–723.)

Cody (1973) found that alcid species (birds of the family Alcidae) differ in their foraging zones at sea and appear to have reduced interspecific competition by this means. Cody studied six species in the North Pacific off the Olympic Peninsula in Washington state and another group of six species on a small island off Northern Iceland in the North Atlantic. At least three of the six coexisting Washington species have similar diets, and all six breed at the same time of the year. A similar pattern appears in the six-species alcid community of Northern Iceland. Inspection of Figure 11–4 will show the reasonably orderly segregation of feeding zones of foraging adults as they collect food for their chicks. Pigeon guillemots feed almost intertidally (mean foraging distance from the breeding rocks was 0.27 km). The common murres feed close by the islands (mean 3.10 km). The tufted puffin is found in abundance further out than the murre (mean 4.67 km), and the rhinoceros auklet ranges more widely yet (mean 5.34 km). Cassin's auklets are only just beginning to become common at the limit of the range of the boat used for the transects (16 km). The distribution of distances of feeding individuals from the breeding rocks gives an excellent indication of the segregation in feeding zones, except for the marbled murrelet whose breeding sites were not known. Murrelets fed within a few kilometers of the shore, however, and probably were about 4.4 km from the nearest breeding colony of other alcids, which puts the species' foraging distance between those of the murre and the puffin.

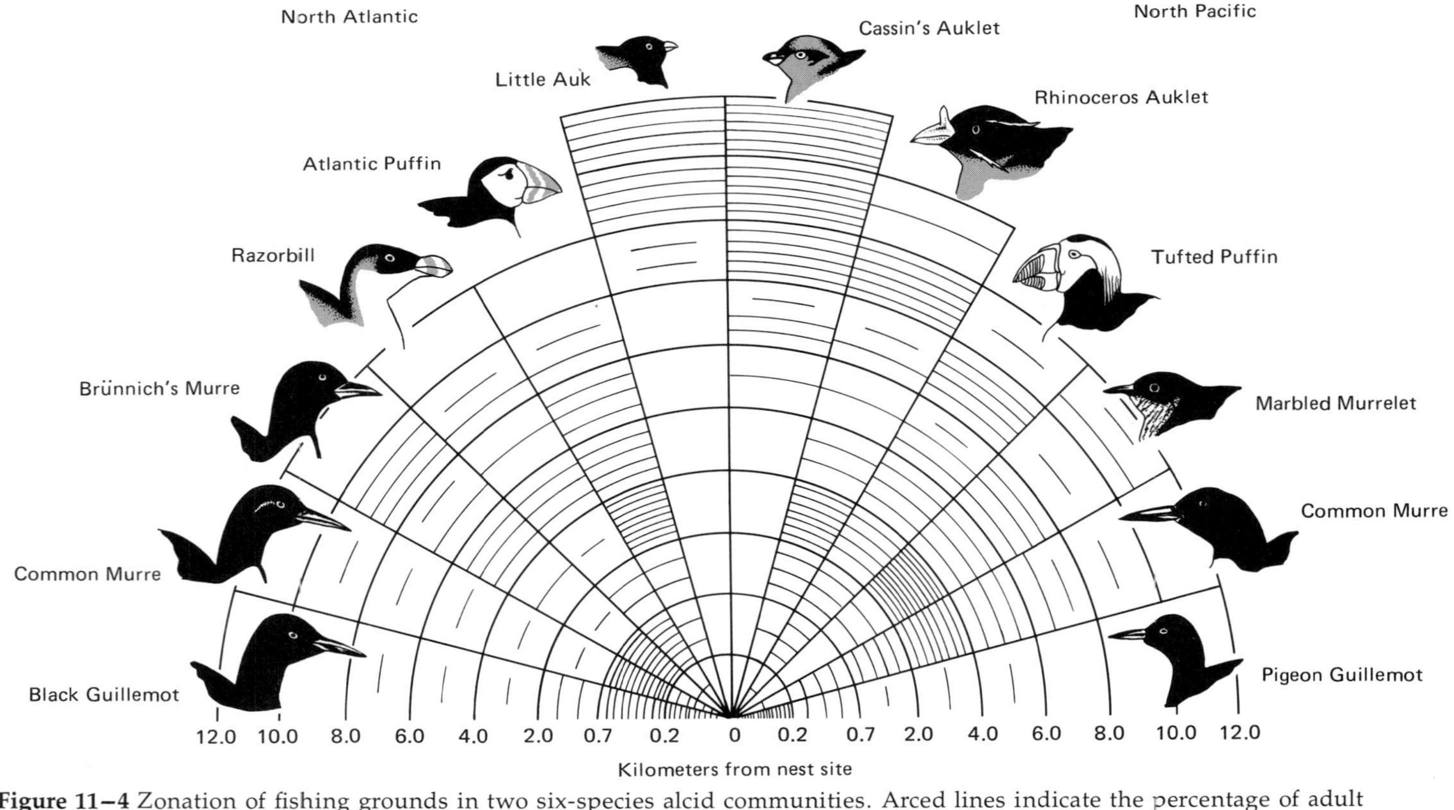

Figure 11–4 Zonation of fishing grounds in two six-species alcid communities. Arced lines indicate the percentage of adult fishing activity spent at various distances from the breeding rock, where each full arc line is 5%. Reading counterclockwise from lower right: pigeon guillemot, common murre, marbled murrelet, tufted puffin, rhinoceros auklet, and Cassin's auklet; reading clockwise from lower left, black guillemot, common murre, Brünnich's murre, razorbill, Atlantic puffin, and little auk. (Source: M. L. Cody, 1973, *Ecology*, *54:* 31–44.)

In interspecific competition, selection will favor the genomes in each species population that confer special adaptation to a particular limited range of the environment, even among parasites in the same host. The European tortoise, *Testudo graeca,* has 10 or more species of colon-inhabiting Oxyuroidea parasites which show considerable niche diversification. Each of these roundworms has a particular distribution in the colon, the physical location of which differs slightly in oxygen and carbon dioxide concentrations or pH (Schad, 1963). Competitive exclusion of one ectoparasite by another is quite common. Carp which are naturally infected with a monogenean trematode, *Dactylogyrus extensus,* may be exposed to a second trematode parasite, *D. vastator,* and in such cases the latter species replaces the former. This occurs because *D. vastator* causes changes in the epithelial lining of the gills which makes the tissues unsuitable substrate for *D. extensus* (Paperna, 1964). A further example of competitive exclusion in endoparasites occurs between two intestinal worms in the three-spined stickleback minnow. When the cestode *Proteocephalus felicollis* and the ancanthocephalan *Neoechinorhynchus ruteli* occur in single-species infections, individuals are widely distributed throughout the gut. In concurrent infections, *P. felicollis* attach themselves more frequently in the anterior intestine, whereas *N. ruteli* attaches itself more frequently in the rectal region (Chappel, 1969).

In small species of mammals, especially rodents (Grant, 1972), the major evidence for interspecific competition is derived from observations of sympatric species. The evidence falls into two categories: (a) two sympatric species will vary inversely in population numbers, and (b) two sympatric species will occupy different parts of the same habitat for no clearly obvious reason associated with variations in the habitat features. In the first category, the population numbers of one species rise while those of the other fall and in some cases, the numerical changes are later reversed, with the common species becoming rare and the rare species becoming common. These changes may reflect the outcome of competitive interaction in a temporarily varying environment. At its extreme, the extinction of one of the species results. Grant (1972) summarizes evidence to show that where species of deer mice (*Peromyscus*) coexist, there is usually a marked difference in body size. The evidence suggested that selection has favored the attributes of the interacting species which render the animal successful in intraspecific aggressive encounters. Because rodents exhibit the same behavioral repertoire in interspecific encounters as in intraspecific encounters, these attributes, largely behavioral and physiological, aid in competitive interactions. The larger species usually has the advantage of free access to food due to dominance, and the ability to deal with large or hard-coated foods such as seeds. The smaller species may gain greater efficiency in energy extraction over most of the size range of foods that the two species exploit, and also be able to avoid predators with greater ease. Thus, selection will favor those attributes of interacting species which renders animals successful in intraspecific aggressive encounters. These evolutionary changes frequently result in sufficient adjustment through competitive exclusion that two species can be generally sympatric within a particular region.

Similar development of size differences has been found between species of starfish in the rocky intertidal zone of the Pacific Coast which compete for food. *Pisaster ochraceus* and *Leptasterias hexactis* show no spatial or temporal subdivision of the habitat while feeding. Although the diets of the two species overlap greatly in both size and types of prey eaten, each obtains most of its energy from a different combination of prey size and species. The larger asteroid *Pisaster* has the primary competitive advantages of being able to reach a larger size (and thus to capture "better" prey) and having highly aggressive behavior. The smaller starfish gains competitive advantage by having a higher rate of energy intake than *Pisaster*. Since this advantage can conceivably give developing *Leptasterias* a great initial size advantage over small *Piaster*, the aggressive behavior of the latter may function to reduce the foraging time of the former and allow coexistence despite active competition (Menge and Menge, 1974).

Frequently in nature we find indirect evidence of competitive exclusion which is later confirmed by unusual circumstances. The two California desert swallowtail butterflies, *Papilio indra fordi* and *Papilio rudkini*, are frequently sympatric in range but mutually exclusive in host plant source. In nondesert areas where a large amount of foodplant is available, a *P. indra* subspecies and another member of *Papilio machaon* complex may fly sympatrically and regularly use the same foodplant. Such a situation exists in the higher elevations of the Sierra Nevada where *P. indra indra* and *P. zelicaon* larvae both feed on the same Umbelliferae (Emmel and Emmel, 1962). In the mountains of San Diego County, California, both *Papilio indra pergamus* and *P. zelicaon* fly together and use another umbellifer as a mutual larval foodplant. In the desert mountain ranges, however, it appears that at certain times foodplant supply becomes a limiting factor for the two desert swallowtail butterflies, *P. i. fordi* and *P. rudkini*. Normally, *P. rudkini* larvae feed on a rutaceous shrub, *Thamnosma*, while *P. indra fordi* feeds on the umbellifer, *Cymopterus*. By feeding on different foodplants in years of normal population levels, larval competition between the species is avoided and many potential foodplant individuals remain unused or damaged only negligibly. Occasionally, optimal rainfall conditions lead to a large simultaneous emergence of the adults of both species from overwintering pupae, synchronizing the growth of the resulting large numbers of larvae. In those rare years when butterfly population peaks of sympatric species are reached simultaneously, the selective advantage of having separate hosts breaks down, and direct competition through foodplant overlap occurs. The larvae that find enough suitable foodplant of any host species to reach pupation stage are the ones to survive (Emmel and Emmel, 1969).

RESOURCE PARTITIONING AND THEORY OF FEEDING STRATEGIES

Observations such as these differences in habitat and foodplant choice are reflected in the more general case of resource partitioning in ecological communities and the theory of feeding strategies. Ecologists have looked at differences in the way species in the same community utilize resources in order to understand the basic role of this division in the natural regulation of species diversity. The major purpose of resource-partitioning studies is to analyze the limits interspecific competition places on the number of species that can coexist in a stable relationship (Schoener, 1974). As Gause (1934) found in laboratory experiments with competition between two species of protozoans, extinction is a common outcome of competitive exclusion if there is insufficient habitat space for niche diversification. Gause cultured two species of *Paramecium* in laboratory tubes of limited size where they competed for food (bacteria), oxygen, and other requirements. When grown alone under carefully controlled conditions and nearly constant food supply, the numbers of *Paramecium caudatum* increased more slowly than *Paramecium aurelia* (Figure 11–5). Eventually, each species reaches the carrying capacity of that environment to support the population, and growth levels off to a population size characteristic of the species. When grown together in the same culture tube, the rate of population growth of each species changes. *Paramecium aurelia* increases more slowly than it did alone and *P. caudatum* grows poorly and eventually becomes extinct, though it is the protozoan with the higher maximal instantaneous rate of increase per individual. The lack of space in these particular environments brought about the outcome of extinction.

As we have seen in preceding examples of competitive exclusion in nature, the effect of one species on another can frequently be demonstrated, but the observations will fail to demonstrate the actual mechanism of the competition. One species of barnacle or rodent may reduce the abundance of a second in a particular habitat by directly depleting its resources, by interfering with its ability to obtain those resources, or by using up in aggressive encounters the energy attained from those resources (Schoener, 1974). In an extensive review of 81 studies bearing on resource partitioning in groups of three or more species, Schoener examined the resource dimensions involved in ecological differences between similar species, including food, microhabitat, food type (e.g., size, texture, hardness, depth in some protective

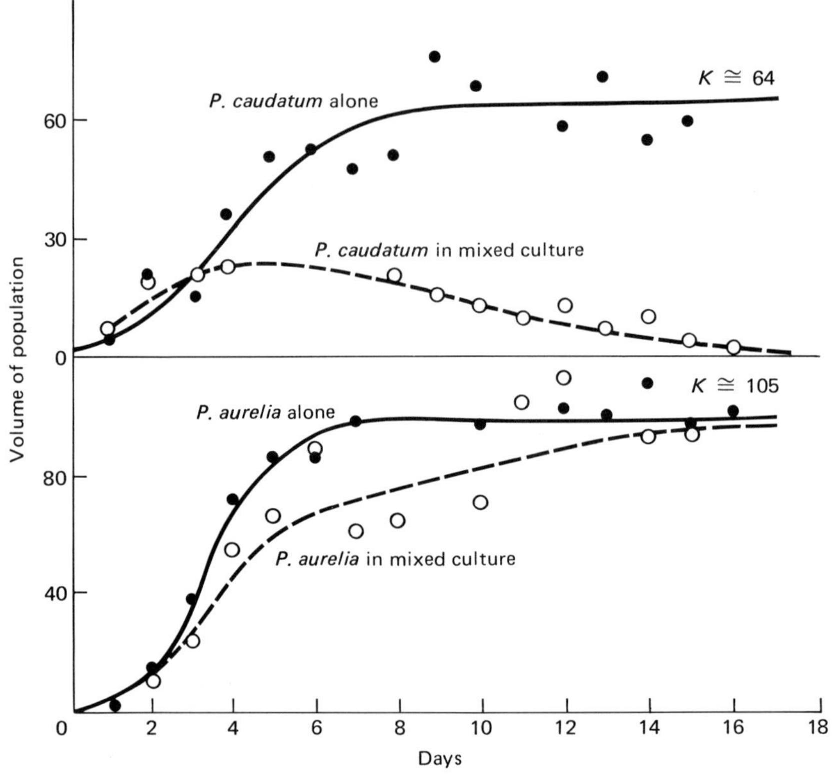

Figure 11–5 Competitive exclusion in a laboratory experiment with two protozoans, *Paramecium caudatum* and *P. aurelia*. (Source: E. R. Pianka, 1974, *Evolutionary Ecology*, Harper & Row, New York, p. 142, From G. F. Gause, 1934, *The Struggle for Existence*, Hafner, New York.)

medium, etc.), and time of day or season of year involved in the competitive interaction. Five major generalizations emerged from his analysis.

Habitat dimensions are important more often than food-type dimensions, which are important more often than temporal dimensions. In 90 percent of the groups of animals examined, the species are separated by habitat. In 78 percent the species are separated by food, and in 41 percent the species are separated by time.

Predators separate more often by being active at different times of the day than do other groups. Thus, within the 63 groups of three or more species in which no differences in daily activity were noted, 49 percent are primarily predators, 27% are primarily herbivores or scavengers, and 24% are primarily omnivores. Out of the 17 groups showing partitioning by daily activity, 82 percent are predators and 12 percent are herbivores. Predators, apparently, will specialize on species by adjusting their daily activity cycle to that of their prey, whereas herbivores can eat plants during the day or night with less difficulty.

Terrestrial poikilotherms relatively often partition food by being active at different times of the day. Being less buffered against external temperature change, these animals showed special sensitivity to diel climatic variation. Thus, of the 23 such groups surveyed, 43 percent segregate by having different periods of daily activity, whereas only 12 percent of other animals do. These differences are most pronounced in terrestrial poikilotherms.

Vertebrates segregate less by seasonal activity than do lower animals. Animals that have relatively long generation times cannot partition the annual cycle as finely as those that mature in shorter time periods. Thus, 44 percent of invertebrates, but only 13 percent of vertebrates, differ in seasonal activity or breeding.

Segregation by food type is more important for animals feeding on food that is large in relation to their own size than it is for animals feeding on relatively small food items. This pattern follows within autotrophic types. Looking at the groups whose food items are relatively large, 71% separate by food type and 33 percent separate by habitat. For the groups whose food items are small, 28 percent separate by food type and 72% by habitat. Many small animals such as herbivorous insects spend most of their lives on a single food item. Animals that eat relatively large prey are often predators that must pursue their prey, and several theoretical models of foraging strategies suggest that pursuers should specialize on food of a limited size range.

The interaction of predation and competition in affecting species differences and diversity is a complex problem that faces present and future ecologists. A comprehensive theory of resource partitioning drawing upon models at the individual and population levels has not yet been developed. Schoener's (1974) article proposes the best current analysis and assemblage of data on this important component of competitive exclusion and interaction at the community level.

We have been looking principally at resource partitioning among diverse groups of competing species. Closely related sympatric species often demonstrate considerable subdivision of habitat space and environmental resources. In lizards of the genus *Ctenotus* in the Australian desert, Pianka (1969) showed that seven sympatric species forage at different times of the day, and in slightly different ways in different microhabitats and often on different food items. Thus, these congeneric species avoid competitive exclusion by fine niche differences. Three distinct components of the overall ecologic distance between these seven sympatric species of *Ctenotus* are diagrammed in Figure 11–6. The first component is the distance (in as many dimensions as appropriate) between the centers of the two niches. The second is the total amount of resource shared by the two species, that is, niche

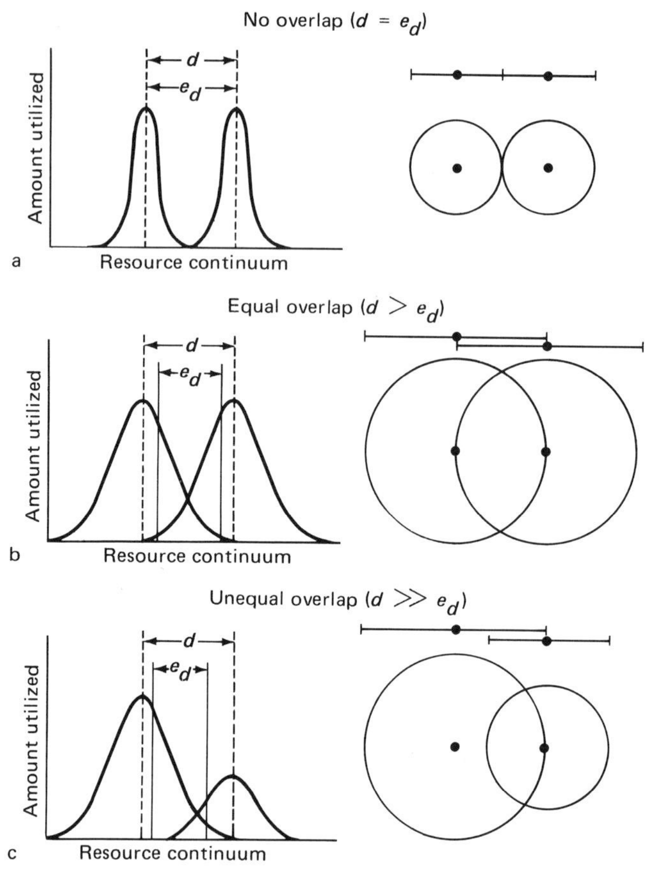

Figure 11–6 Diagrammatic representations of the three components of ecological distance. (a) Abutting niches with no overlap, effective ecologic distance e_d equals the distance d between the centers of the two niches. There is no direct competition in this case. (b) Overlapping niches of equal breadth, e_d is symmetrically less than d. Competition is equal and opposite. (c) Overlapping niches of unequal breadth, e_d is asymmetrically much less than d. Competition is not equal and opposite, but is more intense on the population with the smaller niche. Three different, but equivalent, geometric representations are shown (Source: E. R. Pianka, 1969, *Ecology, 50:* 1012–1030.)

overlap. The third is a competitor's relative usage of the shared resources (Pianka, 1969). The first measure d is normally equal to the effective ecologic distance e_d when there is no niche overlap (Figure 11–6a). The effective ecologic distance between competitors is reduced by niche overlap. This overlap affects two competitors equally only when their niche breadth or population sizes, or both, are identical (Figure 11–6b). Finally, should niche breadths be unequal, the competitor with the narrower niche must share relatively more of its required resources (Figure 11–6c). *Ctenotus* niches usually overlap broadly in the manner shown by Figure 11–6b,c, according to Pianka (1969).

Mixed species flocks in birds demonstrate an interesting foraging strategy to reduce competition. The formation of mixed species flocks occurs in both the Tropics and the Temperate Zones, although it is far commoner in tropical forests. Two or more species band together for a long or short period of foraging. An early evolutionary stage of flock formation is seen in several species of jays in southeastern Arizona (Westcott, 1969). Most jays are gregarious and large flocks are formed by each species during the nonbreeding season, especially during winter. Mexican jays form flocks of 15 to 25 birds, which are often widely dispersed while feeding. This species is very quiet while feeding and moving. They only bunch up when excited, when they are moving fast, or when they are at an abundant food supply. The Scrub jays are not as gregarious and usually travel in flocks of 5 to 15. These are more vociferous birds and are extremely noisy when moving, yet are quite shy and difficult to approach. Steller's jays form large flocks during the winter in the montane coniferous forest, but in the oak grassland habitat where they mix with the other two species, they travel in pairs or in small flocks of up to 48 birds. Pinyon jays are very gregarious, often forming flocks of 50–100 individuals traveling in a tight, noisy group, but this species is rather rare in Arizona. The three overwintering species use the same feeding techniques and concentrate upon the same food sources, mainly acorns, ground dwelling beetles, and lepidopterans. Jays do not normally mix but stay in conspecific flocks. However, when the flocks cross paths, there is a slight tendency toward interspecific flocking. They are attracted by the visual and auditory stimuli of another flock, especially by the noisy Pinyon Jay species, and will travel along together for a while, maintaining a minimal distance of several feet between jays. This flocking association is of a transitory nature, with the response apparently being due to a common "jay pattern": blue color, medium size, and noise.

The benefits of these flocking associations include increased predator perception and food discovery, but the association has not yet advanced to the constant mixed flocking commonly found among temperate fringillids and many tropical passerines. Morse (1970) studied mixed species flocks composed of chickadees, titmice, woodpeckers, nuthatches, creepers, kinglets, and wood warblers in forested areas of Louisiana, Maryland and Maine. Whereas both the size of the flocks and the density of the birds varied with habitat, several interesting relationships were found between population structure and population density in these birds. Flocking conferred improvement in food finding; when one of the many in a spread-out group of birds discovers food, all can share and continue the search in that area. Flocking also minimizes hostilities between species since they are adjusted to foraging in the presence of another species. Dominant species and subordinate species move through parts of the habitat to which they are each best adapted, and thus they obtain a more predictable portion of the food supply by participating in flocks than if they were solitary. The constant movement of these flocks probably insures that birds obtain the most readily available food and do not totally deplete one area. Different species in these mixed feeding flocks have developed similar predator calls, and thus protection against predation may be another advantage of this type of flocking behavior. Overall, however, it seems that the most important explanation for the advantages gained by these flocks is the ability of their component species to exploit available resources in a maximally effective manner and minimize competition in habitat utilization.

Similar types of mixed species foraging flocks have been found in migrant shore birds (Recher, 1966; Recher and Recher, 1969), and in ant-following flocks in the Neotropics (Willis, 1966, 1968; Rettenmeyer, 1963). In the latter case, birds are attracted to the presence of army ant swarms by food items flushed up by these raiding ants, and the flocks break up when the cyclical ants are not swarming. Other mixed feeding flocks in the tropics beat out insects by their group passage through the forest, or they use flocking as a general strategy for locating flower and fruit sources (Davis, 1946; Moynihan, 1962; McClure, 1967). McClure's work (1967) on mixed species flocks in the Old World tropics showed a distinct ecological separation of the species moving in the waves through the forest such that characteristic groupings travelled at different heights. He also found distinct seasonality in flock formation, with the peaks of flocking occurring at the nonbreeding season. Each flock appeared to have a regular route through the forest, moving in a more or less circular pattern. A grouping into feeding flocks in this locality coincided with a reduction of breeding activities of local birds and the influx and breeding presence of northern migrant species. Hence, the flocking behavior probably insured more efficient capture of food resources under conditions of seasonal but strong competition.

CHEMICAL INTERACTIONS BETWEEN SPECIES

In an earlier chapter we looked at the way chemical agents such as plant toxins affect the distribution of organisms. Chemical agents are involved in many types of competitive interactions by which organisms of one species affect the growth, health, behavior or population biologv of organisms of another species; this class of interactions has been termed allelochemical, that is, chemical agents involved in attack, defense, and behavioral responses (Whittaker and Feeny, 1971).

These effects of chemical interactions between species are commonly observed in many plant populations in which the expression of the growth or distribution of certain plants are inhibited by chemicals released from another plant. These poisons are termed allelopathic agents. The effects of such agents were first observed in continuous one-crop agriculture, including cereal crops, grasses, and apple trees. Black walnut trees (*Juglans nigra*) inhibit the growth of grass and alfalfa plants, as well as certain trees, for at least 50 ft away from the walnut trunk. This toxic zone generally is greater than the area covered by the walnut canopy, and Davis (1928) found that a chemical called juglone (5-hydroxy-alpha-napthaquinone) was secreted from the roots and produced allelopathic effects in other plants. This growth inhibitor showed selective effects in that some species, like Kentucky Bluegrass, were not killed.

Allelopathic effects have been observed from many other plants in both temperate and tropical habitats. The substance responsible may be released from the living plant by rain washing over the leaves or root release, whereas other effects are produced by decomposition of litter and the dead remains of roots. Allelopathic materials may also be released by fog-drip from leaf surfaces and glands, by volatilization from leaves, or by other routes. Allelopathic effects have been reported from rainforest trees to desert shrubs (Figure 11–7), and it is reasonable to assume that these effects form an important class of competi-

Figure 11–7 Allochemical interactions and competition for water between root systems space creosote bushes (*Larrea divaricata*) across the Mojave Desert in California. (Photo by T. C. Emmel.)

Table 11–1 Classes of interorganismic chemical effects. Adaptive advantage is indicated by +, detriment by (−), and adaptive indifference by 0, for the releasing organism first and the receiving organism second. The virgule (/) indicates that adaptive advantage or detriment is not specified for one side of the relationship.

I. Allelochemic effects
   A. Allomones (+ /), which give adaptive advantage to the producing organism
      1. *Repellents* (+ /), which provide defense against attack or infection (many secondary plant substances, chemical defenses among animals, probably some toxins of other organisms)
      2. *Escape substances* (+ /) that are not repellents in the usual sense (inks of cephalopods, tension-swimming substances)
      3. *Suppressants* (+ −), which inhibit competitors (antibiotics, possibly some allelopathics and plankton ectocrines)
      4. *Venoms* (+ −), which poison prey organisms (venoms of predatory animals and myxobacteria, aggressins of parasites and pathogens)
      5. *Inductants* (+ /), which modify growth of the second organism (gall, nodule, and mycorrhiza-producing agents)
      6. *Counteractants* (+ /), which neutralize as a defense the effect of a venom or other agent (antibodies, substances inactivating stinging cells, substances protecting parasites against digestive enzymes)
      7. *Attractants* (+ /)
         a. Chemical lures (+ −), which attract prey to a predator (attractants of carnivorous plants and fungi)
         b. Pollination attractants, which are without (+ 0) or with (+ +) advantage to the organism attracted (flower scents)
   B. Kairomones (/ +), which give adaptive advantage to the receiving organism
      1. *Attractants* as food location signals (/ +), which attract the organism to its food source, including (− +) those attracting to a food organism (use of secondary substances as signals by plant consumers, of prey scents by predators or chemical cues by parasites), (+ +) pollination attractants when the attracted organism obtains food, and (0 +) those attracting to nonliving food (response to scent by carrion feeder, chemotactic response by motile bacteria and by fungal hyphae)
      2. *Inductants* (/ +), which stimulate adaptive development in the receiving organism (hyphal loop factor in nematode-trapping fungi, spine-development factor in rotifers)
      3. *Signals* (/ +) that warn of danger or toxicity to receiver [repellent signals (A, 1) that have adaptive advantage to the receiver; scents and flavors that indicate unpalatability of nonliving food, predator scents]
      4. *Stimulants* (/ +), such as hormones, that benefit the second organism by inducing growth
   C. Depressants (0 −), wastes, and so forth, that inhibit or poison the receiver without adaptive advantage to releaser from this effect (some bacterial and parasite toxins, allelopathics that give no competitive advantage, some plankton ectocrines)

II. Intraspecific chemical effects
   A. Autotoxins (− /), repellents, wastes, and so forth, that are toxic or inhibitory to individuals of the releasing populations, with or without selective advantage from detriment to some other species (some bacterial toxins, antibiotics, ectocrines, and accumulated wastes of animals in dense culture)

B. Adaptive autoinhibitors (+ /) that limit the population to numbers that do not destroy the host or produce excessive crowding (staling substance of fungi)
C. Pheromones (+ /), chemical messages between members of a species, that are signals for:
   1. *Reproductive behavior*
   2. *Social regulation and recognition*
   3. *Control of caste differentiation*
   4. *Alarm and defense*
   5. *Territory and trail marking*
   6. *Food location*

SOURCE: Whittaker and Feeny (1971).

tive interactions in organizing communities. In plant succession a dominant species may speed its invasion of a preceding community through allelopathic suppression of growth, and prevent its own replacement by other species through release of alleopathic agents. The known allelopathic substances belong to a few major groups of compounds among the secondary plant substances; these include phenolic acids, flavonoids and other aromatic compounds, terpenoid substances, steroids, alkaloids, and organic cyanides (Whittaker and Feeny, 1971).

Other toxic plant substances affect the palatability of green plant parts to insects or other herbivores. Steroid cardiac glycosides cause convulsive heart attacks in vertebrates eating foxglove (*Digitalis purpurea*), and a single leaf of oleander (*Nerium oleander*) may cause death in man. Cattle may be poisoned by larkspurs (*Delphinium* species) and locoweeds (*Oxytropis*), which contain neurotoxic alkaloids. Certain plants are able to synthesize insect hormones and analogs, such as the molting and juvenile hormones (called phytoecdysones). If a certain critical dose of the phytoecdysones is exceeded, insect metamorphosis is fatally accelerated. A great many plants eaten by insects use poisonous alkaloids, mustard oils and their glycosides, and other secondary substances to serve as repellents or toxins and in some cases, even attractants, for the advantage of the plants. In a biological community the plant species which first successfully synthesizes or utilizes a substance as a toxin, may gain a considerable advantage over its fellows in the competitive race for light and energy and protection against enemies.

Other chemical defenses are used among animals, primarily for defense against predators. Use of such chemicals by an organism confers an advantage in survival compared to closely related species that lack such protection. Many of the species of insects that feed on poisonous plants such as Asclepiadaceae (milkweeds) and Solanaceae (the deadly nightshade family) use the poisonous compounds from their hosts for their own defense against birds and other predators. Predatory bugs may develop toxic salivary secretions which, like snake venoms, consist of a mixture of tissue-destroying enzymes. Species that have such venoms gain a competitive advantage in prey capture over their fellows.

A great many further examples of chemical interactions among species could be cited, including antibiotics secreted by fungi and microbes in the soils, but these examples have served to show the extremely broad use of these compounds in competitive interactions in nature. A summary of the classes of chemical effects, classified by function and adaptive relationship, is given in Table 11–1. As can be seen in a close examination of these many categories, chemicals are useful in both interspecific and intraspecific relationships. We shall examine more detail the evolutionary race between species in utilizing chemical agents in a later section on coevolution. Now let us look at a general model of competitive interactions among natural populations.

A GENERAL MODEL OF COMPETITIVE INTERACTIONS

A general theoretical treatment of competition was first put forth on a firm basis half a century ago by Lotka (1925) and Volterra (1926, 1931). Their mathematical formulas describe relationships between two species utilizing the same resource, and show through their modification of the Verhulst–Pearl logistic equation (Chapter 5) that the primary competitive effect is on the basic population growth curve. Eventually, only one species, the least affected by food shortage or the most adaptable to change in environmental conditions, will survive. These general competition equations indicate the detrimental effect that populations have on each other's abilities to grow. Thus, they involve an effect on the carrying capacities K_1 and K_2 of the two competing species (N_1 and N_2) as well as involving the species' maximal instantaneous rates of increase per individual, r_1 and r_2.

When one of the species, for example species 1, occurs by itself, its growth can be expected to behave approximately in accordance with the logistic equation:

$$\frac{dN_1}{dt} = r_1 N_1 \left(\frac{K_1 - N_1}{K_1}\right)$$

The simultaneous growth of the two competing species occurring together can be described by the Lotka-Volterra competition equations, which are differential logistic equations that are modified to incorporate the effect of the competing species on the logistic equation for a single species (as shown above). These competition equations describe the influence of one species on another:

$$\frac{dN_1}{dt} = r_1N_1\left(\frac{K_1 - N_1 - \alpha N_2}{K_1}\right)$$

and

$$\frac{dN_2}{dt} = r_2N_2\left(\frac{K_2 - N_2 - \beta N_1}{K_2}\right)$$

where

dN_1 is the change in numbers of species 1,
r_1 is the maximal rate of increase of species 1,
N_1 is the number of individuals of species 1, and
K_1 is the carrying capacity for species 1 in the absence of the competing species

and the same symbols with the second subscript are true for species 2. Also,

α is the competition coefficient for species 2 in relation to species 1, and
β is the competition coefficient for species 1 in relation to species 2,

where the competition term α is the relative impact of one individual of species 2 on the growth rate of the population of species 1, and β is the reverse competition coefficient.

If we assume that the outcome of a competitive interaction between two species is not concluded until each population achieves a stable size, even if this requires extinction for one of the populations, then we may use the preceding equations to describe the outcome of competition. The four possible outcomes of interspecific competition for species that are not ecological equivalents are shown in Figure 11–8.

In Case 1, species 1 has driven species 2 to extinction and N_1 equals the carrying capacity K_1. If as in Case 2, species 2 has driven species 1 to extinction, then N_2 equals K_1/α, which means that the density of species 2 in the absence of species 1 depends on the carrying capacity of the environment defined in terms of species 1 modified by the competition coefficient. In Case 3 we have an unstable equilibrium with the eventual winner dependent upon relative population sizes. Case 4 has the outcome with the greatest interest, perhaps; it presents the conditions required for stable equilibrium and coexistence. In order for the two species in Case 4 to live together indefinitely, they must stop increasing before they produce enough individuals to reverse the growth of the competing species. Their own density-dependent controls bring their growth to a halt before they eliminate the competitor. As species 1 increases and species 2 decreases, the two

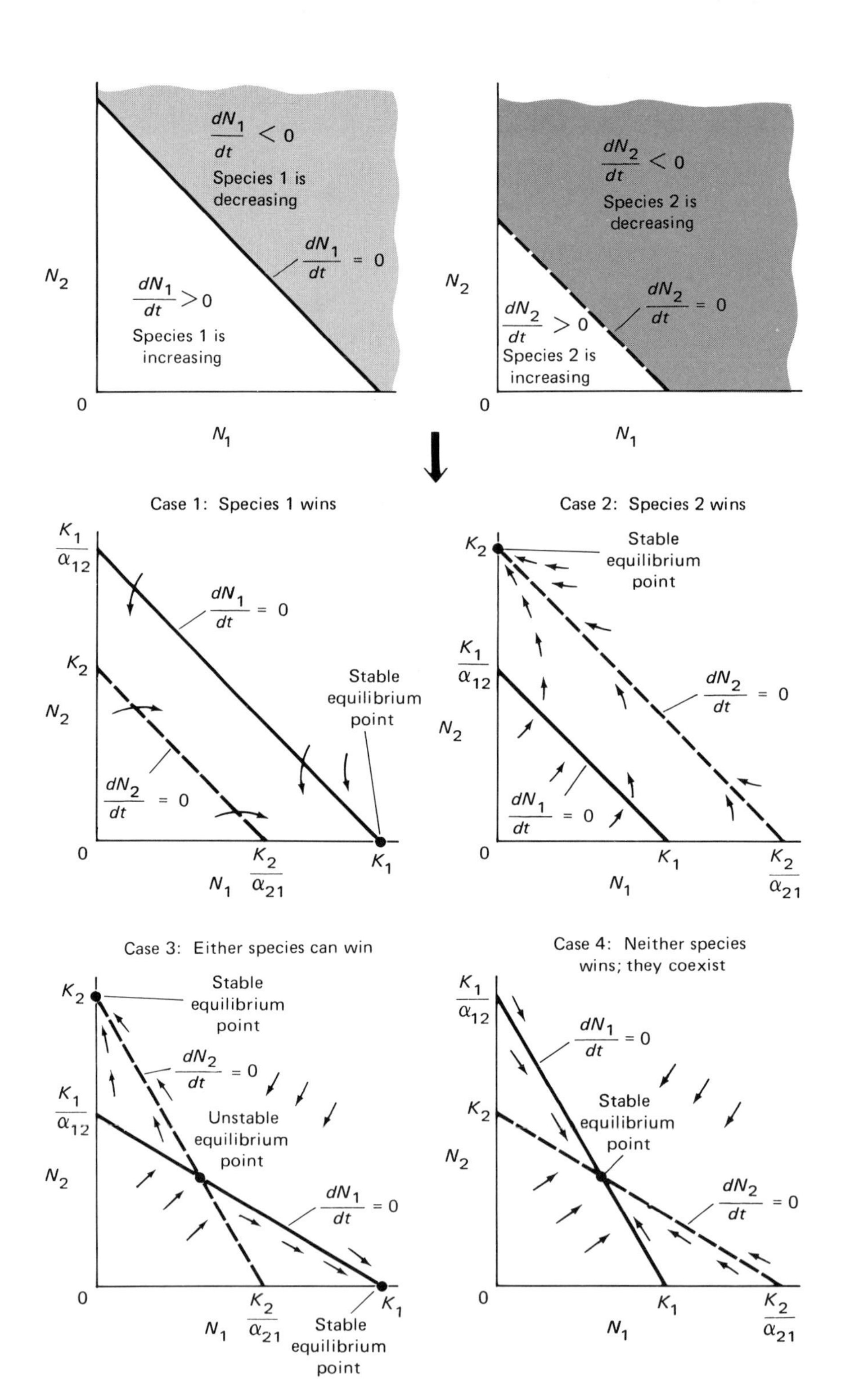

$\frac{dN_1}{dt} < 0$
Species 1 is decreasing
$\frac{dN_1}{dt} = 0$
$\frac{dN_1}{dt} > 0$
Species 1 is increasing
N_2
0
N_1
$\frac{dN_2}{dt} < 0$
Species 2 is decreasing
$\frac{dN_2}{dt} = 0$
$\frac{dN_2}{dt} > 0$
Species 2 is increasing
Case 1: Species 1 wins
$\frac{K_1}{\alpha_{12}}$
K_2
$\frac{dN_1}{dt} = 0$
$\frac{dN_2}{dt} = 0$
Stable equilibrium point
$\frac{K_2}{\alpha_{21}}$
K_1
Case 2: Species 2 wins
K_2
Stable equilibrium point
$\frac{K_1}{\alpha_{12}}$
$\frac{dN_2}{dt} = 0$
$\frac{dN_1}{dt} = 0$
K_1
$\frac{K_2}{\alpha_{21}}$
Case 3: Either species can win
K_2
Stable equilibrium point
$\frac{dN_2}{dt} = 0$
$\frac{K_1}{\alpha_{12}}$
Unstable equilibrium point
$\frac{dN_1}{dt} = 0$
$\frac{K_2}{\alpha_{21}}$
K_1
Stable equilibrium point
Case 4: Neither species wins; they coexist
$\frac{K_1}{\alpha_{12}}$
$\frac{dN_1}{dt} = 0$
K_2
Stable equilibrium point
$\frac{dN_2}{dt} = 0$
K_1
$\frac{K_2}{\alpha_{21}}$

must eventually reach a point, the stable equilibrium point, where this process is reversed. Here species 1 is so abundant that its own density-dependent controls halt its growth, but because its niche is sufficiently different from that of species 2, it does not become abundant enough that it can stop species 2 from beginning to increase. Species 2 likewise moves toward this equilibrium point and eventually the joint abundance at this level allows no further change through time (Wilson and Bossert, 1971).

Although these equations probably do not match the situation in many natural populations precisely, they provide an important theoretical framework for further exploration of competitive phenomena (for example, see Pianka, 1974; Levins, 1968; MacArthur, 1968, 1972; Vandermeer, 1970, 1972).

PREDATOR–PREY INTERACTIONS

We have already looked at some of the effects of predators on prey populations in Chapters 3 and 5. In contrast to competition in which both species are adversely affected in their growth patterns, and a selective premium is placed on minimizing competitive interactions, predation is a directional selective phenomenon in which one member of the pair (the predator) benefits from the association, while the other (the prey) is affected adversely in at least a proximate sense. Ultimately, prey populations benefit from predation operating as a density-dependent regulating mechanism. Natural selection tends to favor predators that are adept at finding, capturing, and handling prey efficiently. Prey that have adaptations to more readily escape predators tend to be at a selective advantage. Hence, there is a kind of coevolutionary race between the predator and the prey which results in an inherently oscillatory system. The prey and predator populations are jointly adapted to each other, and an evolutionary advance in one causes a selective response in the other. That is, when prey evolve new means of avoiding their predator, that predaceous species has a high selective premium placed on variants which are able to find or handle the new prey variants. As we have seen in Chapter 5, oscillations also occur on a short-term basis in the sense of time lags in population size. Predator response to an increase in prey population size lagged by the length of time needed for young predators to grow up,

---

Figure 11–8 Competition between two species: the four possible outcomes. In the top figures, the zero-growth curve for species 1, labeled $dN_1/dt = 0$, falls outside the zero-growth curve for species 2, labeled $dN_2/dt = 0$. As a consequence, changes in joint abundance through time, indicated by the arrows, always favor species 1 at the expense of species 2 and lead ultimately to the extinction of species 2. (After E. O. Wilson and W. H. Bossert, 1971, *A Primer of Population Biology*, Sinauer, Stamford, Conn., pp. 160–161.)

and as a result the predators produce enough offspring that eventually they will overeat their food supply and the prey species will then precipitously decline, with the predator suffering considerable mortality (or at least lower fucundity), and thus also decline. Expressing these relationships in a theoretical model has been less successful than in competition theory, but again the pioneers in this field as with competition were Lotka (1925) and Volterra (1926, 1931). These two mathematicians separately proposed formulas to express the relationship between prey and predator population. As we have seen in Chapter 5, the following simple pair of predation equations shows that as the predator population increases, the prey decreases to a point where the trend is reversed and oscillations are produced:

$$\frac{dN_1}{dt} = r_1N_1 - p_1N_1N_2$$

$$\frac{dN_2}{dt} = p_2N_1N_2 - d_2N_2$$

where

N_1 = the prey population density,
N_2 = the predator population density,
r_1 = the instantaneous rate of increase of the prey population (per individual),
d_2 = the death rate of the predator population (per individual), and
p_1 and p_2 = predation constants.

This pair of differential equations produces an oscillating predator-prey interaction (Figure 11–9), with the population densities of both prey and predator changing cyclically and out of phase over time (see Wilson and Bossert, 1971; see also Chapter 5).

Predation includes foraging activities of organisms that eat all or parts of other live organisms as energy sources. Therefore, under this definition, herbivores as well as carnivores are predators. Parasites are not included because they have usually adjusted their behaviors sufficiently to allow their host to remain alive and continue to provide them with energy. Evolutionary interactions among different kinds of organisms—plants and herbivores, models and mimics in mimicry systems, carnivorous predators and prey, parasites and hosts, etc.—involve a continual flux of adaptive genetic changes which has been called coevolution (Ehrlich and Raven, 1965). Many of these interactions involve chemical defenses and we have looked at some of these earlier in the present chapter. The delicately balanced interactions that are frequently present in such associations may have ramifications on populations beyond those of primary concern. For instance, biochemical changes that affect the balance between a butterfly with

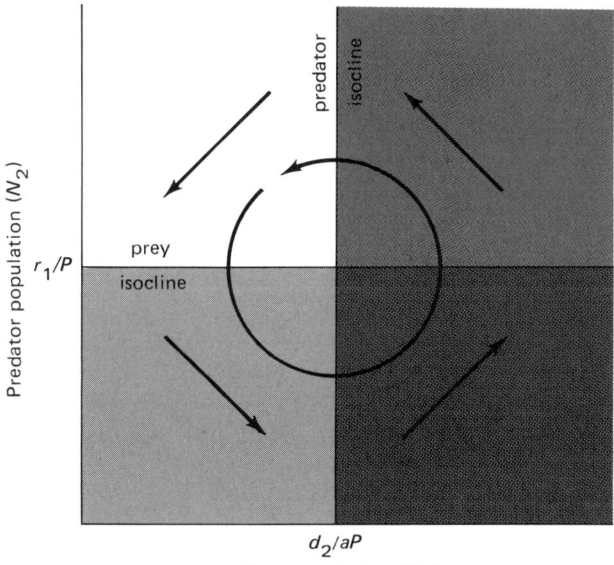

Figure 11–9 Representation of the Lotka-Volterra predator–prey model on a population graph. The area in which the prey population can increase is indicated at upper left, and the area in which the predator population can increase is indicated in the lower half. The trajectories of the populations show that the predator and prey will continually oscillate out of phase with each other (full circle). (Source: R. E. Ricklefs, 1973, *Ecology,* Chiron Press, Newton, Mass., p. 538.)

distasteful tissues and a palatable species that mimics it, will have ramifications on the bird populations feeding on these butterflies and therefore on still other species which are ecologically or coevolutionarily associated with the birds. Escape from predation by use of external chemical defenses, reducing the palatability of prey to predators, has also been surveyed previously (Table 11–1). Many beetles, mammals such as skunks, millipedes and other potential prey items, have defensive secretions which can be sprayed at predators to actively ward them off before injurious attack on the prey can occur.

Mimetic assemblages of species of toxic and nontoxic organisms coevolve with complexes of predators and foodplants. Many insects derive their toxic or noxious compounds from their foodplants and utilize them as antipredator devices. Such insects will generally evolve *warning* or *aposematic* coloration to provide externally apparent signals to potential predators that they are distasteful (Figure 11–10). These insects frequently become models in *mimicry* complexes, as do other poisonous animals. The two classic kinds of mimicry—Batesian mimicry and Mullerian mimicry—were named after the nineteenth century naturalists who first recognized and described these phenomena. The

Figure 11–10 Warning coloration: this bright red poison-arrow frog, *Dendrobates grandulifera,* contains at least two toxic steroidal alkaloids that cause nerve and muscle paralysis and, in minimum lethal dose, are four times as poisonous as sodium cyanide. (Photo by C. L. Hogue.)

coevolutionary systems in such mimicry complexes involve food plants, insects, and visual predators such as birds.

A *Batesian* mimicry complex includes a distasteful or dangerous species called the *model* and one or more similarly appearing edible species, the *mimics,* which gain protection from predation by resembling the model. Abundant experimental evidence (summarized in Ford, 1965, 1972) has shown that naive vertebrate predators inexperienced with the model species learn after one or two trials to avoid that species on all future occasions, and birds and mammal predators such as monkeys generalize from a bad experience such that they avoid all prey items that resemble the object which created the original adverse reaction. The Monarch butterfly (*Danaus plexippus*) feeds on poisonous milkweeds over most of its North American range and is mimicked by the similarly colored non-noxious Viceroy (*Limenitis archippus*) whose larvae feed on willows (Figure 11–11). The ground-running clubionid spider *Castianeira rica* is a benign species resembling stinging myrmicine and ponerine ants. In this complicated Batesian mimicry relationship in Costa Rica, Reiskind (1970) found that the multiple mimetic forms of the spider result from (1) sexual dimorphism (two forms), (2) color variation in the adult female (an additional form), and (3) developmental changes in the preadult spider instars (at least two ad-

Figure 11–11 Batesian mimicry. The monarch (*Danaus plexippus*) carries cardiac glycosides from larval feeding on milkweeds, hence is toxic to vertebrate predators. It serves as a model for the harmless mimic Viceroy (*Limenitis archippus*) over much of North America. (Photo by T. C. Emmel.)

ditional forms). The mimics and model ants occur in the same microhabitat—heavily shaded leaf litter—and the spiders have a general antlike behavior (running, jerky gait, with the first pair of legs raised like the searching antennae of ants) in addition to an antlike physical appearance. Some fish, such as the poison-fang blenny (*Meiacanthus atrodorsalis*), have large canine teeth capable of imparting a toxic bite to potential predators (Losey, 1972). Morphologically and behaviorally similar species have been shown to enjoy predator protection through Batesian mimicry.

In *Mullerian mimicry* all potential prey species in a mimicry complex are distasteful models and share a common warning color pattern and behavior (Figure 11–12). This becomes selectively advantageous to each of the prey species because the feedback for the predator from the experience of capturing a prey item is always negative. The local predators will learn to associate distastefulness with that pattern without trying all the species. With more models evolving to join the Mullerian complex, the possible number of negative rewards increases. The obvious advantages to the prey from sharing one aposematic pattern instead of many include minimization of handling by predators and thus reduction of the chance of prey injury while the predator is discovering its noxious qualities. Mullerian mimicry complexes tend to include mostly members of the same phylogenetic line (genus or family), probably because the ability of the model to store or produce a noxious chemical is not a simple genetic adjustment. Once it has arisen in an evolutionary line, it likely becomes a standard feature of that major taxon, such as the butterfly family Ithomiidae, which has achieved the ability to sequester poisonous alkaloids from the so-

Figure 11–12 A Mullerian mimicry complex of distasteful ithomiine butterflies from Limoncocha, Ecuador: *Mechanitis lysimnia elisa, M. mazaeus mazaeus* (left); *Forbestra equicola* nr. *equicoloides,* and *Mechanitis isthmis eurydice* (right). (Photo by T. C. Emmel.)

lanum plants. In Batesian mimicry, however, it is much more common for the model and mimic members of the complex to be from widely different evolutionary lines and convergent only in external phenotype and behavior.

Hespenheide (1973) has shown that a striking color pattern which mimics medium-sized to large flies occurs among more than 60 species of Central America beetles of at least five families. Both flies and beetles with this pattern tend to be abundant and to perch on relatively isolated and exposed tree boles. The advantage to a beetle of looking like a fly is hypothesized by Hespenheide to rest in the difficulty that birds and other visually-hunting predators have in capturing flies of the size and type mimicked by the beetles. Because most of the beetles are also very quick and elusive, it is suggested that the mimicry system is functionally Mullerian, in which both flies and beetles are avoided by birds because of the high energetic cost to the bird in their capture, rather than a chemical unpalatability.

Several other classes of mimetic relationships have been proposed. *Mertensian mimicry,* described by Wolfgang Wickler (1968) and named after the German herpetologist Robert Mertens for his studies of the complexes of South American coral snakes and their mimics, involves an interesting twist: the most offensive species is not the model because it is *too* offensive to allow the signal-receiver (predator) to learn. Instead, it is the mimic of a less offensive model species, which has venom poisonous enough to sicken the potential predator but not to kill him. Harmless snakes of nonpoisonous families can also be involved in Mertensian mimicry complexes, mimicking the moderately poisonous species.

Brower and his co-workers (1970) found that not all monarch butterflies were Batesian models. Some fed on species or forms of milkweeds that did not contain poisonous cardiac glycosides, and indeed he found they were perfectly edible to predators when he fed them to naive laboratory-reared birds. It appears that in the eastern United States the monarch populations are composed of partly noxious and partly palatable butterflies. The natural predators would not be expected to detect this *automimicry,* or models and mimics within a single prey species, unless the ratio of noxious to palatable individuals favored the latter category excessively.

Another more ecologically intricate interaction between predator and prey occurs in *aggressive mimicry.* In this predator strategy, the predator deceives the prey by evolving coloration, physical characteristics, and behavior, that give it the appearance of a nonpredator. Thus ambush bugs (members of the insect order Hemiptera) are colored and shaped to resemble flower parts in composite blossoms, and their concealment makes it relatively easy for them to attack unsuspecting bees and flies coming for nectar. The aggressive mimicry shown by predaceous fireflies of the genus *Photinus* (Lloyd, 1965) represents an even more elaborate way to decrease the escape response by prey. Male fireflies of a related genus, *Photurus,* signal with light flashes from the top of blades of grass at night and call in females of the same species for mating (Figure 11–13). However, if a female *Photinus* firefly spots the flash of the calling male *Photurus,* she will imitate the normal flashing response of a *Photurus* female and be allowed by the male *Photurus* to fly in and land next to him. At this point, the predaceous female *Photinus* drops its mimic role of potential mate and seizes the male *Photurus* to eat him (Figure 11–13b).

Cryptic coloration involves a resemblance of the prey to some nonedible or background object in its environment, thereby deceiving potential predators. Camouflage is the commonest form of protective coloration (Figure 11–14). It has profound ecological as well as evolutionary implications, for many more prey species can be packed into a community if each is protectively colored in some way, reducing the impact of predation upon the individual species. Examples seem particularly abundant in the tropics, but occur in every community, from arctic to desert to ocean habitats. Disruptive coloration in the form of body stripes can conceal even large mammals like the boldly striped zebra or gazelle. Many insects resemble sticks, bark,

a

rocks, leaves, and bird droppings. Prey species frequently have bizarre or unusual morphology which disrupts the shape of the organisms such that predators are less likely to recognize them as prey.

Several other widely observed aspects of animal coloration have arisen through natural selection in response to predator–prey interactions. In addition to sharp claws, tearing teeth, rapid behavioral reactions, and other prey-capturing adaptations, many vertebrate predators have color lines leading forward from the eye which apparently are aiming sights (Figure 11–15). These circles, eye lines, and grooves are found about the eyes of predatory birds, reptiles, mammals, amphibians, and fish (Ficken et al., 1971). In addition to their role as lines of sight in tracking and capturing prey, dark eye marks also reduce glare in bright open habitats. Light circles around the eyes probably function as light-gathering devices. Other facial patches and stripes on prey species may serve to hide the eyes and further decrease visibility to potential predators by interrupting the animal's outline (Figure 11–16). Many markings, of course, probably serve primarily in intra- and interspecific recognition; facial coloration in the American redstart (*Setophaga ruticilia*), for instance, functions for recognition of sex, age, and species, as well as for sighting during feeding (Ficken et al., 1971).

b

Figure 11–13 Aggressive mimicry. (a) A *Photinus* firefly in normal position for evening flashing. (b) A female *Photinus* firefly has successfully attracted and is eating a *Photurus* male by imitating the flash pattern of a *Photorus* female. (Photos by James E. Lloyd, Univ. of Florida, Gainesville.)

Erratic display can often serve as a specific device against predators. Humphries and Driver (1967) present considerable evidence that prey animals in many taxonomic groups behave erratically when attacked by predators. Such protean displays function to confuse and disorient the predator and to increase its reaction time. Thus, the survival of the prey is assisted, and the selective advantage is created whereby such erratic behavioral patterns of the prey animals may have evolved.

Selective forces shaping predator–prey relationships intimately affect the social system of the predator species, as we have seen to some degree in the evolution of flocking behavior in birds. Large carnivores such as the big cats exhibit an interesting array of social behavior patterns related to their predatory activities. Some are solitary, whereas

a

b

Figure 11–14 Examples of protective coloration. (a) Disruptive coloration of stripes running off the body assists in concealing the body outlines of these zebras in Kruger National Park. (b) The dusty tan coats and vertical rump stripes of these impala serve as cryptic coloration in the African brush. (Photos by T. C. Emmel.)

Figure 11–15 Eye lines serve as sighting and aiming devices for predators in catching prey. (a) Simple eye line of the partially insectivorous blue tit (*Parus caeruleus*). This is the most common type of eye line in vertebrates. The eye line in this species is wider than it is in other avian examples. (b) Combined eye circle and eye line of the yellow-throated vireo (*Vireo flavifrons*). (c) Raised yellow feathers above the eye line in a white-throated sparrow (*Zonotrichia albicollis*); these may cast light along the line of sight. (d) Red-necked grebe (*Podiceps griseigena*) showing eye line slanting downward. Such a line also occurs in some other fish-eating birds. (e) Long-billed curlew (*Numenius americanus*) showing direction of eye line forward of center of pupil to bill tip. (f) Teardrop mark of the pickerel (*Esox americanus*) associated with downward dashes at prey. (g) Rearward pointed eye line of the European woodcock (*Scolopax rusticola*), probably used to sight predators coming from behind. Associated with 360° vision plus front and back binocularity. (h) Head of the arboreal vine snake (*Oxybelis aeneus*) showing eye line and groove. (Source: R. W. Ficken et al., 1971, *Science, 173:* 936–939. Copyright by the American Association for the Advancement of Science.)

others are quite social. The lion lives in prides consisting of one or more males, several females, and cubs, numbering anywhere from four to 30 or more individuals. Such prides generally remain constant in composition for several years except for births and deaths and the emigration of some subadults. Schaller (1972) has pointed out that lion society has probably been shaped and maintained by several interacting selective forces. First, the social lion lives on the open sa-

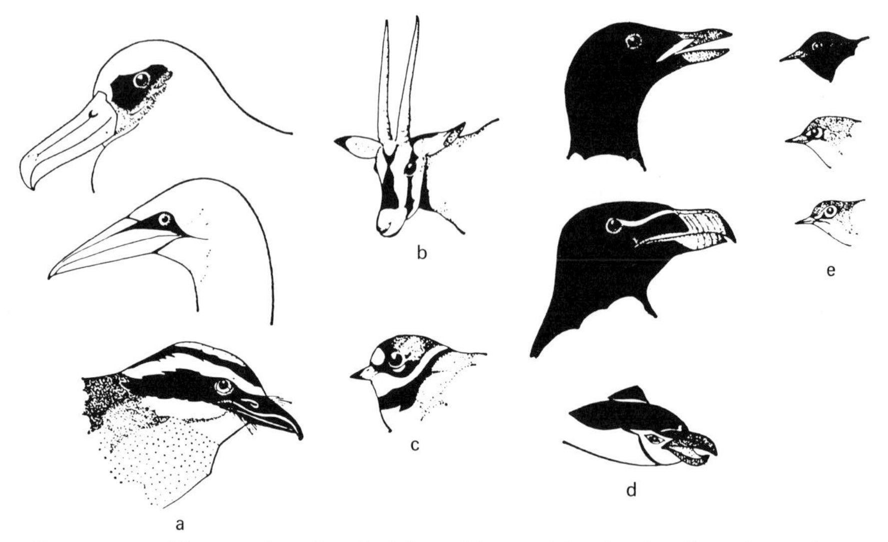

Figure 11–16 Figures showing facial markings with adaptive functions other than predatory sighting alone. (a) Facial patches and stripes associated with reduction of glare in very bright habitats. (b) Here eye lines may hide eyes and further decrease visibility to potential predators by interrupting the beast's outline; also, the pattern may decrease glare on the line of sight toward clumps of grass in the animal's usually bright open habitat, and the pattern may be important as a social signal. (c) Here there is no sighting line; stripes are probably mainly for disruption of the head pattern (concealing the eye) and for aiding species recognition. (d) Alcids and penguins are very highly specialized for catching fish. The lines about the eyes and heads of virtually all birds of these two groups could not be for sighting of prey; they probably serve in intra- and interspecific recognition. (e) Male (top), female (middle), and first-year male (bottom) American redstart (*Setophago ruticilia*), showing development of differences for recognition of sex, species, and age as well as a complex eye line and eye circle associated with feeding. (Source: R. W. Ficken et al., 1971, *Science, 173:* 936–939. Copyright by the American Association for the Advancement of Science.)

vanna whereas the more solitary cat species tend to live in closed forest environments. Group cohesion is easier to maintain in an area of good visibility and high game concentrations. Cooperation enables lions to catch large prey like wildebeest and zebra with considerable success, whereas solitary lions cannot utilize prey of such size. A division of labor is possible in a group, such that one lion may guard the cubs while the others hunt for meat for all, guard the carcass, or seek shade (Figure 11–17). Tigers, on the other hand, essentially lead a solitary existence although several individuals may meet and share a kill of a large water buffalo, then part again. A tiger's nocturnal habits and forest habitat, where the prey species are widely scattered, probably are not conducive to long-term groupings, although the species formally occupied grassland areas in parts of its wide geographic distri-

Figure 11–17 A pair of African lions in Ngorongoro Crater, Tanzania. An average lion pride consists of four to six individuals, the appropriate number to enable each individual to gorge itself on a wildebeest, zebra, topi, or similar prey. Several lions stalking together employ such cooperative methods as encircling their quarry and hence are generally twice as successful in catching their prey as a solitary lion (Schaller, 1972). Young lions typically remain dependent on the predatory activity of the pride until they are at least $2\frac{1}{2}$ years old. (Photo by T. C. Emmel.)

bution. Schaller (1967) also notes that much of the tiger's food consists of relatively small mammals such as pigs and deer, which, if shared with several other adults in a social group, would not provide enough meat for them all. The nocturnal leopard is essentially a solitary species in East Africa (Schaller, 1972), hunting in the riverine bush, whereas the diurnal cheetah are solitary also, although they hunt in the open plains. Several males may occasionally associate as companions (Figure 11–18). The cheetah's method of hunting involves running down its prey in a short, extremely rapid chase (Eaton, 1974); the leopard usually waits along game trails and pounces from a tree onto its prey. Small antelope such as Thomson's gazelle and impala are the usual prey for cheetah and leopard. Both types of hunting strategies work better for solitary predators than for a social group. The encircling method of the lion pride enables these larger cats to take heavier game species.

Many lizards are predaceous on insects and other small arthropod prey. Even the large herbivorous tree iguana, *Iguana iguana,* which sometimes reach six feet in length from head to end of tail, and the

Figure 11–18 Cheetahs are usually solitary hunters, although several males may associate as companions like this pair at Samburu in northern Kenya. Cheetah young become fully independent at the age of about 16 months, apparently not requiring the extra months of dependence which the lion cubs experience in the pride in order to learn cooperative group hunting. (Photo by T. C. Emmel.)

ground iguanas such as *Ctenosaura similans* in Mexico and Central America, have insectivorous young. This predaceous stage lasts for only the first year or so of growth, for beyond about a foot in length the immature iguanas switch to a grazing diet of foliage (Heath, 1963). In the rain forests and savannas of six Indonesian islands, however, the largest lizards in the world track down their animal prey, dead or alive. The formidable Komodo dragon, *Varanus komodensis,* reaches a length of 10 ft and a weight of 200 lb, and is capable of bringing down a large mammal as a kill. The thorough studies of the population biology of this giant monitor lizard by Auffenberg (1972; in press) offer exceptionally interesting comparisons with the predator-prey systems we have seen among the big cats. A highly opportunistic carnivore, the adult lizard roams the monsoon forest in search of prey or carrion (Figure 11–19). In the first year of growth, young monitors remain in trees, feeding on small animals such as geckos. At about 3 ft in length and one year of age, the lizards begin scavenging as well as hunting larger prey. Komodo monitors can successfully attack virtually every animal—from insects to water buffaloes—found on the islands of Komodo, Flores, Rintja, Padar, Owadi Sami, and Gili Moto. Goats stocked on neighboring islets by natives will be attacked by monitors

Figure 11–19 The formidable Komodo dragon (*Varanus komodoensis*), largest lizard in the world, roams the monsoon forests and open savanna on Komodo Island in Indonesia in search of prey or carrion. Its predatory behavior is remarkably flexible for a reptile, and ranges from setting an ambush along game trails or smelling carrion on the wind to actively hunting a learned route daily for resting deer. (Photo by Walter Auffenberg, New York Zoological Society.)

swimming 1300 ft of swift tidal currents to temporarily take up residence there. Grasshoppers, rats, birds, goats, hogs and deer are the most common prey, depending upon the size of the monitor. Large monitors do most of their serious hunting by lying in wait along game trails descending to valleys. Deer move along these trails in late morning to reach the shade of the thicker vegetation on the valley floors. Lizards that are unsuccessful in morning ambushes along the trails follow the deer slowly and once reaching the thickets, they begin an afternoon hunt in which they attempt to capture sleeping deer. Monitors attack water buffaloes weighing as much as 1300 lb by creeping close to and biting a leg with their sharp, serrated teeth. With their vise-like grip, the Achilles tendon is usually severed and the buffalo is forced to drop to the ground where it is rapidly eviserated. Over a period of several days, an increasing number of monitors will slowly gather to feed at the rotting carcass. However, actual hunting for prey and carrion is essentially a solitary endeavour, and the small, casual

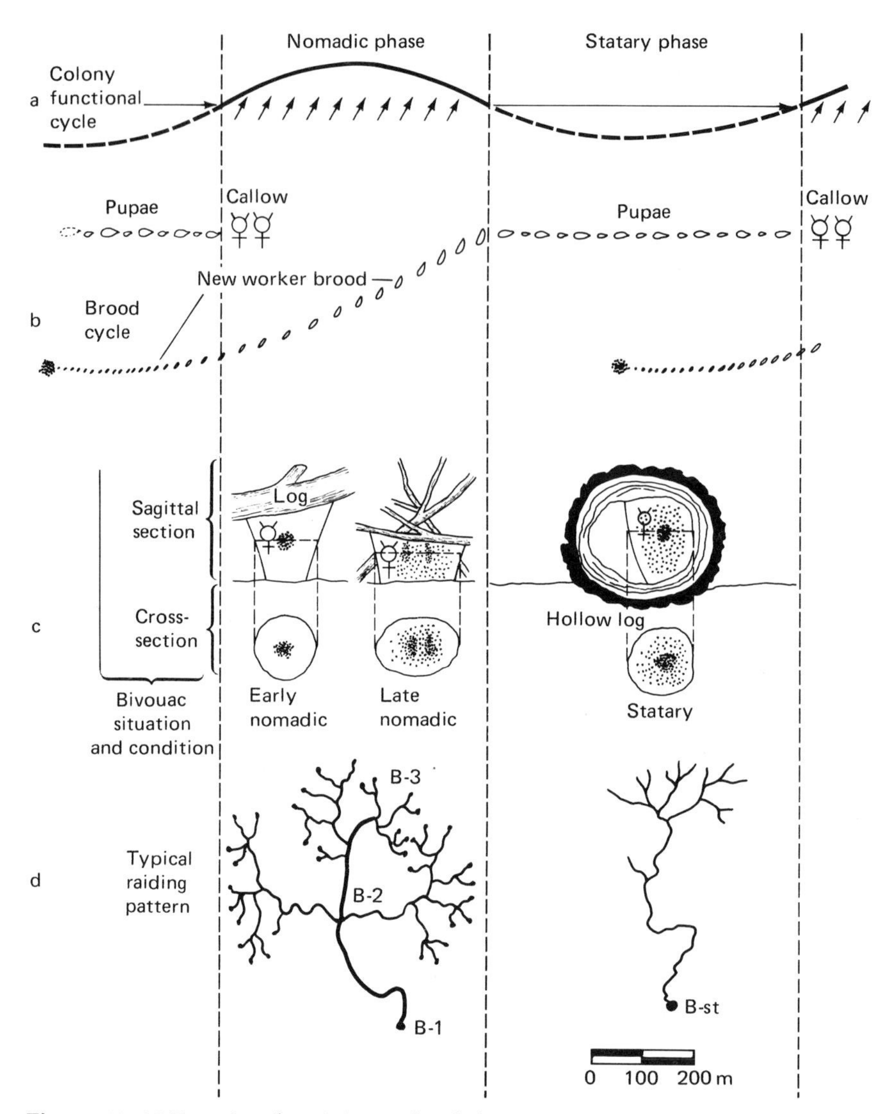

Figure 11–20 Functional activity cycle of the army ant *Eciton hamatum* in the tropical rain forest of Panama. (a) Phases in the cycle are indicated by a sine curve; arrows indicate large daily raids and nightly emigrations. (b) The concurrent development of successive all-worker broods, from eggs (at left) to larvae, pupae, and finally to freshly emerged (callow) workers (modified female signs, at right). (c) Types of bivouac in each functional phase, indicating typical placement of brood in each. (d) Patterns of raiding typical of the functional phases. B-1, B-2, and B-3 are successive bivouac sites, whereas B-st indicates a statary bivouac site. (Source: T. C. Schneirla, 1971, *Army Ants: A Study in Social Organization,* Freeman, San Francisco, p. 2.)

aggregations at a dead animal last no longer than one or two days and involve little true coordinated behavior of any sort.

The army ants, representing more than 200 species of the subfamily Dorylinae distributed through the tropics and subtropics of the world, exemplify an extraordinary development of an exclusively predatory habit and carnivorous diet in animals with huge colony populations. They share much the same way of life, including (1) massive predatory raids for insects and other prey, (2) great nomadic movements of the colonies based on the ability to form and readily abandon temporary nests, and (3) the ability of all colonies to change more or less regularly from a high level of activity and reproduction to a low level (Schneirla, 1971). The functional cycle of the Neotropical army ant *Eciton hamatum* is depicted in Figure 11–20. A colony typically consists of 150,000 to 500,000 workers, a large brood of developing individuals, and a single queen. The worker population is polymorphic in overall size and structural proportions; the individuals with the largest jaws are commonly called "soldiers" (Figure 11–21b). Workers carry out all tasks except reproduction. During the nomadic phase of the colony's cycle (Figure 11–20), large fan-shaped swarm raids of workers sweep through the forest daily (Figure 11–21c), securing prey for the adults and the developing worker brood (Figure 11–21a). The queen stays in a diurnal bivouac surrounded by masses of workers (Figure 11–21d), and the colony moves nightly in a mass emigration. This phase lasts for about two weeks. When the brood passes into its nonfeeding pupal stage, the colony enters the stationary phase and establishes a stable bivouac with small daily raids. The queen lays great numbers of new eggs at this time. When the pupae of the older brood mature after several weeks, the brood emerges as a great population of new workers (called callows) and the colony enters a new nomadic phase of activity. The energizing or stimulative effect of the brood condition on the colony is a major factor in controlling the adaptive patterns of predatory activity and movement of all army ants. The great sweeping predator raids through the rain forest, then, periodically affect prey populations according to the developmental schedule of the army ant brood.

The effects of predators on biological communities may be shown in other, less dramatic but nevertheless profound ways. Estes and Palmisano (1974) found in a comparison of western Aleutian islands supporting and lacking sea otter populations that this predator species is important in determining littoral and sublittoral marine community structure. Sea otters control herbivorous invertebrate populations such as sea urchins, which are voracious algal and kelp grazers. Removal of sea otters by Russian fur traders in the Near Islands during the eighteenth century caused great increases in herbivore populations and ultimately resulted in the destruction of macrophyte kelp associations.

a

c

Figure 11–21 The fierce swarm-raiding army ant, *Eciton hamatum*, on the Osa Peninsula in Costa Rica. (a) The workers capture and dismember a katydid during a swarm raid. (b) The giant mandibles of a soldier-caste worker. (c) Part of a swarm raid; in an emigration column, the workers would be carrying brood. (d) The bivouac, with a mass of living workers surrounding the queen and brood. The workers hook into each other with tarsal claws on the legs. (Photos: a, b, Charles L. Hogue; c, Charles L. Hogue and Julian Donahue; d, Julian Donahue.)

b

d

Because these plants are of considerable importance to nearshore productivity in temperate waters, many faunal species such as seals and bald eagles are lacking in the Near Islands though they are abundant in the Aleutian islands where sea otters occur. Estes and Palmisano suggest that the current reestablishment of sea otters along the Pacific Coast of North America will have profound ecological effects, probably decreasing invertebrate populations and increasing vegetational biomass. In this case, then, the predaceous sea otter becomes a keystone species in the sense of Paine (1969) and may be an evolutionary component essential to the integrity and stability of the Pacific Coast ecosystem.

SYMBIOSIS

Symbiosis refers to a long-term interspecific relationship in which two species live together in more or less intimate association. This is not a social system like those on the intraspecific population level but an ecological and behavioral association involving some adaptive benefit or transfer of energy. Symbiosis represents an evolutionary resolution of the problems of competition and predation through coevolution of a pair of species towards a more or less satisfactory accommodation enabling the two to live together. Even in parasitic symbioses, a good parasite avoids killing its host until the host as well as the parasite can reproduce. In other types of symbiosis, when a species receives direct or indirect aid and protection from another species, part of the energy normally required for maintenance activities and defense can be put into reproduction or other outlets such as social behavior and group organization. Phenotypes that might be normally eliminated from a population can survive and indeed even be advantageous in a symbiotic relationship, thus providing a means of retaining genetic diversity in species.

Symbiosis in the broadest sense includes four types of relationships: phoresis, commensalism, mutualism, and parasitism. *Phoresis* involves an association where two species live together but are not dependent upon each other (especially physiologically). In fact, they are perfectly capable of living as independent organisms. Along the shores of many tropical beaches and mangrove swamps live medium-sized to large land crabs. These crustaceans excavate rather deep burrows in which they live much of the time, coming out principally to forage. Land crabs show behavioral specialization in their burrow structure, nocturnal or crepuscular habits, brackish habitat preference, and the use of the burrow for seclusion during ecysis. Hogue and Bright (in literature) have emphasized the frequent symbiotic use of the crab's burrow by other organisms (Table 11–2). Phoresis or association relationships include the utilization of the burrow water and

Table 11–2 Summary of the known ecological structure of the crabhole community (arthropod fraction).

LEVEL OF INTERSPECIES REACTION	NICHE		MAJOR EXAMPLES
	PLACE	FUNCTIONAL	
Phoresis or association	Burrow water	Developing, breeding, feeding, etc.	Immature mosquitoes Diving beetles (*Ridessus*) *Cyclops*
	Burrow chambers	Resting, mating, etc.	Adult mosquitoes
	Surface of burrow water		Adult biting gnats (*Culicoides*)
Commensalism	Peribuccal cavity and renal grooves of *Gecarcinus ruricola*	Attaching to host and feeding on food debris	*Drosophila carcinophila*
Parasitism	Gill chamber of *Gecarcinus lateralis*	Attaching and feeding	Mite (*Laelaps cancer*)

SOURCE: Charles L. Hogue and Donald Bright.

chamber by mosquitoes, aquatic beetles, and gnats as breeding, feeding, and resting sites. These organisms are phoronts in the host's burrow.

Commensalism occurs when one species benefits from the symbiotic association but the other species is not significantly affected. Thus many bromeliads, orchids, ferns, and other epiphytes live on trees, especially in areas of high rainfall, and these commensals gain a place in which to grow and reproduce above ground-level vegetative competition (Figure 11–22). In the crabhole community (Table 11–2), the larvae of the fruit fly *Drosphila carcinophila* live in the peribuccal cavity and renal grooves of the host crab (*Gercarcinus ruricola*), feeding on food debris.

In *mutualism,* both species in the relationship benefit and a mutual functional dependency exists between the members of the association (Figure 11–23). Often, mutualistic symbionts evolve highly specialized, species-specific responses to chemical and physical stimuli which serve to bring the partners together and maintain the association. Large flightless weevils feeding on leaves of woody plants in high-altitude moss forests of New Guinea have extensive plant growth on their backs (Gressitt et al., 1965). Fungi, algae, lichens and liverworts of at least 12 families live on these heavily sclerotized, slow-moving, and longlived beetles (some of the plant growth appears to have required three to five years to develop). The fungi and lichens

Figure 11–22 Epiphytic bromeliads growing on cypress trunks in the Florida Everglades, an example of commensalism. (Photo by T. C. Emmel.)

are inhabited by orbatid mites which may transfer the plant spores from beetle to beetle. A thick waxy secretion from glands and the presence of specialized scales or hairs in roughly sculptured depressions on the newly emerged weevils represent modifications to encourage the initial growth of the plants. The mutualistic nature of this symbiotic relationship seems to be clearly indicated. Presumably, the flightless slow-moving weevils are concealed by the plant growth from predators, while the plants find less competition and fewer enemies growing on their moving environment than in the normal habitat of bark and leaf surfaces.

Ambrosia beetles cultivate one or more species of ambrosia fungi in their wooden tunnels and periodically remove some of the spore-bearing conidia for larval and adult food. Female beetles also carry the fungi to new tunnels for propagation. Batra (1966) has suggested that this mutualism probably originated by fungi being fortuitously carried into the tunnels of bark- and wood-inhabiting beetles. Initially the fungi established themselves beneath the bark, ramifying into the frass-filled tunnels. Fungi were occasionally consumed by the bark-feeding beetles. Eventually some beetles abandoned the eating of bark and wood, became wholly adapted to mycetophagy, and developed mutualism with ambrosia fungi to the degree we see today, where the fungus is required by the beetles for viable eggs to be produced (Norris and Baker, 1967). Similar specificity of mutualistic relationships is known from leaf-cutter ants that grow fungi in underground galleries (Weber, 1966), termites of the Old World tropics that cultivate fungi (Batra and Batra, 1966), fig wasps and fig trees of the genus *Ficus* Ramírez, 1969), orchids and male Euglossini bees (Dodson, 1962), the yucca moth (*Tegeticula*) and the yucca plant (Powell and Mackie, 1966), cleaning symbioses in fish, shrimp, and crabs (McCutcheon and McCutcheon, 1964), and many other associations (see reviews in Davenport, 1955, and Henry, 1967). It has been suggested by B. H. Howard (in Henry, 1967) that herbivore vertebrates such as ruminants may have evolved only because of their mutualistic association with fiber-digesting intestinal microorganisms.

The evolution of pollination mechanisms in the flowering plants represents a classic series of varying degrees of mutualistic relationships between pollinator and plant (Baker, 1963). In fact, the remarkable adaptive radiation of the angiosperms since their origin in Cretaceous times around 130 million years ago or earlier has been attributed to mutualistic pollination associations with early insect groups (see review by Baker and Hurd, 1968). Flowers are pollinated by beetles, flies, hummingbirds, hawk moths, small moths, butterflies, bats, and even occasionally by arboreal mammals such as possums and monkeys. Ants are usually considered villains in floral interactions because they are common flower visitors and often take nectar. But

a

b

c

Figure 11–23 Mutualistic symbiosis. (a) Cattle egret and African buffalo on the floor of Ngorongoro Crater in East Africa. (b) An ithomiine butterfly, *Ceratinia poecilia*, carries pollen between composite flowers in Ecuador. (c) Leaf-cutting ants (*Atta* species) in Ecuador carry these pieces of leaves underground and use them as a culture medium for fungi whose bromatia bodies are eaten by the ants. (Photos: a, T. C. Emmel; b, c, Boyce A. Drummond.)

Hickman (1974) has shown that inconspicuously flowered desert and Mediterranean-climate annuals are ant-pollinated.

The coevolution of mutualism between the swollen-thorn acacias and their ant inhabitants in the New World tropics is one of the more thoroughly studied plant–insect mutualistic systems. Janzen (1966) has shown that the ant is dependent upon the acacia for food and a place to live, and the acacia is dependent upon the ant for protection from phytophagous insects and neighboring competitor plants. The swollen-thorn acacias comprise less than 10 percent of the genus *Acacia* in Central America and have evolved specialized features for symbiosis with ants (Table 11–3). These include (1) enlarged hollow stipular thorns which the ants tenant; (2) enlarged foliar nectaries to supply the ants with nectar (Figure 11–24); (3) modified leaflet tips called Beltian bodies, which are high in protein content and are eaten by the ants; and (4) nearly year-round leaf production and maintenance of nectar and food biosynthesis even in areas with a distinct dry season. The obligate acacia ants in the genus *Pseudomyrmex* patrol the acacia plant and aggressively ward off insects and other herbivores by biting and stinging. They also attack by mauling with mandibles any living foreign plants which touch the swollen-thorn acacia's foliage or grow in a basal circle area 10 to 150 cm in diameter below the acacia. Thus, the acacia grows in a vertical cylindrical space free of other plants, such as clinging vines. The cleared area also serves to protect the host tree as a firebreak during dry-season grass fires.

Table 11–3 Acacia traits related to the ant-acacia coevolution.

A. General features of acacias of importance to the interaction	B. Specialized features of swollen-thorn acacias (coevolved traits)
1. Woody shrub or tree life form	1. Woody but with very high growth rate
2. Reproduce from suckers	2. Rapid and year-round sucker production
3. Moderate seedling and sucker mortality	3. Very high unoccupied seedling and sucker mortality
4. Plants of dry areas	4. Plants of moister areas
5. Ecologically widely distributed	5. Ecologically very widely distributed
6. Leaves shed during dry season	6. Year-round leaf production[a]
7. Shade-intolerant, sometimes covered by vines	7. Shade-intolerant and free of vines
8. Stipules often persistent	8. Stipules longer persistent, woody with soft pith[a]
9. Bitter-tasting foliage	9. Bland-tasting foliage
10. Each species with a group of relatively host-specific phytophagous insects, able to feed in the presence of the physical and chemical properties of the acacia	10. Each species with a few host-specific phytophagous insects, able to feed in the presence of the ants
11. Foliar nectaries	11. Very enlarged foliar nectaries[a]
12. Compound unmodified leaves	12. Leaflets with tips modified into Beltian bodies[a]
13 . Flowers insect-pollinated, outcrossing	13. Same as A 13
14. Seeds dispersed by water, gravity, and rodents	14. Seeds dispersed by birds
15. Lengthy seed maturation period	15. Same as A 15
16. Not dependent upon another species for survival	16. Dependent upon another species for survival

SOURCE: Janzen (1966).
[a] Essential to the interaction.

The seeds of many plants are dispersed by animals, as we have seen earlier in this book. In the Galapagos Islands, the giant tortoises (Figure 11–25) not only serve as a dispersal agent for seeds of native tomato species, but are an important natural agent in breaking seed dormancy. Rick and Bowman (1961) found that Galapagos tomato seeds will not germinate unless they are passed through the animal gut for one to three weeks. The advantages for dispersal of seeds inherent in such long passages are obvious when one considers the extensive distances these tortoises roam while feeding and searching for mates.

Parasitism, the fourth category of symbiotic relationship, involves one species (the parasite), which is benefited by the association while

Figure 11–24 A *Pseudomyrmex* ant feeding at one of the foliar nectaries at the base of a swollen-thorn acacia leaf in Costa Rica. (Photo by T. C. Emmel.)

Figure 11–25 The Hood Island race of the Galapagos tortoise, *Geochelone elephantopus hoodensis.* This female shows the saddle-backed character of the carapace in this race, apparently a selective response to arid conditions on this low-lying island where tortoises must graze on hanging *Opuntia* cactus pads. (Photo by T. C. Emmel.)

the other species (the host) is harmed. Generally, the parasite is smaller than its partner and is metabolically dependent upon the host. Organisms that spread parasites from host to host are known as vectors. The mosquito *Culex aikenii,* for instance, is an efficient natural vector which transmits the virus parasite causing Venezuelan equine encephalitis in man and horses in the Panama Canal Zone (Galindo and Grayson, 1971).

Parasitism is not inevitably disadvantageous to the host. In brood parasitism among birds, a parasitic species like cuckoos, African honey-guides, some finches, and certain members of the blackbird family (Icteridae) places its eggs in the nest of a host species. The hosts incubate the similarly-appearing eggs and raise the parasites when they hatch, along with their own offspring. Often, the parasite nestling will eliminate its competitors and end up being the only product of its host's nesting efforts that season. But under certain conditions the presence of the parasite may actually enhance the survival of the host nestlings. Thus, Smith (1968) found in Panama that a greater number of host offspring in oropendola and cacique colonies reach breeding age from cowbird-parasitized nests than from unparasitized nests. The chief source of mortality among the oropendola and cacique nestlings came from an invertebrate parasite—botflies of the genus *Philornis.* These conspicuous noisy flies place their eggs or living larvae on the chicks. The larvae burrow into the chick's body to feed, and eventually crawl out to pupate on the bottom of the nest. Chicks with more than seven larvae usually die. Smith found host chicks from exposed treetop nests containing one or two parasite chicks of the giant cowbird were almost never parasitized by these botflies. In contrast, host chicks in exposed nests lacking cowbird chicks were heavily parasitized and suffered a high mortality. Smith discovered that in nests with a mixture of several host and cowbird chicks, the giant cowbird chicks actually preened their nestmates, removing any eggs and larvae of the botfly. Host chicks did not reciprocate. The aggressive parasitic nestlings provided protection against botflies for all the offspring, and one or two host chicks could be raised to maturity along with one or two cowbird chicks in the same nest. Female oropendolas and caciques are capable of feeding and raising three or even four offspring, but because of the adaptive advantage of also maintaining one or two parasitic cowbird nestlings in the nest, the host species normally lays clutches of only two eggs in this area, for these broods are the most successful in surviving to reproductive maturity.

References

Abrahamson, W. G., and M. Gadgil. 1973. Growth form and reproductive effort in goldenrods (Solidago, Compositae). *American Naturalist, 107:* 651–661.

Aiken, D. E. 1969. Photoperiod, endocrinology and the crustacean molt cycle. *Science, 164:* 149–155.

Alcala, A. C., and W. C. Brown. 1967. Population ecology of the tropical scincoid lizard, *Emoia atrocostata,* in the Philippines. *Copeia, 1967* (3): 596–604.

Allison, A. C. 1954. Notes on sickle cell polymorphism. *Annals of Human Genetics, 19:* 39–57.

Anderson, E., and L. Hubricht. 1940. A method for describing and comparing blooming-seasons. *Bulletin of the Torrey Botanical Club, 67:* 639–648.

Anderson, P. K. 1964. Lethal alleles in *Mus musculus:* local distribution and evidence for isolation of demes. *Science, 145:* 177–178.

Andrewartha, H. G. 1961. *Introduction to the Study of Animal Populations.* University of Chicago Press, 281 pages.

Asplund, K. K. 1967. Ecology of lizards in the relictual Cape Flora, Baja California. *American Midland Naturalist, 77:* 462–475.

Atsatt, P. R. 1965. Angiosperm parasite and host: coordinated dispersal. *Science, 149:* 1389–1390.

Auffenberg, W. 1972. Komodo dragons. *Natural History, 81*(4): 52–59.

Ayala, F. J. 1971. Competition between species: frequency dependence. *Science, 171:* 820–824.

Bailey, H. P. 1964. Toward a unified concept of the temperate climate. *Geographical Review, 54*(4): 211–220.

Baker, H. B., and P. D. Hurd, Jr. 1968. Intrafloral ecology. *Annual Review of Entomology, 13:* 385–414.

Baker, H. G. 1963. Evolutionary mechanisms in pollination biology. *Science, 139:* 877–883.

Bastock, M. 1967. *Courtship: An Ethological Study*. Chicago: Aldine, 220 pages.

Bastock, M., and A. Manning. 1955. The courtship of *Drosophila melanogaster. Behaviour, 8:* 85–111.

Batra, L. R. 1966. Ambrosia fungi: extent of specificity to ambrosia beetles. *Science, 153:* 193–195.

Batra, L. R., and S. W. T. Batra. 1966. Fungus-growing termites of tropical India and associated fungi. *Journal of the Kansas Entomological Society, 39:* 725–738.

Bell, R. H. V. 1971. A grazing ecosystem in the Serengeti. *Scientific American, 225*(1): 86–93.

Benson, W. W. 1972. Natural selection for Müllerian mimicry in *Heliconius erato* in Costa Rica. *Science, 176:* 936–939.

Benson, W. W., and T. C. Emmel. 1973. Demography of gregariously roosting populations of the nymphaline butterfly *Marpesia berania* in Costa Rica. *Ecology, 54*(2): 326–335.

Betts, M. M. 1955. The food of titmice in oak woodland. *Journal of Animal Ecology, 24*(2): 282–323.

Birch, L. C. 1960. The genetic factor in population ecology. *American Naturalist, 94:* 5–24.

Bishop, J. A., and J. S. Bradley. 1972. Taxi-cabs as subjects for a population study. *Journal of Biological Education, 6:* 227–231.

Bishop, J. A., and P. M. Sheppard. 1973. An evaluation of two capture-recapture models using the technique of computer simulation. *In* M. S. Bartlett and R. W. Hiorns, eds., *The Mathematical Theory of the Dynamics of Biological Populations.* New York: Academic Press, pp. 235–252.

Bonnell, M. L., and R. K. Selander. 1974. Elephant seals: genetic variation and near extinction. *Science, 184:* 908–909.

Boulière, F., and M. Hadley. 1970. The ecology of tropical savannas. *Annual Review of Ecology and Systematics, 1:* 125–152.

Boyd, W. C. 1964. Modern ideas on race, in the light of our knowledge of blood groups and other characters with known mode of inheritance. In C. A. Leone, ed., *Taxonomic Biochemistry and Serology,* New York: Ronald Press, pages 119–169.

Brattstrom, B. H. 1962. Homing in the giant toad, *Bufo marinus. Herpetologica, 18:* 176–180.

Breedlove, D. E., and P. R. Ehrlich. 1972. Coevolution: patterns of legume predation by a lycaenid butterfly. *Oecologia (Berlin), 10:* 99–104.

Brewer, R. 1963. Stability in bird populations. *Occasional Papers of the C. C. Adams Center for Ecological Studies,* No. 7, 12 pages.

Brower, L. P., F. H. Pough, and H. R. Meck. 1970. Theoretical investigations of automimicry. I. Single trial learning. *Proceedings of the National Academy of Science, 66:* 1059–1066.

Brussard, P. F., P. R. Ehrlich, and M. C. Singer. 1974. Adult movements and population structure in *Euphydryas editha. Evolution, 28:* 408–415.

Bumpus, H. C. 1896. The variations and mutations of the introduced sparrow, *Passer domesticus. Biological Lectures, Marine Biology Laboratory, Woods Hole* (1896–1897), 1–15.

Burns, J. M., and F. M. Johnson. 1967. Esterase polymorphism in natural populations of a sulfur butterfly, *Colias eurytheme. Science, 156:* 93–96.
Bush, G. L. 1969. Sympatric host race formation and speciation in frugivorous flies of the genus *Rhagoletis* (Diptera, Tephritidae). *Evolution, 23:* 237–251.
Carlquist, S. 1965. *Island Life.* New York: Natural History Press, 451 pages.
Carlquist, S. 1966. The biota of long-distance dispersal. IV. Genetic systems in the floras of oceanic islands. *Evolution, 20*(4): 433–455.
Carlquist, S. 1974. *Island Biology.* New York: Columbia University Press, 660 pages.
Carpenter, C. C. 1966a. Comparative behavior of the Galápagos lava lizards (*Tropidurus*). *In* R. I. Bowman, ed., *The Galápagos.* Berkeley: University of California Press, pp. 269–273.
Carpenter, C. C. 1966b. The marine iguana of the Galápagos Islands, its behavior and ecology. *Proceedings of the California Academy of Sciences* (4th series), *34*(6): 329–376.
Carr, A. 1962. Orientation problems in the high seas travel and terrestrial movements of marine turtles. *American Scientist, 50:* 359–374.
Carr, A. 1964. Transoceanic migrations of the green turtle. *Bioscience, 14:* 49–52.
Carr, A., and L. Ogren. 1960. The ecology and migration of sea turtles. IV. The green turtle in the Caribbean Sea. *Bulletin of the American Museum of Natural History, 121:* 1–48.
Carson, H. L. 1975. The genetics of speciation at the diploid level. *American Naturalist, 109:* 83–92.
Cesnola, A. P. di. 1904. A first study of natural selection in *Mantis religiosa. Biometrika, 3:* 58–59.
Chanter, D. O., and D. F. Owen. 1972. The inheritance and population genetics of sex ratio in the butterfly *Acraea encedon. Journal of Zoology (London), 166:* 363–383.
Chapman, D. G. 1954. The estimation of biological populations. *Annals of Mathematical Statistics, 25:* 1–15.
Chappell, L. H. 1969. Competitive exclusion between two intestinal parasites of the three-spined stickleback, *Gasterosteus aculeatus* L. *Journal of Parasitology, 55:* 775–778.
Chitty, D. 1952. Mortality among voles (*Microtus agrestis*) at Lake Vyrnwy, Montgomeryshire in 1936–39. *Philosophical Transactions of the Royal Society of London, Ser. B, 236:* 505–552.
Chitty, D. 1954. Tuberculosis among wild voles: with a discussion of other pathological conditions among certain mammals and birds. *Ecology, 35:* 227–237.
Chitty, D. 1957. Self-regulation of numbers through changes in viability. *Cold Spring Harbor Symposia on Quantitative Biology, 22:* 277–280.
Chitty, D. 1960. Population processes in the vole and their relevance to general theory. *Canadian Journal of Zoology, 38:* 99–113.
Christian, J. J., and D. E. Davis. 1964. Endocrines, behavior, and population. *Science, 146:* 1550–1560.
Clark, L. R., P. W. Geier, R. D. Hughes, and R. F. Morris. 1967. *The Ecology of Insect Populations in Theory and Practice.* London: Methuen, 232 pages.
Clarke, B. 1962. The evidence for apostatic selection. *Heredity, 24:* 347–352.
Clarke, B., and J. Murray. 1969. Ecological genetics and speciation in land snails of the genus *Partula. Biological Journal of Linnaean Society, 1:* 31–42.
Clough, G. C. 1965. Lemmings and population problems. *American Scientist, 53:* 199–212.
Cody, M. L. 1966. A general theory of clutch size. *Evolution, 20*(2): 174–184.

Cody, M. L. 1971. Ecological aspects of reproduction. *In* D. S. Farner and J. R. King, eds., *Avian Biology,* New York: Academic Press, Volume 1, Pp. 461–512.

Cody, M. L. 1973. Coexistence, coevolution and convergent evolution in seabird communities. *Ecology, 54:* 31–44.

Cohen, B. H. 1964. Family patterns of mortality and life span. *Quarterly Review of Biology, 39:* 130–181.

Cole, L. C. 1951. Population cycles and random oscillations. *Journal of Wildlife Management, 15:* 233–252.

Cole, L. C. 1954. The population consequences of life history phenomena. *Quarterly Revue of Biology, 29:* 103–137.

Cole, L. C. 1957. Sketches of general and comparative demography. *Cold Spring Harbor Symposia on Quantitative Biology, 22:* 1–15.

Cole, L. C. 1958. Population fluctuations. *Proceedings of the Tenth International Congress on Entomology, 2* (1956): 639–647.

Colwell, R. K., and D. J. Futuyma. 1971. On the measurement of niche breadth and overlap. *Ecology, 52*(4): 567–576.

Connell, J. H. 1961. The influence of interspecific competition and other factors on the distribution of the barnacle *Chthamalus stellatus. Ecology, 42:* 710–723.

Cook, L. M., K. Frank, and L. P. Brower. 1971. Experiments on the demography of tropical butterflies. I. Survival rate and density in two species of *Parides. Biotropica, 3:* 17–20.

Cooper, R. A., and J. R. Uzmann. 1971. Migrations and growth of deep-sea lobsters, *Homarus americanus. Science, 171:* 288–290.

Craig, C. C. 1953. On the utilisation of marked specimens in estimating populations of flying insects. *Biometrika, 40:* 170–176.

Crow, J. F., and M. Kimura. 1970. *An Introduction to Population Genetics Theory.* New York: Harper & Row, 591 pages.

Cunha, A. B. da. 1949. Genetic analysis of the polymorphism of color pattern in *Drosophila polymorpha. Evolution, 3:* 239–251.

Darwin, C. 1859. *On the Origin of Species.* Cambridge: Harvard University Press, (facsimile of the 1st edition, 1964), 502 pages.

Darwin, C. 1871. *The Descent of Man and Selection in Relation to Sex.* Englewood Cliffs, N. J.: Prentice-Hall, 2 volumes.

Davenport, D. 1955. Specificity and behavior in symbioses. *Quarterly Review of Biology, 30*(1): 29–46.

Davidson, J., and H. G. Andrewartha. 1948a. Annual trends in a natural population of *Thrips imaginis* (Thysanoptera). *Journal of Animal Ecology, 17:* 193–199.

Davidson, J., and H. G. Andrewartha. 1948b. The influence of rainfall, evaporation and atmospheric temperature on fluctuations in the size of a natural population of *Thrips imaginis* (Thysanoptera). *Journal of Animal Ecology, 17:* 200–222.

Davis, D. E. 1946. A seasonal analysis of mixed flocks of birds in Brazil. *Ecology, 27:* 168–181.

Davis, D. E. 1957. The existence of cycles. *Ecology, 38*(1): 163–164.

Davis, E. F. 1928. The toxic principle of *Juglans nigra* as identified with synthetic juglone, and its toxic effects on tomato and alfalfa plants. *American Journal of Botany, 15:* 620.

Day, J. C. L., and W. H. Dowdeswell. 1968. Natural selection in *Cepaea* on Portland Bill. *Heredity, 23:* 169–188.

DeLury, D. B. 1947. On the estimation of biological populations. *Biometrika, 3:* 147–167.

Dinsmore, J. J. 1970. History and natural history of *Paradisaea apoda* on Little Tobago Island, West Indies. *Caribbean Journal of Science, 10:* 93–100.

Dobzhansky, T. 1951. *Genetics and the Origin of Species.* New York: Columbia University Press, 364 pages.

Dobzhansky, T., H. Burla, and A. B. da Cunha. 1950. A comparative study of chromosomal polymorphism in sibling species of the *willistoni* group of Drosophila. *American Naturalist, 84:* 229–246.

Dobzhansky, T. and C. Epling. 1944. Taxonomy, geographic distribution, and ecology of *Drosophila pseudoobscura* and its relatives. *In* T. Dobzhansky and C. Epling, *Contributions to the Genetics, Taxonomy, and Ecology of Drosophila pseudoobscura and its Relatives.* Washington, D. C.: Carnegie Institution of Washington, Publication 554, pp. 1–46.

Dodson, C. H. 1962. The importance of pollination in the evolution of the orchids of tropical America. *American Orchid Society Bulletin, 31:* 525–534; 641–646; 731–735.

Dowdeswell, W. H., R. A. Fisher, and E. B. Ford. 1940. The quantitative study of populations in the Lepidoptera. I. *Polyommatus icarus* (Rott.) *Annals of Eugenics, 10:* 123–136.

Dowdeswell, W. H., and K. G. McWhirter. 1967. Stability of spot-distribution in *Maniola jurtina* throughout its range. *Heredity, 22:* 187–210.

Dublin, L. L., and A. J. Lotka. 1925. On the true rate of natural increase as exemplified by the population of the United States, 1920. *Journal of American Statistics A, 20:* 305–339.

Dustan, G. G. 1965. Tagging the Oriental Fruit Moth, *Grapholitha molesta* (Busck) with radioactive phosphorus for flight and dispersal studies. *Canadian Entomologist, 97:* 810–816.

Eaton, R. L. 1974a. Second cat conference (Report on the Second International Symposium on the Ecology, Behavior, and Conservation of the World's Cats). *Science, 185:* 284–285.

Eaton, R. L. 1974b. *The Cheetah: The Biology, Ecology, and Behavior of an Endangered Species.* New York: Van Nostrand, 178 pages.

Ehrlich, P. R. 1961. Intrinsic barriers to dispersal in checkerspot butterfly. *Science, 134:* 108–109.

Ehrlich, P. R. 1965. The population biology of the butterfly, *Euphydryas editha.* II. The structure of the Jasper Ridge colony. *Evolution, 19:* 327–336.

Ehrlich, P. R. 1970. Coevolution and the biology of communities. *In* K. L. Chambers, ed., *Biochemical Coevolution.* Corvallis: Oregon State University Press, pp. 1–11.

Ehrlich, P. R., and L. C. Birch. 1967. The "balance of nature" and "population control." *American Naturalist, 101:* 97–107.

Ehrlich, P. R., D. E. Breedlove, P. F. Brussard, and M. A. Sharp. 1972. Weather and the "regulation" of subalpine populations. *Ecology, 53*(2): 243–247.

Ehrlich, P. R., and L. E. Gilbert. 1973. Population structure and dynamics of the tropical butterfly *Heliconius ethilla. Biotropica, 5:* 69–82.

Ehrlich, P. R., and R. W. Holm. 1962. Patterns and populations. *Science, 137:* 652–657.

Ehrlich, P. R., and R. W. Holm. 1963. *The Process of Evolution.* New York: McGraw-Hill, 347 pages.

Ehrlich, P. R., and P. H. Raven. 1969. Differentiation of populations. *Science, 165:* 1228–1232.

Ehrlich, P. R., R. R. White, M. C. Singer, S. W. McKechnie, and L. E. Gilbert. 1975. Checkerspot butterflies: a historical perspective. *Science, 188:* 221–228.

Ehrman, L. 1969. The sensory basis of mate selection in *Drosophila*. *Evolution, 23:* 59–64.

Eisenberg, J. F. 1966. The social organization of mammals. In *Handbuch der Zoologie*. Berlin: Walter de Gruyter, Volume 8(39), pp. 1–92.

Eisenberg, J. F., and R. W. Thorington, Jr. 1973. A preliminary analysis of a neotropical mammal fauna. *Biotropica, 5:* 150–161.

Elton, C. S. 1924. Periodic fluctuations in the numbers of animals: their causes and effects. *British Journal of Experimental Biology, 2:* 119–163.

Elton, C. S. 1927. *Animal Ecology*. London: Sidgewick & Jackson, 204 pages.

Elton, C. S. 1933. *The Ecology of Animals*. London: Methuen, 226 pages.

Elton, C. S. 1942. *Voles, Mice and Lemmings: Problems in Population Dynamics*. New York: Oxford University Press, 496 pages.

Emmel, T. C. 1964. The ecology and distribution of butterflies in a montane community near Florissant, Colorado. *American Midland Naturalist, 72:* 358–373.

Emmel, T. C. 1972. Mate selection and balanced polymorphism in the tropical nymphalid butterfly, *Anartia fatima*. *Evolution, 26:* 96–107.

Emmel, T. C. 1973. On the nature of the polymorphism and mate selection phenomena in *Anartia fatima* (Lepidoptera: Nymphalidae). *Evolution, 27:* 164–165.

Emmel, T. C., and J. F. Emmel. 1962. Ecological studies of Rhopalocera in a High Sierran community—Donner Pass, California. I. Butterfly associations and distributional factors. *Journal of the Lepidopterists' Society, 16:* 23–44.

Emmel, T. C., and J. F. Emmel. 1963. Ecological studies of Rhopalocera in a High Sierran community—Donner Pass, California. II. Meteorologic influence on flight activity. *Journal of the Lepidopterists' Society, 17:* 7–20.

Emmel, T. C., and J. F. Emmel. 1969. Selection and host plant overlap in two desert *Papilio* butterflies. *Ecology, 50*(1): 158–159.

Emmel, T. C., and C. F. Leck. 1970. Seasonal changes in organization of tropical rain forest butterfly populations in Panama. *Journal of Research on the Lepidoptera, 8*(4): 133–152.

Endler, J. A. 1973. Gene flow and population differentiation. *Science, 179:* 243–250.

Estes, J. A., and J. F. Palmisano. 1974. Sea otters: their role in structuring nearshore communities. *Science, 185:* 1058–1060.

Ficken, R. W., P. E. Matthiae, and R. Horwich. 1971. Eye marks in vertebrates: aids to vision. *Science, 173:* 936–939.

Fisher, R. A., and E. B. Ford. 1947. The spread of a gene in natural conditions in a colony of the moth *Panaxia dominula* (L.). *Heredity (London), 1:* 143–174.

Fleming, T. H. 1971. Artibeus jamaicensis: delayed embryonic development in a neotropical bat. *Science, 171:* 402–404.

Ford, E. B. 1940. Polymorphism and taxonomy. *In* Julian Huxley, ed., *The New Systematics*. New York: Oxford University Press (Clarendon Press), pp. 493–513.

Ford, E. B. 1964. *Ecological Genetics*. London: Methuen, 325 pages.

Fox, R. M., A. W. Linsey, Jr., H. K. Clench, and L. D. Miller. 1965. The Butterflies of Liberia. *Memoirs of the American Entomological Society, 19,* 1–438.

Gadgil, M. D., and W. H. Bossert. 1970. Life historical consequences of natural selection. *American Naturalist, 104:* 1–24.

Gadgil, M. D., and O. T. Solbrig. 1972. The concept of *r*- and *K*-selection: evidence from wild flowers and some theoretical considerations. *American Naturalist, 106:* 14–31.

Galindo, P., and M. A. Grayson. 1971. *Culex (Melanoconion) aikenii*: natural vector in Panama of endemic Venezuelan encephalitis. *Science, 172:* 594–595.

Gause, G. F. 1934. *The Struggle for Existence.* New York: Macmillan (Hafner Press).

Geist, V. 1971. *Mountain Sheep: A Study in Behavior and Evolution.* University of Chicago Press, 383 pages.

Goddard, J. 1967. Home range, behaviour, and recruitment rates of two black rhinoceros populations. *East African Wildlife Journal, 5:* 133–150.

Golley, F. B. 1960. Energy dynamics of a food chain of an old-field community. *Ecological Monographs, 30:* 187–206.

Gordon, H., and M. Gordon. 1957. Maintenance of polymorphism by potentially injurious genes in eight natural populations of the platyfish, *Xiphophorus maculatus. Journal of Genetics, 55:* 1–44.

Gordon, M. 1947. Speciation in fishes. Distribution in time and space of seven dominant multiple alleles in *Platypoecilus maculatus. Advances in Genetics, 1:* 95–132.

Grant, P. R. 1972. Interspecific competition among rodents. *Annual Review of Ecology and Systematics, 3:* 79–106.

Gressitt, J. L., J. Sedlacek, and J. J. H. Szent-Ivany. 1965. Flora and fauna on backs of large Papuan moss-forest weevils. *Science, 150:* 1833–1835.

Grinnell, J. 1924. Geography and evolution. *Ecology, 5:* 225–229.

Grinnell, J. 1928. Presence and absence of animals. *University of California Chronicles, 30:* 429–450.

Gwynne, M. D., and R. H. V. Bell. 1968. Selection of vegetation components by grazing ungulates in the Serengeti National Park. *Nature (London), 220:* 390–393.

Haldane, J. B. S. 1954. The measurement of natural selection. *Proceedings of the Ninth International Congress of Genetics,* pp. 480–487.

Hanson, W. R., and W. Hovanitz. 1968. Trials of several density estimators on a butterfly population. *Journal of Research on the Lepidoptera, 7:* 35–49.

Harper, J. L., P. H. Lovell, and K. G. Moore. 1970. The shapes and sizes of seeds. *Annual Review of Ecology and Systematics, 1:* 327–356.

Harper, J. L., and G. R. Sagar. 1953. Some aspects of the ecology of buttercups in permanent grassland. *Proceedings of the First British Weed Control Conference,* 256–265.

Henry, S. M. editor. 1967. *Symbiosis. Vol. 2: Associations of Invertebrates, Birds, Ruminants, and Other Biota.* New York: Academic Press, 461 pages.

Herrnkind, W. F. 1969. Queuing behavior of spiny lobsters. *Science, 164:* 1425–1427.

Herrnkind, W. F. 1970. Migration of the spiny lobster. *Natural History, 79*(5): 36–43.

Hespenheide, H. A. 1973. A novel mimicry complex: beetles and flies. *Journal of Entomology (A), 48*(1): 49–56.

Hiatt, R. W., and D. W. Strasburg. 1960. Ecological relationships of the fish fauna on coral reefs of the Marshall Islands. *Ecological Monographs, 30:* 65–127.

Hickman, J. C. 1974. Pollination by ants: a low-energy system. *Science, 184:* 1290–1292.

Hirth, H. F. 1963. The ecology of two lizards on a tropical beach. *Ecological Monographs, 33:* 83–112.

Holling, C. S. 1963. An experimental component analysis of population processes. *Memoirs of the Entomological Society of Canada, 32:* 22–32.

Holling, C. S. 1965. The functional response of predators to prey density and its role in mimicry and population regulation. *Memoirs of the Entomological Society of Canada, 45:* 1–60.

Holling, C. S. 1966. The functional reponse of invertebrate predators to prey density. *Memoirs of the Entomological Society of Canada, 48:* 1–86.

Hovanitz, W. 1948. Differences in the field activity of two female color phases of *Colias* butterflies of various times of the day. *Contributions of the Laboratory for Vertebrate Biology, Michigan, 41:* 1–37.

Hovanitz, W. 1950. The biology of *Colias* butterflies. II. Parallel geographical variation of dimorphic color phases in North American species. *Wasmann Journal of Biology, 8:* 197–219.

Howard, L. O., and W. F. Fiske. 1911. The importation into the United States of the parasites of the Gypsy Moth and Brown-tailed Moth. *Bureau of Entomology Bulletin, 91,* 344.

Howard, W. E. 1960. Innate and environmental dispersal of individual vertebrates. *American Midland Naturalist, 63:* 152–161.

Humphries, D. A., and P. M. Driver. 1967. Erratic display as a device against predators. *Science, 156:* 1767–1768.

Hutchinson, G. E. 1957. Concluding remarks. *Cold Spring Harbor Symposia on Quantitative Biology, 22:* 415–427.

Inger, R. F., and B. Greenberg. 1966. Annual reproductive patterns of lizards from a Bornean rain forest. *Ecology, 47:* 1007–1021.

Istock, C. A. 1967. The evolution of complex life cycle phenomena: an ecological perspective. *Evolution, 21:* 592–605.

Jackson, C. H. N. 1933. On the true density of tsetse flies. *Journal of Animal Ecology, 2:* 204–209.

Janzen, D. H. 1967. Synchronization of sexual reproduction of trees within the dry season in Central America. *Evolution, 21:* 620–637.

Janzen, D. H. 1969. Seed-eaters versus seed size, number, toxicity and dispersal. *Evolution, 23:* 1–27.

Janzen, D. H. 1971. Seed predation by animals. *Annual Review of Ecology and Systematics, 2:* 465–492.

Janzen, D. H., and T. Schoener. 1967. Dry season X community structure. Unpublished field problem report. San Jose, Costa Rica: Organization for Tropical Studies, pp. 10–11.

Johnson, C. 1964a. The evolution of territoriality in the Odonata. *Evolution, 18:* 89–92.

Johnson, C. 1964b. Mating expectancies and sex ratio in the damselfly, *Enallagma praevarum* (Odonata: Coenagrionidae). *Southwestern Naturalist, 9:* 297–304.

Johnson, C. 1968. Seasonal ecology of the dragonfly *Oplonaeschna armata* Hagen (Odonata: Aeshnidae). *American Midland Naturalist, 80*(2): 449–457.

Johnson, C. G. 1969. *Migration and Disperal of Insects by Flight.* London: Methuen, 763 pages.

Jolly, G. M. 1965. Explicit estimates from capture-recapture data with both death and immigration—stochastic model. *Biometrika, 52:* 225–247.

Jolly, A. 1972. *The Evolution of Primate Behavior.* New York: Macmillan, 397 pages.

Kalela, O. 1961. Seasonal change of habitat in the Norwegian lemming (*Lemmus lemmus*). *Annales Academiae Scientiarum Fennicae A IV, 55:* 1–72.

Keith, L. B. 1963. *Wildlife's Ten-year Cycle.* Madison: University of Wisconsin Press, 201 pages.

Kelker, G. H. 1940. Estimating deer population by a differential hunting loss in the sexes. *Proceedings of the Utah Academy of Science, Arts, and Letters, 17:* 65–69.

Kendeigh, S. C. 1961. *Animal Ecology*. Englewood Cliffs, N. J.: Prentice-Hall, 468 pages.

Kenyon, K. W., and V. B. Shaffer. 1954. A population study of the Alaska fur-seal herd. U. S. Fish and Wildlife Service, Special Scientific Report: Wildlife No. 12, 77 pages.

Kettlewell, B. 1973. *The Evolution of Melanism*. Oxford University Press (Clarendon Press), 423 pages.

Kettlewell, H. B. D. 1956. Further selection experiments on industrial melanism in the Lepidoptera. *Heredity, 10:* 287–301.

Kettlewell, H. B. D., and R. J. Berry. 1969. Gene flow in a cline: *Amathes glareosa* Esp. and its melanic f. *edda* Staud. (Lep.) in Shetland. *Heredity, 24:* 1–14.

King, H. D. 1939. Life processes in grey Norway rats during fourteen years in captivity. *American Anatomy Memoirs, 17:* 1–72.

King, J. A. 1955. Social behavior, social organization, and population dynamics in a black-tailed prairiedog town in the Black Hills of South Dakota. *Contributions of the Laboratory of Vertebrate Biology, University of Michigan, 67:* 1–123.

Krebs, C. J. 1964. The lemming cycle at Baker Lake, Northwest Territories, during 1959–62: Discussion. *Arctic Institute of North America Technical Paper, 15:* 50–67.

Krebs, C. J. 1972. *Ecology: the Experimental Analysis of Distribution and Abundance*. New York: Harper & Row, 694 pages.

Krebs, C. J., M. S. Gaines, B. L. Keller, J. H. Myers, and R. H. Tamarin. 1973. Population cycles in small rodents. *Science, 179:* 35–41.

Labine, P. 1964. Population biology of the butterfly, *Euphydryas editha*. I. Barriers to multiple inseminations. *Evolution, 18:* 335–336.

Lack, D. 1954. *The Natural Regulation of Animal Numbers*. New York: Oxford University Press, 343 pages.

Lack, D. 1968. *Ecological Adaptations for Breeding in Birds*. London: Methuen, 409 pages.

Levin, D. A. 1973. The age structure of a hybrid swarm in *Liatris* (Compositae). *Evolution, 27:* 532–535.

Levins, R. 1968. *Evolution in Changing Environments*. Princeton University Press, 120 pages.

Levins, R., and R. H. MacArthur. 1966. Maintenance of genetic polymorphism in a heterogenous environment: variations on a theme by Howard Levene. *American Naturalist, 100:* 585–590.

Lewis, H. 1953. The mechanism of evolution in the genus *Clarkia*. *Evolution, 7:* 1–20.

Lewontin, R. C. 1957. The adaptations of populations to varying environments. *Cold Spring Harbor Symposia on Quantitative Biology, 22:* 395–408.

Li, C. C. 1955. *Population Genetics*. University of Chicago Press, 366 pages.

Lidicker, W. Z. 1962. Emigration as a possible mechanism permitting the regulation of population density below carrying capacity. *American Naturalist, 96:* 29–33.

Likens, G. E., and F. H. Bormann. 1972. Nutrient cycling in ecosystems. *In* Wiens, J. A., ed., *Ecosystem Structure and Function*, Corvallis: Oregon State University Press, pp. 25–67.

Lindburg, D. G. 1969. Rhesus monkeys: mating season mobility of adult males. *Science, 166:* 1176–1178.

Littlejohn, M. J. 1969. The systematic significance of isolating mechanisms. *In* Charles Sibley, ed., *Systematic Biology*, Publication 1692. Washington, D. C.: National Academy of Sciences, pp. 459–482.

Lloyd, J. E. 1965. Aggressive mimicry in *Photuris:* firefly femmes fatales. *Science, 149:* 653–654.

Lloyd, M., and H. S. Dybas. 1966. The periodical cicada problem. II. Evolution. *Evolution, 20:* 466–505.

Losey, G. S. 1972. Predation protection in the poison-fang blenny, *Meiacanthus atrodorsalis,* and its mimics, *Ecsenius bicolor* and *Runula laudandus* (Blenniidae). *Pacific Science, 26*(2): 129–139.

Lotka, A. J. 1925. *Elements of Physical Biology.* Baltimore: Williams & Wilkins, 460 pages.

Lowther, J. K. 1961. Polymorphism in the white-throated sparrow, *Zonotrichia albicollis* (Gmelin). *Canadian Journal of Zoology, 39:* 281–292.

MacArthur, R. H. 1965. Ecological consequences of natural selection. *In* T. H. Waterman and H. J. Morowitz, eds., *Theoretical and Mathematical Biology.* New York: Blaisdell, pp. 388–397.

MacArthur, R. H. 1968. The theory of the niche. *In* R. C. Lewontin, ed., *Population Biology and Evolution.* Syracuse University Press, pp. 159–176.

MacArthur, R. H. 1972. *Geographical Ecology: Patterns in the Distribution of Species.* New York: Harper & Row, 269 pages.

MacArthur, R. H., and R. Levins. 1964. Competition, habitat selection and character displacement in a patchy environment. *Proceedings of the National Academy of Sciences U.S., 51:* 1207–1210.

MacArthur, R. H., and J. W. MacArthur. 1961. On species diversity. *Ecology, 42:* 594–598.

MacArthur, R. H., and E. O. Wilson. 1967. *The Theory of Island Biogeography.* Princeton University Press, 203 pages.

McClure, H. E. 1967. The composition of mixed species flocks in lowland and sub-montane forests of Malaya. *Wilson Bulletin, 79:* 131–154.

McCutcheon, F. H., and A. E. McCutcheon. 1964. Symbiotic behavior among fishes from temperate ocean waters. *Science, 145:* 948–949.

McFarland, W. N., and S. A. Moss. 1967. Internal behavior in fish schools. *Science, 156:* 260–262.

McNaughton, S. J., and L. L. Wolf. 1973. *General Ecology.* New York: Holt, Rinehart and Winston, 710 pages.

Manly, B. F. J., and M. J. Parr. 1968. A new method of estimating population size survivorship and birth rate from capture-recapture data. *Transactions of the Society of British Entomologists, 18:* 81–89.

Marler, P. 1969. Colobus guereza: territoriality and group composition. *Science, 163:* 93–95.

Mason, L. G. 1972. Natural insect populations and assortative mating. *American Midland Naturalist, 88:* 150–157.

Mather, K. 1941. Variation and selection of polygenic characters. *Journal of Genetics, 41:* 159–193.

Mather, K. 1943. Polygenic inheritance and natural selection. *Biological Reviews, 18:* 32–62.

Menge, J. L., and B. A. Menge. 1974. Role of resource allocation, aggression and spatial heterogeniety in coexistence of two competing intertidal starfish. *Ecological Monographs, 44:* 189–209.

Mertens, R. 1947. Studien zur Eidonomie and Taxonomie de Ringelnatter (*Natrix natrix*). *Abhandl. senckenberg. naturforsch. Ges., 476:* 1–38.

Mettler, L. E., and T. G. Gregg. 1969. *Population Genetics and Evolution.* Englewood Cliffs, N. J.: Prentice-Hall, 212 pages.

Miller, A. H. 1963. Seasonal activity and ecology of the avifauna of an American equatorial cloud forest. *University of California Publications in Zoology, 66*(1): 1–78.

Milne, A. 1957. Theories of natural control of insect populations. *Cold Spring Harbor Symposia on Quantitative Zoology, 22:* 253–271.
Milne, A. 1958. Perfect and imperfect density dependence in population dynamics. *Nature (London), 182:* 1251–1252.
Milne, A. 1961. Definition of competition among animals. *Symposia of the Society of Experimental Biology, 15:* 40–61.
Milne, A. 1962. On a theory of natural control of insect population. *Journal of Theoretical Biology, 3:* 19–50.
Moreau, R. E. 1950. The breeding seasons of African birds. 1. Land birds. *Ibis, 92:* 223–267.
Morris, D. J. 1956. The function and causation of courtship ceremonies. *In* P. P. Gross, ed., *L'Instinct dans le compartement des animaux et de l'Homme.* Paris: Masson, pp. 261–287.
Morse, D. H. 1970. Ecological aspects of some mixed-species foraging flocks of birds. *Ecological Monographs, 40:* 119–168.
Morton, E. S. 1971. Nest predation affecting the breeding season of the clay-colored robin, a tropical song bird. *Science, 171:* 920–921.
Moynihan, M. 1962. The organization and probable evolution of some mixed species flocks of Neotropical birds. *Smithsonian Miscellaneous Collections, 143*(7): 140 pp. (Publication 4473.)
Murie, A. 1944. The wolves of Mount McKinley. *U. S. Department of the Interior, National Park Service, Fauna Series No. 5,* 238 pages.
Murphy, G. I. 1968. Pattern in life history and the environment. *American Naturalist, 102:* 391–403.
Myers, J. H., and C. J. Krebs. 1971. Genetic, behavioral, and reproductive attributes of dispersing field voles *Microtus pennsylvanicus* and *Microtus ochrogaster. Ecological Monographs, 41*(1): 53–78.
Neel, J. V. 1951. The population genetics of two inherited blood dyscrasias in man. *Cold Spring Harbor Symposia on Quantitative Biology, 15:* 141–158.
Nevo, E. 1969. Mole rate *Spalax ehrenbergi:* mating behavior and its evolutionary significance. *Science, 163:* 484–486.
Nicholson, A. J. 1933. The balance of animal populations. *Journal of Animal Ecology, 2:* 132–178.
Nicholson, A. J. 1954. An outline of the dynamics of animal populations. *Australian Journal of Zoology, 2:* 9–65.
Nicholson, A. J. 1957. The self-adjustment of populations to change. *Cold Spring Harbor Symposia on Quantitative Biology, 22:* 153–173.
Nishida, T., and H. A. Bess. 1950. Applied ecology in Melon Fly control. *Journal of Economic Entomology, 43:* 877–883.
Norris, D. M., and J. K. Baker. 1967. Symbiosis: effects of a mutualistic fungus upon the growth and reproduction of *Xyleborus ferrugineus. Science, 156:* 1120–1122.
Odum, H. T. 1957. Trophic structure and productivity of Silver Springs, Florida. *Ecological Monographs, 27:* 55–112.
O'Meara, G. F., and D. G. Evans. 1973. Blood-feeding requirements of the mosquito: geographical variation in Aedes taeniorhynchus. *Science, 180:* 1291–1293.
Orians, G. H. 1969. On the evolution of mating systems in birds and mammals. *American Nationalist, 103:* 589–603.
Owen, D. F. 1963. Polymorphism and population density in the African land snail, *Limicolaria martensiana. Science, 140:* 666–667.
Owen, D. F. 1966a. Polymorphism in Pleistocene land snails. *Science, 152:* 71–72.

Owen, D. F. 1966b. Predominantly female populations of an African butterfly. *Heredity, 21:* 443–451.

Owen, D. F. 1966c. *Animal Ecology in Tropical Africa.* San Francisco: Freeman, 122 pages.

Owen, D. F. 1970. Inheritance of sex ratio in the butterfly *Acraea encedon. Nature (London), 225:* 662–663.

Owen, D. F., J. Owen, and D. O. Chanter. 1973a. Low mating frequencies in an African butterfly. *Nature (London)*, 244: 116–117.

Owen, D. F., J. Owen, and D. O. Chanter. 1973b. Low mating frequency in predominantly female populations of the butterfly, *Acraea encedon* (L.) (Lep.). *Entomologica Scandinavia, 4:* 155–160.

Paine, R. T. 1969. A note on trophic complexity and community stability. *American Naturalist, 103:* 91–93.

Paperna, I. 1964. Competitive exclusion of *Dactylogyrus extensus* by *Dactylogyrus vastator* (Trematoda, Monogenea) on the gills of reared carp. *Journal of Parasitology, 50:* 94–98.

Parsons, P. A. 1962. The initial increase of a new gene under positive assortative mating. *Heredity, 17:* 267–276.

Pearson, O. P. 1966. The prey of carnivores during one cycle of mouse abundance. *Journal of Animal Ecology, 35:* 217–233.

Pianka, E. R. 1970. On r- and K-selection. *American Naturalist, 104:* 592–597.

Pianka, E. R. 1974. *Evolutionary Ecology.* New York: Harper & Row, 356 pages.

Piearce, T. G. 1972. The calcium relations of selected Lumbricidae. *Journal of Animal Ecology, 41:* 167–188.

Pielou, E. C. 1972. Niche width and niche overlap: a method for measuring them. *Ecology, 53:* 687–692.

Pipkin, S. B. 1953. Fluctuations in *Drosophila* populations in a tropical area. *American Naturalist, 86:* 317–322.

Pitelka, F. A. 1959. Population studies of lemmings and lemming predators in northern Alaska. *XVth International Congress on Zoology Section X*, Paper 5, 3 pages.

Pitelka, F. A. 1964. The nutrient-recovery hypothesis for arctic microtine cycles: Introduction. *In* D. T., Crisp, ed., *Grazing in Terrestrial and Marine Environments, A Symposium of the British Ecological Society.* Oxford: Blackwell Scientific Publications.

Pitelka, F. A., P. Q. Tomich, and G. W. Treichel. 1955. Ecological relations of jaegers and owls as lemming predators near Barrow, Alaska. *Ecological Monographs, 25:* 85–117.

Powell, J. A., and R. A. Mackie. 1966. Biological interrelationships of moths and *Yucca whipplei* (Lepidoptera: Gelechiidae, Blastobasidae, Prodoxidae). *University of California Publications in Entomology, 42:* 1–46.

Ramirez, W. 1969. Fig wasps: mechanism of pollen transfer. *Science, 163:* 580–581.

Reiskind, J. 1970. Multiple mimetic forms in an ant-mimicking clubionid spider. *Science, 169:* 587–588.

Rendel, J. M. 1951. Mating of ebony vestigial and wild type *Drosophila melanogaster* in light and dark. *Evolution, 5:* 226–230.

Rick, C. M., and R. I. Bowman. 1961. Galápagos tomatoes and tortoises. *Evolution, 15:* 407–417.

Royama, T. 1971. Evolutionary significance of predators' response to local differences in prey density: A theoretical study. *In* P. J. Boer and G. R. Gradwell, eds., *Dynamics of Populations.* Wageningen: Center of Agricultural Publication and Documentation, pp. 344–357.

Salisbury, E. J. 1942. *The Reproductive Capacity of Plants. Studies in Quantitative Biology.* London: Bell, 285 pages.

Salt, G. W. 1974. Predator and prey densities as controls of the rate of capture by the predator *Didinium nasutum*. *Ecology, 55:* 434–439.
Schad, G. A. 1963. Niche diversification in a parasite species flock. *Nature (London), 198:* 404–406.
Schaffer, W. M. 1974. Optimal reproductive effort in fluctuating environments. *American Naturalist, 108:* 783–790.
Schaller, G. B. 1967. *The Deer and the Tiger: A Study of Wildlife in India.* University of Chicago Press, 370 pages.
Schaller, G. B. 1972. *The Serengeti Lion: A Study of Predator-Prey Relations.* University of Chicago Press, 480 pages.
Schneirla, T. C. 1971. *Army Ants: A Study in Social Organization.* San Francisco: Freeman, 349 pages.
Schoener, T. W. 1974. Resource partitioning in ecological communities. *Science, 185:* 27–39.
Schultz, A. M. 1962. The nutrient-recovery hypothesis for arctic microtine cycles: ecosystem variables in relation to arctic microtine cycles. *In* D. T. Crisp. ed. *Grazing in Terrestrial and Marine Environments, A Symposium of the British Ecological Society.* Oxford: Blackwell Scientific Publications. 111–121.
Sexton, O. J. 1960. Some aspects of the behavior and of the territory of a dendrobatid frog, *Prostherapis trinitatis*. *Ecology, 41:* 107–115.
Sexton, O. J. 1967. Population changes in a tropical lizard *Anolis limifrons* on Barro Colorado Island, Panama Canal Zone. *Copeia, 1967:* 219–222.
Sharp, M. A., D. R. Parks, and P. R. Ehrlich. 1974. Plant resources and butterfly habitat selection. *Ecology, 55:* 870–875.
Sharpe, F. R., and A. J. Lotka. 1911. A problem in age-distribution. *Philosophical Magazine, 21:* 435.
Sheppard, P. M. 1952a. Polymorphism and population studies. *Symposia of the Society for Experimental Biology, 7:* 274–289.
Sheppard, P. M. 1952b. A note on non-random mating in the moth *Panaxia dominula* (L). *Heredity, 6:* 239–241.
Sheppard, P. M., and J. A. Bishop. 1974. The study of populations of Lepidoptera by capture-recapture methods. *Journal of Research on the Lepidoptera, 12:* 135–144.
Shoop, C. R. 1965. Orientation of *Amblystoma maculatum:* movements to and from breeding ponds. *Science, 149:* 558–559.
Simpson, G. G. 1953. *The Major Features of Evolution.* New York: Columbia University Press, 434 pages.
Singer, M. C. 1972. Complex components of habitat suitability within a butterfly colony. *Science, 176:* 75–77.
Skutch, A. F. 1950. The nesting seasons of Central American birds in relation to climate and food supply. *Ibis, 92:* 185–222.
Skutch, A. F. 1961. Helpers among birds. *Condor, 63:* 198–226.
Skutch, A. F. 1967. Adaptive limitation of the reproductive rate of birds. *Ibis, 109:* 579–599.
Slobodkin, L. B. 1961. *Growth and Regulation of Animal Populations.* New York: Holt, Rinehart and Winston, 184 pages.
Smith, N. G. 1968. The advantage of being parasitized. *Nature (London), 219:* 690–694.
Smith, R. L. 1974. *Ecology and Field Biology.* New York: Harper & Row, 850 pages.
Snedecor, G. W. 1956. *Statistical Methods,* 5th ed. Ames: Iowa State College Press, 534 pages.
Snow, D. W. 1966. A possible selective factor in the evolution of fruiting seasons in tropical forest. *Oikos, 15*(2): 274–281.

Snow, D. W., and B. K. Snow. 1964. Breeding seasons and annual cycles of Trinidad land-birds. *Zoologica, 49*(1): 1–39.
Solomon, M. E. 1949. The natural control of animal populations. *Journal of Animal Ecology, 18:* 1–35.
Southwick, C. H. 1955. Regulatory mechanisms of house mouse populations: social behavior affecting litter survival. *Ecology, 36*(4): 627–634.
Southwood, T. R. E. 1966. *Ecological Methods.* London: Methuen, 391 pages.
Southwood, T. R. E., R. M. May, M. P. Hassell, and G. R. Conway. 1974. Ecological strategies and population parameters. *American Naturalist, 108:* 791–804.
Spieth, H. T. 1968. Evolutionary implications of sexual behavior in Drosophila. *Evolutionary Biology, 2:* 157–193.
Spieth, P. T. 1974. Theoretical considerations of unequal sex ratios. *American Naturalist, 108:* 837–849.
Stebbings, G. L., Jr. 1950. *Variation and Evolution in Plants.* New York: Columbia University Press, 643 pages.
Street, J. M. 1969. An evaluation of the concept of carrying capacity. *Professional Geographer, 21*(2): 104–107.
Taylor, O. R. 1973. A non-genetic "polymorphism" in *Anartia fatima* (Lepidoptera: Nymphalidae). *Evolution, 27:* 161–164.
Thoday, J. M., and J. B. Gibson. 1962. Isolation by disruptive selection. *Nature (London), 193:* 1164–1166.
Thompson, D. Q. 1955. The role of food and cover in population fluctuations of the brown lemming at Point Barrow, Alaska. *Transactions of the Wildlife Conference, 20:* 166–176.
Thompson, E. Y., J. Bell, and K. Pearson. 1911. A third cooperative study of *Vespa vulgaris.* Comparison of queens of a single nest with queens of the general autumn population. *Biometrika, 8:* 1–12.
Thompson, W. R. 1939. Biological control and the theories of the interactions of populations. *Parasitology, 31:* 299–388.
Thorneycroft, H. B. 1966. Chromosomal polymorphism in the white-throated sparrow, *Zonotrichia albicollis* (Gmelin). *Science, 154:* 1571–1572.
Tinbergen, L. 1949. Bosvogels en insecton. *Nederlands Bosch Tijdschrift, 4:* 91–105.
Tinkle, D. W. 1967. Home range, density, dynamics, and structure of a Texas population of the lizard *Uta stansburiana. In* W. W. Milstead, ed. *Lizard Ecology: A Symposium.* Columbia: University of Missouri Press, pp. 5–29.
Tinkle, D. W., H. M. Wilbur, and S. G. Tilley. 1970. Evolutionary strategies in lizard reproduction. *Evolution, 24:* 55–74.
Tothill, J. D. 1922. The natural control of the fall webworm (*Hyphentia cunea* Drury) in Canada. *Bulletin of the Canadian Department of Agriculture, 3*(n.s.) (*Entomology Bulletin, 19*): 1–107.
Turner, J. R. G. 1971. Experiments on the demography of tropical butterflies. II. Longevity and home-range behaviour in *Heliconius erato. Biotropica, 3:* 21–31.
Twitty, V. C. 1966. *Of Scientists and Salamanders.* San Francisco: Freeman, 178 pages.
Vandermeer, J. H. 1970. The community matrix and the number of species in a community. *American Naturalist, 104:* 73–83.
Vandermeer, J. H. 1972. Niche theory. *Annual Review of Ecology and Systematics, 3:* 107–132.
Verner, J., and M. F. Willson. 1966. The influence of habitats on mating systems of North American passerine birds. *Ecology, 47:* 143–147.

Volterra, V. 1926. Fluctuations in the abundance of a species considered mathematically. *Nature (London), 118:* 558–560.

Volterra, V. 1931. Variation and fluctuations of the number of individuals in animal species living together. Appendix (pp. 409–448). *In* R. N. Chapman (1939), *Animal Ecology.* New York: McGraw-Hill.

Waddington, C. H. 1957. *The Strategy of the Genes.* London: Allen & Unwin.

Wallace, B. 1968. *Topics in Population Genetics.* New York: Norton, 481 pages.

Wassersug, R. J. 1974. Evolution of anuran life cycles. *Science, 185:* 377–378.

Weber, N. A. 1966. Fungus-growing ants. *Science,* 153: 587–604.

Wehner, R., and R. Menzel. 1969. Homing in the ant *Cataglyphis bicolor. Science, 164:* 192–194.

Weir, B. J., and I. W. Rowlands. 1973. Reproductive strategies of mammals. *Annual Review of Ecology and Systematics, 4:* 139–163.

Westcott, P. W. 1969. Relationships among three species of jays wintering in southeastern Arizona. *Condor, 71:* 353–359.

Whittaker, R. H., and P. P. Feeny. 1971. Allelochemics: chemical interactions between species. *Science, 171:* 757–770.

Wickler, W. 1968. *Mimicry in Plants and Animals.* New York: World University Library (McGraw-Hill), 255 pages.

Wiegert, R. G. 1974. Competition: A theory based on realistic, general equations of population growth. *Science, 185:* 539–542.

Wiegert, R. G., and F. C. Evans. 1967. *In* K. Petrusewicz, ed., *Secondary Productivity of Terrestrial Ecosystems.* Warsaw and Krakow: Polish Academy of Science, pp. 499–518.

Wilbur, H. M., D. W. Tinkle, and J. P. Collins. 1974. Environmental certainty, trophic level, and resource availability in life history evolution. *American Naturalist, 108:* 805–817.

Williams, A. B. 1947. Census No. 19. Climax beech-maple forest with some hemlock (fifteen year summary). *Audubon Field Notes, 2:* 231.

Williams, G. C. 1971. *Group Selection.* Chicago: Aldine–Atherton, 210 pages.

Williams, G. R. 1963. A four-year population cycle in California quail, *Lophortyx californicus* (Shaw) in the South Island of New Zealand. *Journal of Animal Ecology, 32:* 441–459.

Willis, E. O. 1966. Interspecific competition and the foraging behavior of Plain-Brown Woodcreepers. *Ecology, 47:* 667–672.

Willis, E. O. 1968. Studies of the behavior of Lunulated and Salvin's antbirds. *Condor, 70:* 128–148.

Willis, E. O. 1974. Populations and local extinctions of birds on Barro Colorado Island, Panama. *Ecological Monographs, 44:* 153–169.

Wilson, E. O. 1971. *The Insect Societies.* Cambridge: Harvard University Press, (Belknap Press), 548 pages.

Wilson, E. O. 1975. *Sociobiology: The New Synthesis.* Cambridge: Harvard University Press, (Belknap Press), 697 pages.

Wilson, E. O., and W. H. Bossert. 1971. *A Primer of Population Biology.* Stamford, Connecticut: Sinauer, 192 pages.

Wright, S. 1964. The distribution of self-incompatibility alleles in populations. *Evolution, 18:* 609–619.

Wynne-Edwards, V. C. 1962. *Animal Dispersion in Relation to Social Behavior.* Edinburgh: Oliver & Boyd, 653 pages.

Young, A. M., and J. H. Thomason. 1974. The demography of a confined population of the butterfly *Morpho peleides* during a tropical dry season. *Studies on the Neotropical Fauna, 9:* 1–34.

Index

Illustrated terms are identified by *italicized* numbers.

76 77 78 79 9 8 7 6 5 4 3 2 1